Green Energy and Technology

Climate change, environmental impact and the limited natural resources urge scientific research and novel technical solutions. The monograph series Green Energy and Technology serves as a publishing platform for scientific and technological approaches to "green"—i.e. environmentally friendly and sustainable—technologies. While a focus lies on energy and power supply, it also covers "green" solutions in industrial engineering and engineering design. Green Energy and Technology addresses researchers, advanced students, technical consultants as well as decision makers in industries and politics. Hence, the level of presentation spans from instructional to highly technical.

Indexed in Scopus.

Indexed in Ei Compendex.

Firyal Latrache · Fatima Zahra Ouariach ·
Soufiane Ouariach · Najwa Jbira
Editors

Digital Technology for an Innovative Energy Transition

Perspectives and Opportunities

Springer

Editors
Firyal Latrache
Mechanics, Energetics, Systems and Signals Team
Laboratory of Modelization and Scientific Calculus
National School of Applied Sciences
Mohammed First University
Oujda, Morocco

Soufiane Ouariach
Research Team in Computer Science and University Pedagogical Engineering (S2IPU)
Higher Normal School of Tetouan
Abdelmalek Essaâdi University
Tétouan, Morocco

Fatima Zahra Ouariach
Research Team in Educational Science
Language Didactics
Human and Social Sciences
Higher Normal School of Tetouan
Abdelmalek Essaâdi University
Tétouan, Morocco

Najwa Jbira
Multidisciplinary Laboratory of Research and Innovation (LAMRI)
Energy, Materials, Atomics and Information Fusion Team
Department of Physics
Polydisciplinary Faculty of Khouribga
Sultan Moulay Slimane University
Beni Mellal, Morocco

ISSN 1865-3529 ISSN 1865-3537 (electronic)
Green Energy and Technology
ISBN 978-3-031-89656-9 ISBN 978-3-031-89657-6 (eBook)
https://doi.org/10.1007/978-3-031-89657-6

This Springer imprint is published by the registered company Springer Nature Switzerland AG
The registered company address is: Gewerbestrasse 11, 6330 Cham, Switzerland

To my father ***Farid****, my mother* ***Fatima*** *and my brother* ***Simohammed*** *for their endless support*

To Prof. ***Mohamed Aklalouch*** *from Cadi Ayyad University, Marrakesh, Morocco, who helped me to rediscover an extraordinary science—Chemistry—after 8 years when I chose physics as a field of study.*

I dedicate this book.

It was 8 years ago that I decided to continue my studies in physics to discover the contributions of Isaac Newton, Ludwig Boltzmann and Albert Einstein to the scientific community. I also decided to study the Field Theory, the Standard Model and string theory, while totally ignoring the issue of climate change and the need to contribute to the implementation of the energy transition. Today, with the help of this Professor, I have gone back to chemistry to discover the widespread use of lithium-ion batteries in electric vehicles to achieve decarbonization by 2050.

Firyal Latrache

I dedicate this book to my parents and brothers for their unconditional love and unwavering support.
A special thanks to my friend ***Dr. Rida Benmehdi****, whose friendship and constant encouragement have been an invaluable source of motivation. Thank you all for being there throughout this journey.*

Fatima Zahra Ouariach

I dedicate this book to my parents, my brother, and my sister for their love and constant support.
A special thanks to ***Dr. Inam Ullah Khan****, whose friendship and encouragement have been essential throughout this project. Thank you all for your presence and inspiration.*

Soufiane Ouariach

I dedicate this work to my parents, who have always been my greatest source of strength and inspiration. Their unfailing love and unfailing support have given me the courage to pursue my goals with determination. To my brothers and sisters, whose constant presence and encouragement have accompanied me at every stage of my life, I express my gratitude for their unwavering support. To my true friends, who have provided moments of comfort and serenity, and whose sincere friendship has enriched me along the way. This dedication is for you, as a token of my gratitude and deep affection.

Najwa Jbira

Preface

We are the four editors of the edited book volume *Digital Technology for an Innovative Energy Transition: Perspectives and Opportunities*: Firyal Latrache, Fatima Zahra Ouariach, Soufiane Ouariach and Najwa Jbira and we are delighted to write the Preface of this book.

What is the Preface? What is the purpose of writing it?

As editors, we constantly focus on presenting scientific information in the most impressive form. As this is our first experience, we looked for the optimal methodology for writing this preface and found that the approach differs depending on whether the book belongs to the scientific field or not. As usual, the preface of the scientific books is based on describing the problematic and the objectives and presenting the contributions in the field concerned. But we wondered, if "the preface" is not our gateway to our dear readers and to share this experience with them, what is the objective of having an introduction in the book at all? Of course, without philosophizing on the subject and at the same time without steering away from it.

But why edit a book on the energy transition?

On February 2024, Firyal Latrache asked Fatima Zahra Ouariach to publish together a book. Afterward, Fatima Zahra Ouariach invited his brother Soufiane Ouariach and Firyal Latrache invited her close friend Najwa Jbira to create a working group to launch this famous project. The subject of energy transition was chosen after a working meeting, with the aim of contributing to the global energy transition and subsequently to confronting climate change and achieving decarbonization. In view of the global wave of digitalization, the editors wanted to exploit this perspective to achieve sustainability while at the same time enabling the optimal utilization and deployment of smart energy systems. Finally, the title was fixed on *Digital Technology for an Innovative Energy Transition: Perspectives and Opportunities*. In fact, the motivation for bringing this idea of publishing an edited book goes back to Professor and Doctor Saad Motahhir from Sidi Mohamed Ben Abdellah University Fez Morocco for his continuous support and encouragement of junior's researchers and for his outstanding expertise in the field of renewable energies.

Lavoisier said "*Nothing is lost, nothing is created, all are transformed*". However, the waste are lost everywhere and factories emit emissions polluting the environment, thereby creating a warning signal about the increasing degree of climate change. Perhaps the Lavoisier quote reflects the creating of dangerous emissions in the atmosphere through the waste, but which type of transformation was intended? In addition to the frightening transformation of the unexplained weather conditions (the temperature rise in autumn and the warm winter in particular in the Mediterranean countries and again the natural disasters like the Tropical storm Milton 2024 of Florida, g), the current transformation is about digitalization, where everyone digitalizes. The last investigations illustrates the lifestyle facilitation through adopting this "digitalization", which includes facilitating administrative services, deepening knowledge and improving the quality of production. As long as the current global objective is to reduce the carbon footprint, why not exploit digitalization to both confront climate change and guarantee energy security, as well as eliminating energy poverty? In view that the reports launched since the adoption of the concept of energy transition present prospects for starting its circulation and implementation, we (the editors) take into consideration the efforts made at this stage and already assume that energy transition has been implemented. With this book, we move on to the second stage involving the realization of the principles and objectives of the energy transition, but this time in light of digital transformation.

This book exploits the digital technologies to further implement the energy transition and achieve the decarbonization. It encourages using the renewable resources such as the photovoltaic energy PV to increase this share in energy mix and benefit from the sustainable materials like the phase change material PCM to perform optimal thermal management. Furthermore, the book uses the AI algorithms to develop and optimize the energy consumption. In global, the book exploits several approaches including modeling, experimental as well as reviews to analyze the current contributions and introduce new ones. In fact, the experimental approaches concerns the development of the building materials sector and the PV energy to promote the sustainability. As a sustainable material, the modeling of incorporating PCM in buildings through entropy equations supports the decreasing of the carbon footprint. To both mitigate climate change and encourage the digital transformation, the book analyzes the climate change impact and illustrates the energy transition promotion through a robust review on the current contributions in this area. Other reviews were performed on financing innovations models for clean digital energies and on the contributions of the 17 Sustainable Development Goals and AI in developing the Moroccan administrations. Not to mention the major contribution of lithium-ion batteries in smart mobility and then in promoting energy transition and the book reviews this technology and encourage the researchers to more implement it in further applications. To apply digital technologies in order to promote decarbonization both now and by 2050, the book illustrates ethics and principles of digital transformation in energy sector and highlights opportunities and perspectives for promoting smart energy transition by 2050.

To collect several contributions and invite powerful researchers in this field, it was very hard in particular that we must satisfy the requirements of book scope.

Through LinkedIn, as a professional network, we succeeded in inviting leading researchers and spreading the idea of exploiting digital technologies in implementing energy transition. In this context, we would appreciate of the effort of the greater researchers as contributors in our book including Fatima Zahra Ouariach, Ayoub Ouchlif, Karri Pallavi, Hamza El Kezazy, Hicham Kaddouri, Chouaib Benqlilou, Mumtaz Ahmad Qaisrani, Akhil Nigam, Mohamed Aklalouch, Othmane Horma, Boutahar Laaouar, Mohammed Remaidi and Firyal Latrache. We appreciate their effort and their approaches to develop our book. Furthermore, we appreciate the effort of their co-authors and their role in developing the book outlooks.

If Prof. Saad Motahhir was motivated us to edit a book and the greater researchers cited above have contributed with vigorous chapters, Prof. Mohamed Saidi hassani Alaoui has the major role in revising the content of this book to satisfy the scientific criteria. Not to mention his ongoing motivation for the editors and his positivity and generosity.

Through our book *Digital Technology for an Innovative Energy Transition: Perspectives and Opportunities* we are no longer just desiring to implement energy transition and mitigate climate change and then perform the decarbonization but we would emphasize the implementation of a smart energy transition to both contribute in developing the digital technologies and achieving the sustainability. The book highlights opportunities and perspectives to encourage the researchers and fields for starting work and research contributions with passion and enthusiasm.

As editors, we propose further perspectives in editing and publishing books to contribute in depth in the major role of digital technologies in the energy sector and highlights the new exigence to implement energy transition in this era of digitalization. We invite the energy passionate to become a key player in decarbonization and energy transition to reduce the carbon footprint and provide more sustainable word.

Oujda, Morocco — Firyal Latrache
Tétouan, Morocco — Fatima Zahra Ouariach
Tétouan, Morocco — Soufiane Ouariach
Beni Mellal, Morocco — Najwa Jbira

Introduction

It seems that the urgent need to restore the natural balance has motivated the political organisms and the actors to perform a transition in the consumption and supply of energy by adopting the energy transition concept as a political and technical objective. The adoption of the energy transition is not only about protecting biodiversity and reducing the impact of climate change, but it is about recognizing the best way for proceeding forward to the sustainability with the incursion in the actual wave of digital transformation. In fact, achieving the energy transition in light of this digital transformation depends on several factors such as the exploitation of the digital technologies to enhance the electricity grids and optimize the energy storage instruments and then to get access to the developed energy tools. There are then a variety of opportunities and perspectives aiming to achieve an innovative energy transition and performing a global developed economy.

The most recent statistics of the World Meteorological Organization illustrate that the average temperature of the earth surface reaches 1.45 °C in 2023 with fears of worsening the death rate due to the climate change of about 250,000 deaths by year by 2030 (*Climate Change Indicators Reached Record Levels in 2023*, 2024). Indeed, if the climate change does not cause a high mortality rate, it affects numerous parts like the human health, the water and food quality, the environment and the infrastructure. In this context, the glacial melt in western North America and Europe, which has not been recorded since 1950 (*Climate Change Impacts|National Oceanic and Atmospheric Administration*, s. d.), causes the rise of ocean and sea levels. In addition, the people who suffered from the insecure food is amounted by 333 million people since 2023. It is clear that climate change causes several dangerous effect and has a several reasons including the air pollution, where it is responsible of 45% of the actual global warming. This air pollution is resulting from different sectors such as energy, industry, transport, agriculture and cooking (*Climate Change Impacts|National Oceanic and Atmospheric Administration*, s. d.). In the energy sector, burning the fossil fuels (coal, oil and natural gas) is considered the most cause of air pollution with 85% of the emissions (*Climate Explainer*, s. d.). In spite of their harmful effect on the biodiversity, the fossil fuels still contribute in the energy mix, which the statistics of the International Energy Agency IEA of 2022 indicates the generation of 35.7% of electricity from coal, followed by natural gas (22.3%) and hydropower (15.3%) (*Climate Explainer*, s. d.). On the other hands, there are efforts and measures to reduce the part of these fossil fuels and the air pollution especially that the last statistics indicates a reduction of 6.1% in the global gross domestic product GDP (*Climate Explainer*, s. d.). Therefore, the effective measure is about performing an innovative energy transition from the fossil resources (coil, natural gas and oil) to the renewable ones (wind, solar and hydropower) while reducing the climate change degree.

The adoption of the energy transition concept is intended to provide the energy security (*Energy Security in Energy Transitions—World Energy Outlook 2022—Analysis*, s. d.) and decrease the energy poverty (*Energy poverty*, s. d.). In fact, the energy security cover the energy supply with a competitive price, which is considered as a vulnerable issue toward the wars and the geopolitical conflicts (*Energy Security in Energy Transitions—World Energy Outlook 2022—Analysis*, s. d.). The energy poverty may seems that is refers to the depletion of energetic resources as a disastrous consequence of these wars (*Energy poverty*, s. d.). However, it includes several dimensions such as the policies of residential sector, the infrastructure, the geographical location and the social background of the habitats. It is clear that providing the energy security and decreasing the energy poverty are from the specialties of the political organizations, whom they should establish the energy transition strategies by promoting peace and security to both decrease the climate-damaging remnants of war and providing more energy supply (*Geopolitics of the Energy Transition*, 2024). They can also provide vulnerable populations with assistance to improve the quality of life through benefiting of energy resources. As well as their assistance, the political organizations help this category by allowing tax exemption in terms of energy consumption to contribute to the decarbonization by 2050. Without forgetting the major impact of the political organization measures to develop the global economy by providing energy supply, in particular resolving the global gross domestic product GDP decreasing issue. Within this context, the energy transition do not include only the decrease of the climate change degree, but providing the energy security and eliminating the energy poverty while developing the global economy and reducing the carbon footprint.

In light of the current digitalization wave, the providing of more energy security and low energy poverty with the protection of the natural resources requires the understanding and the applications of the digital technologies and approaches for exploiting renewable resources, analyzing their consumption and their potential of achieving the energy transition outlook. The artificial intelligence AI, digital twin, internet of thing IoT, and blockchain are all supporting the energy transition by providing a strong analyst for the data of the intelligent energy systems and introducing the smart climate concept to encourage the net zero energy and low-carbon. For illustration, AI predicts energy demand and consumption in real time to test the flexibility of the energy systems, and can even automate systems for optimal operation. In this context, the approach based on the machine learning algorithms such as LightGBM and random forest helps in the energy demand estimation. The internet of thing IoT helps to the data management and analyze the energy systems to predict their maintenance through connected sensors. Without forgetting the major role of blockchain in encouraging the decentralized energy systems. In view of the existing approaches and digital technologies, the literature and the scientific community need an understandable opportunities and perspectives for implementing an innovative energy transition in light of the digital transformation (Ferdaus et al. 2024).

As presented above, the energy transition requires the exploitation of sustainable resources (solar, wind and water) and even use sustainable materials in particular the bio-based material for promoting sustainability. As an inexhaustible resource, the solar energy is exploited through the photovoltaic panel to generate the electricity and again through a variety of solar concentrators' geometries to produce a steam then the electricity using turbines. Afterward, the generated electricity is stored in batteries for later use and the lithium-ion battery is the optimal technology chosen to store the electricity as they has a high energy density and reduced self-discharge. The electric vehicles and the mobiles phones are among the most the most widely used applications. As well as the current outlook of the energy transition requires the electric vehicles as a key factor for achieving the decarbonization, the Direct Torque Control must control their motors for efficient operation. On the other side, reaching the energy transition not only means using renewable resources, but at the same time optimizing the building materials industry by incorporating more additives to ensure efficient thermal building management and to contribute to the zero-energy building concept. The energy transition includes too the use of the pyro-piezoelectric materials in intelligent buildings for thermal control and as an autonomous sensor for surveillance and energy recuperation.

So, why we launch our book *Digital Technology for an Innovative Energy Transition: Perspectives and Opportunities*?

As the Paris Agreement of 2015 (*The Paris Agreement|UNFCCC*, s. d.) has a charter aiming to limit the average temperature of the earth's surface at 1.5 °C and as the Sustainable Development Agenda of 2018 set 17 goals for sustainability and climate change mitigation (*THE 17 GOALS|Sustainable Development*, s. d.), the achievement of an innovative energy transition in light of the digital transformation requires a book for motivating the researchers and the experts to initiate the necessary measures and actions. While the current contributions in energy transition in recently published books are interested in the acceleration of this energy transition or in its geopolitical and democratic dimensions, our book presupposes the transition in the energy sector and presents measures to ensure its successful implementation in light of the digital transformation. Relating to this context, our book provide the artificial intelligence AI and deep learning algorithms to enhance the exploitation of the renewable resources. It presents the numerical software, which they are considered as perfect tools to simulate the scientific phenomena before the AI era. The numerical software are currently being developed to satisfy industrial demands to not be late for the AI algorithms. Therefore, this book exploits the digital technologies, the numerical software and the AI algorithms to develop and optimize the energy transition key factors (presented above) such as bio-based materials, PV energy, building and the pyro-piezoelectric materials. In addition, the book proposes measures to mitigate the climate change as a mandatory action for achieving the energy transition and provides the valuation of Sustainable Development Goals in the administrative sector. The contribution of lithium-ion batteries and the building materials industry are also presented and their major roles in promoting the energy transition are explained. Indeed the future digital technologies for green energy need investment and funds and the present book contributes in the optimization of the financing innovations models. In global and to get on the path of the digitalization, this book resumes the opportunities and perspectives to achieve a smart energy transition by 2050. It is not just about implementing only an energy transition by mitigating the climate change and exploiting the renewable resources, but it is about tracking the way of digital transformation and adopting digital technologies to achieve a smart energy transition based on both sustainability and smart systems. By considering that each measure and action require principles and by going from the AI ethics (transparency, impartiality, accountability, reliability and security), the book reveals the principles and the ethics of implementing the energy transition in light of the digital transformation.

The present book is composed of 13 chapters organized as follows:

Chapter "Principles and Ethics for the Implementation of Digital Transformation on the Energy Transition" (the corresponding author is **Fatima Zahra Ouariach**) presents the principles and ethics of the implementation of digital transformation on the energy transition.

Chapter "A Systematic Literature Review on Modernizing Land Acquisition Process for Smart Decision Making in Urban Planning" (the corresponding author is **Ayoub Ouchlif**) shows the potential of AI in Moroccan administrative expertise commission process and investigate a review on modernizing real estate valuation for Sustainable Development Goals SDG.

Chapter "Optimization Financing Innovations Models for Next-Generation Sustainable Clean Energy Technologies" (the corresponding author is **Karri Pallavi**) optimizes the financing innovations models for next generation sustainable clean energy technologies.

Chapter "Promoting the Energy Transition Throughout Dealing with the Climate Change Issue" (the corresponding author is **Hamza El Kezazy**) analyzes the climate change issue as a key factor for promoting the energy transition.

Chapter "Energy, Economic, and Thermal Assessment of Insulating a Moroccan Building Envelope with a Biobased Material for Promoting Sustainability" (the corresponding author is **Hicham Kaddouri**) evaluates the energy, economy, and thermal of insulating a Moroccan Building envelope with a biobased material for promoting sustainability.

Chapter "Deep Learning Method for Evaluating Morocco's Urban Photovoltaic Potential" (the corresponding author is **Chouaib Benqlilou**) proposes the deep learning method for evaluating Morocco's urban photovoltaic potential.

Chapter "Implementation of AI Techniques on Efficiency Enhancement of Solar PV Relating to Performance Parameter" (the corresponding author is **Mumtaz Ahmad Qaisrani**) implements the AI techniques on efficiency enhancement of solar PV relating to performance parameter.

Chapter "The Study of Direct Torque Controller (DTC) With the Proportional Integral (PI) Controller to Promote the Energy Transition" (the corresponding author is **Akhil Nigam**) studies the Direct Torque Controller (DTC) with the Proportional Integral (PI) Controller to promote the energy transition.

Chapter "The Evolution of the Lithium-Ion Batteries LIBs Technologies and An Overview of Their Contribution in the Global Energy Transition" (the corresponding author is **Mohamed Aklalouch**) analyses the Evolution of the Lithium-Ion Batteries LIBs technologies and an overview of their contribution in the global energy transition.

Chapter "Infusing Sustainability: Mint Tea By-Products as Thermal Insulating Mortar Additives for Energy Transition in the Construction Industry" (the corresponding author is **Othmane Horma**) contributes in the construction industry by adding the mint tea by-products as thermal insulating mortar additives for promoting energy transition and sustainability.

Chapter "Evaluating Moroccan Soils for Sustainable Building Applications: An Experimental and Computational Analyses" (the corresponding author is **Boutahar Laaouar**) evaluates the Moroccan soils for sustainable building applications through experimental and computational analyses.

Chapter "Enhancing Energy Harvesting Efficiency in Autonomous Devices Using Dual Pyro-Piezoelectric Materials" (the corresponding author is **Mohammed Remaidi**) enhances the energy harvesting efficiency in autonomous devices using dual pyro-piezoelectric materials.

Chapter "Opportunities and Perspectives of the Implementation of Smart Energy Transition by 2050" (the corresponding author is **Firyal Latrache**) presents the opportunities and perspectives of the implementation of smart energy transition by 2050.

References

Energy Poverty (n.d.) Consulté 8 octobre 2024, à l'adresse. https://energy.ec.europa.eu/topics/markets-and-consumers/energy-consumers-and-prosumers/energy-poverty_en

Ferdaus MM, Dam T, Anavatti S, Das S (2024) Digital technologies for a net-zero energy future: a comprehensive review. Renew Sustain Energy Rev 202:114681. https://doi.org/10.1016/j.rser.2024.114681

Geopolitics of the energy transition: energy security (2024) https://www.irena.org/Publications/2024/Apr/Geopolitics-of-the-energy-transition-Energy-security

IEA (n.d.) Energy security in energy transitions—World Energy Outlook 2022—Analysis. Consulté 8 octobre 2024, à l'adresse. https://www.iea.org/reports/world-energy-outlook-2022/energy-security-in-energy-transitions

National Oceanic and Atmospheric Administration (n.d.) Climate change impacts. Consulté 7 octobre 2024, à l'adresse. https://www.noaa.gov/education/resource-collections/climate/climate-change-impacts

Sustainable Development (n.d.) The 17 Goals. Consulté 11 octobre 2024, à l'adresse. https://sdgs.un.org/goals

UNFCCC (n.d.) The Paris Agreement. Consulté 11 octobre 2024, à l'adresse. https://unfccc.int/process-and-meetings/the-paris-agreement

WMO (2024) Climate change indicators reached record levels in 2023. World Meteorological Organization. https://wmo.int/news/media-centre/climate-change-indicators-reached-record-levels-2023-wmo

World Bank (n.d.) Climate explainer: climate change and air pollution. Consulté 7 octobre 2024, à l'adresse. https://www.worldbank.org/en/news/feature/2022/09/01/what-you-need-to-know-about-climate-change-and-air-pollution

Contents

Editors and Contributors

About the Editors

Firyal Latrache is a Ph.D. in mechanical and energy engineering. She works in solar thermal and thermodynamic energies, green hydrogen and smart cities. Her research focuses on the mathematical modeling of solar thermal energy applications, as well as the dimensioning and study of the technical feasibility of thermodynamic solar power plants. Firyal is a major player in the energy transition and sustainable development, and in achieving decarbonization by 2050. She uses the electrical energy generated in thermodynamic solar power plants to generate green hydrogen. Its research focuses on developing mathematical and numerical models for heat exchangers and solar concentrators and on analyzing the thermomechanical behavior of materials to predict their industrial maintenance for Industry 4.0. She is currently working on books on the energy transition and smart cities. Firyal has reviewed numerous articles in international journals and conferences (Springer and Scientific.net). She is an Arabic-language poet, portraitist and painter of abstract art.

Fatima Zahra Ouariach is a doctoral researcher specializing in Educational Technology at Abdelmalek Essaadi University. She is currently working on her Ph.D., focusing on the management of communication tools in e-learning through an LMS platform. Her research aims to design an efficient architecture for communication tools in the specific context of online learning using an LMS platform. In addition to her academic work, she has gained experience as a reviewer for prestigious publishing houses such as Hindawi, Inderscience and IGI Global. Furthermore, she has made valuable contributions as a member of the jury for oral interviews at Abdelmalek Essaadi University, highlighting her dedication and contribution to the advancement of research and higher education.

Soufiane Ouariach is a doctor specializing in Computer Science and University Pedagogical Engineering. Focusing on the innovative approach of flipped classroom models using SPOC (Small Private Online Courses) as well as distance learning design and implementation. His research aims to break new ground in these exciting fields of educational technology. In addition to his academic work, he has gained experience as a reviewer for prestigious publishing houses such as IGI Global and Inderscience. Furthermore, he has served as a reviewer for several international conferences, highlighting his dedication and contribution to the advancement of research and higher education. His academic prowess extends to distance learning design and implementation, where he's keen on pushing the boundaries of innovation in educational technology.

Najwa Jbira is a Ph.D. in energy efficiency, specializing in energy integration within the industrial sector. Her work focuses on two advanced optimization methods: pinch analysis and exergy analysis, with a particular emphasis on the dairy industry. Her research aims to improve energy efficiency and sustainability in this key sector, contributing to reducing its environmental footprint. As a reviewer, she evaluates articles for several scientific journals and international conferences, regularly contributing to the enhancement of publications in these fields. Najwa Jbira is also a highly active figure in academic circles, having organized and participated in numerous doctoral symposiums and workshops aimed at fostering collaboration and innovation among researchers. A member of the "Green Brains" association, dedicated to innovation and sustainable development. Professionally, Najwa works as a renewable energy specialist at the Ministry of the Interior, where she is involved in projects aimed at promoting energy efficiency, renewable energy and environmental initiatives.

Contributors

Faheem Abbas Institute of Mechanical and Manufacturing Engineering, Khwaja Fareed University of Engineering and Information Technology, Rahim Yar Khan, Pakistan

Abderrahim Abidouche Innovative Technologies Laboratory, Sidi Mohammed Ben Abdellah University, High school of technology, Fez, Morocco

Mohamed Aymen Ben Achour Polytechnic University of Hauts-de-France, Valenciennes, France

Naveed Ahmed U.S.-Pakistan Center for Advanced Studies in Energy National Univsersity of Sciences and Technology, Islamabad, Pakistan

Kenza Aitelkadi RU19-Geospatial Technologies for Intelligent Decision Making, Hassan II Agronomic and Veterinary Institute, Rabat, Morocco

Mohamed Aklalouch Chemistry of Materials & Environment Team, IMED-Laboratory, Department of Chemistry Sciences, Faculty of Sciences and Technologies, University Cadi Ayyad, Marrakesh, Morocco

Mohamed Saidi Hassani Alaoui Experimentation and Modelling Team in Mechanics and Energy Systems, National School of Applied Sciences, Abdelmalek Essaadi University, Al Hoceima, Morocco

Benaissa Bellach Mechanics, Energetics, Systems and Signals Team, Laboratory of Modelization and Scientific Calculus, National School of Applied Sciences, University Mohammed I, Oujda, Morocco

Chouaib Benqlilou Ecole Nationale Supérieure des Mines de Rabat, Rabat, Morocco

Ranjit Bindal EED, Chandigarh University, Mohali, Punjab, India

Meryiem Derraz Laboratory of Engineering Sciences for Energy, National School of Applied Sciences, Chouaib Doukkali University, El Jadida, Morocco

Ismael Driouch Experimentation and Modelling Team in Mechanics and Energy Systems, Abdelmalek Essaadi University, National School of Applied Sciences, Al-Hoceima, Morocco

Abderrahim Ed-Daoui Laboratory of Physics of Matter and Radiations, Mohamed First University, Faculty of Sciences, Oujda, Morocco

Aboubakr El Hammouti Mechanics and Energy Laboratory, Mohammed First University, Oujda, Morocco;
Institut de Recherche de la Construction, ESTP, Cachan, France

Sara El Hassani Laboratory of Energetics, Abdelmalek Essaâdi University, Tetouan, Morocco

Abdesadeq Elbayad Ecole Nationale Supérieure des Mines de Rabat, Rabat, Morocco

Abdelhafid Elouassidi Experimentation and Modelling Team in Mechanics and Energy Systems, Abdelmalek Essaadi University, National School of Applied Sciences, Al-Hoceima, Morocco

Chouaib Ennawaoui Laboratory of Engineering Sciences for Energy, National School of Applied Sciences, Chouaib Doukkali University, El Jadida, Morocco

Muhammad Umer Farooq Institute of Mechanical and Manufacturing Engineering, Khwaja Fareed University of Engineering and Information Technology, Rahim Yar Khan, Pakistan

Syed Saad Farooq Institute of Mechanical and Manufacturing Engineering, Khwaja Fareed University of Engineering and Information Technology, Rahim Yar Khan, Pakistan

Abdelowahed Hajjaji Laboratory of Engineering Sciences for Energy, National School of Applied Sciences, Chouaib Doukkali University, El Jadida, Morocco

Hicham Hajji RU19-Geospatial Technologies for Intelligent Decision Making, Hassan II Agronomic and Veterinary Institute, Rabat, Morocco

Yassine Hilmi Laboratory of Studies and Research in Economic Sciences and Management National School of Commerce and Management, Chouaib Doukkali University, El Jadida, Morocco

Othmane Horma Mechanics and Energy Laboratory, Mohammed First University, Oujda, Morocco

Asma Ihechmyen Chemistry of Materials & Environment Team, IMED-Laboratory, Department of Chemistry Sciences, Faculty of Sciences and Technologies, University Cadi Ayyad, Marrakesh, Morocco

Najwa Jbira Multidisciplinary Laboratory of Research and Innovation (LAMRI), Energy, Materials, Atomics and Information Fusion Team, Department of Physics,

Polydisciplinary Faculty of Khouribga, Sultan Moulay Slimane University of Beni Mellal, Khouribga, Morocco

Hicham Kaddouri Experimentation and Modelling Team in Mechanics and Energy Systems, National School of Applied Sciences, Abdelmalek Essaadi University, Al Hoceima, Morocco

Sarah Kahlaoui Ecole Nationale Supérieure des Mines de Rabat, Rabat, Morocco

Amnah Kamran Institute of Mechanical and Manufacturing Engineering, Khwaja Fareed University of Engineering and Information Technology, Rahim Yar Khan, Pakistan

Hamza El Kezazy Economie, Management et Finance des Organisations (EEMFO), Université Sultane Moulay Slimane (USMS), Beni Mellal, Morocco

Boutahar Laaouar Mechanics and Energy Laboratory, Mohammed First University, Oujda, Morocco

Firyal Latrache Mechanics, Energetics, Systems and Signals Team, Laboratory of Modelization and Scientific Calculus, National School of Applied Sciences, University Mohammed I, Oujda, Morocco

Youssouf Maazou Manomi Ecole Nationale Supérieure des Mines de Rabat, Rabat, Morocco

Ahmed Mezrhab Mechanics and Energy Laboratory, Mohammed First University, Oujda, Morocco

Mohammed Amine Moussaoui Mechanics and Energy Laboratory, Mohammed First University, Oujda, Morocco

Akhil Nigam EED, Chandigarh University, Mohali, Punjab, India

Fatima Zahra Ouariach Research Team in Educational Science, Language Didactics, Human and Social Sciences, Higher Normal School of Tetouan, Abdelmalek Essaadi University, Tetouan, Morocco

Soufiane Ouariach Research team in Computer Science and University Pedagogical Engineering (S2IPU), Higher Normal School of Tetouan, Abdelmalek Essaadi University, Morocco, Tetouan, Morocco

Ayoub Ouchlif RU19-Geospatial Technologies for Intelligent Decision Making, Hassan II Agronomic and Veterinary Institute, Rabat, Morocco

Karri Pallavi VIT Bhopal University, Sehore, Madhya Pradesh, India

Mumtaz A. Qaisrani Institute of Mechanical and Manufacturing Engineering, Khwaja Fareed University of Engineering and Information Technology, Rahim Yar Khan, Pakistan

Hassan Radoine Mohammed VI Polytechnic University, Marrakesh, Morocco

Mohammed Remaidi Laboratory of Engineering Sciences for Energy, National School of Applied Sciences, Chouaib Doukkali University, El Jadida, Morocco

Sonali Srivastava Jaypee Business School, Jaypee Institute of Information Technology, Noida, India

Nabila Zrira Ecole Nationale Supérieure des Mines de Rabat, Rabat, Morocco

Principles and Ethics for the Implementation of Digital Transformation on the Energy Transition

Fatima Zahra Ouariach, Soufiane Ouariach, Firyal Latrache, and Najwa Jbira

Abstract This chapter explores the intersection of energy transition and digital transformation, two major challenges of the twenty-first century. It examines how digital technologies can optimize the transition from fossil fuels to renewable energies. The analysis focuses on the opportunities and challenges of this transformation, including issues of ethics, sustainability, and equity. The chapter addresses four axes: the purpose of digitization, its implementation, ethical considerations, and a case study on Energy 4.0. The aim is to develop an ethical framework to guide the digitization of the energy sector, balancing efficiency, environmental sustainability, and social justice. This exploration aims to contribute to a sustainable, efficient, and human-centered energy future.

Keywords Digital transformation · Energy transition · Artificial intelligence · Ethics

1 Introduction

At the dawn of the twenty-first century, humanity faces two major and interconnected challenges that will shape our collective future: the energy transition and the digital transformation. While these phenomena are distinct, they are deeply intertwined and present unprecedented opportunities and complex challenges, necessitating deep reflection on the principles and ethics that should guide their implementation. The global energy landscape has long been dominated by fossil fuels. Coal, oil, and

F. Z. Ouariach (✉) · S. Ouariach
Higher Normal School of Tetouan, Abdelmalek Essaadi University, Tétouan, Morocco
e-mail: fatimazahra.ouariach@etu.uae.ac.ma

F. Latrache
Laboratory of Modelization and Scientific Calculus, National School of Applied Sciences, University Moham-Med I, Oujda, Morocco

N. Jbira
Department of Physics, Polydisciplinary Faculty of Khouribga, Sultan Moulay Slimane University of Beni Mellal, Khouribga, Morocco

F. Latrache et al. (eds.), *Digital Technology for an Innovative Energy Transition*, Green Energy and Technology, https://doi.org/10.1007/978-3-031-89657-6_1

natural gas fueled the Industrial Revolution and continue to play a dominant role in our energy mix [1]. However, the devastating environmental consequences of their use, particularly climate change and air pollution, have highlighted the urgent need for a transition to more sustainable energy sources [2].

In this context, the energy transition has emerged as a global imperative. This process involves not only a shift from fossil fuels to renewable energy but also a fundamental transformation of our energy production, distribution, and consumption systems [3]. Renewable energies, such as wind, solar, hydro, and geothermal, promise a cleaner and more sustainable energy future [4]. However, their large-scale integration into our existing energy networks poses significant technical and economic challenges [5]. This is where digital transformation comes into play. This technological revolution, characterized by the widespread adoption of technologies such as the Internet of Things (IoT), artificial intelligence (AI), big data, and machine learning, has the potential to revolutionize the energy sector [6]. Digitalization provides powerful tools to optimize renewable energy production, improve energy efficiency, manage real-time demand, and facilitate the integration of decentralized energy sources into the grid [7].

Research on digital transformation in the energy sector has highlighted its transformative potential. Studies have shown how smart grids can enhance the stability and resilience of the power network [8], how AI can optimize renewable energy production by predicting weather conditions [9], and how blockchain technologies can facilitate peer-to-peer energy transactions [10]. However, this research has also revealed significant gaps. First, the ethical and societal implications of these technologies in the energy context remain largely unexplored [11]. Additionally, the accessibility of these technologies in developing countries raises questions about equity and energy justice [12]. Finally, challenges related to privacy protection and cybersecurity in an increasingly connected energy system require careful attention [13]. Our approach in this chapter aims to address these gaps by critically examining the principles and ethics that should guide the implementation of digital transformation in the energy transition.

This chapter explores the principles and ethics of implementing digital transformation in the energy transition. It is structured around four main axes, providing a comprehensive perspective on this crucial intersection. We will begin by exploring the purpose of digital transformation, focusing on the fundamental goals of this revolution and its potential in the energy sector. Next, we will discuss the practical implementation of digital transformation in the energy sector, examining the real-world application of digital technologies in this industry. The third axis will take us from AI ethics to digital transformation ethics, where we will develop an ethical framework specific to the context of the energy transition. Finally, we will conclude with an analysis of the Energy 4.0 case study, examining a concrete example of digital transformation in the energy sector, highlighting its principles, ethical considerations, and limitations.

Through this exploration, we aim to provide a nuanced and critical understanding of how digital transformation can serve the energy transition while remaining true to strong ethical principles. Our goal is to help shape an energy future that is not only sustainable and efficient but also equitable and human-centered.

2 The Purpose of Digital Transformation

Organizational transformation through digital technologies is a holistic process that companies undertake to enhance their existing operations and explore innovative digital solutions to revamp their business models [14]. This transformation is primarily driven by the adoption of emerging digital technologies, such as mobile, artificial intelligence, cloud computing, blockchain, and the Internet of Things (IoT), which enable businesses to significantly improve customer experience, operational efficiency, and develop novel business models [15]. The core aim of digital transformation is to evolve a company's business model by leveraging digital technologies, leading to changes in products, organizational structure, and process automation, ultimately striving to optimize financial resource allocation and increase profit opportunities [16].

Digital transformation is fundamentally an investment in developing new technologies and business models to create value for both companies and consumers within the evolving digital economy [17]. It encompasses the innovative deployment of advanced digital technologies to establish and sustain a value network for various stakeholders, cultivating organizational identity through widespread changes in activities, structure, approaches, and mindsets [18]. Furthermore, digital transformation strategies are designed to generate value for organizations, encouraging them to expand beyond their traditional boundaries and drive purposeful digital transformations [19].

The process of digital transformation is closely intertwined with the creation and diffusion of various advanced technologies, such as blockchain, artificial intelligence, machine learning, the Internet of Things, and virtual/augmented reality, collectively driving organizational change and innovation [20]. In the context of the retail sector, digitalization represents a significant ongoing transformation that is crucial for the industry [21]. The overarching purpose of digital transformation can be categorized into two patterns: excellence and renewal, emphasizing the importance of continuous improvement and innovation [22].

Digital transformation empowers small and medium enterprises (SMEs) to reimagine their decision-making processes and technology applications in meaningful and sustainable ways, particularly highlighted during events like the COVID-19 pandemic [23]. This transformation often involves visible and dramatic changes, including deliberate repositioning and restructuring of organizations and industries [24]. Furthermore, digital transformation plays a vital role in managing transaction relationships within and between companies, especially as intercompany dependencies and environmental uncertainties increase [25].

Digital transformation acts as a catalyst for organizational change, driven by the integration of advanced digital technologies to enhance processes, drive innovation, and create value for stakeholders. It involves a strategic shift toward leveraging digital tools and technologies to optimize operations, improve customer experiences, and foster business growth in the digital age.

Digital transformation is a key driver in the energy sector, leading to advancements in energy production, transmission, and consumption processes [26]. By integrating digital technologies, energy companies can automate the development of new technologies such as smart grids, energy storage systems, and renewable energy solutions, enhancing operational efficiency, and promoting the adoption of clean energy sources [27].

The role of digitalization in the global energy transition is crucial, enabling the energy sector to utilize information and communication technologies to facilitate the shift toward clean energy sources [28]. Through the deployment of digital technologies, energy companies can optimize supply chain efficiency, create new production models, and improve overall operational performance [29–31]. Additionally, digital transformation in the energy sector not only enhances energy efficiency but also actively influences the dynamics of energy transition and emission reduction [32]. Digital transformation within the energy sector has the potential to reduce production costs, increase energy savings, enhance productivity, and unlock innovation [33]. By embracing digitization, energy companies can pursue sustainable development through green innovation and low-carbon transition, contributing to environmental sustainability [29–31]. Furthermore, digital technologies empower energy firms to progress toward carbon neutrality by improving energy conservation practices and diminishing carbon emissions [34].

The impact of digital transformation on the energy transition extends beyond operational efficiency to encompass broader environmental and economic implications. Digital technologies have the potential to reshape the energy industry by restructuring energy consumption patterns, enhancing energy efficiency, and decreasing pollution emissions [35]. Moreover, the rapid development of the digital economy has facilitated more sustainable manufacturing methods, leading to lessened energy usage and environmental pollution in the manufacturing industry [36].

Digital governance plays a pivotal role in driving carbon emission performance, allowing cities and regions to achieve higher value generation while minimizing energy consumption and carbon emissions [37]. While the scale effect of digital transformation may initially lead to an increase in energy demand, the widespread adoption of digital technologies ultimately enhances energy utilization efficiency and reconfigures the energy supply–demand structure [38]. This shift is instrumental in transitioning energy companies from traditional high-energy-consuming practices to green, low-carbon, and sustainable development models [29–31].

Digital transformation within the energy industry serves as a catalytic force propelling the energy transition toward a more eco-friendly and efficient future. Through incorporating digital technologies, energy firms can streamline functions, diminish expenses, boost productivity, and hasten embracing clean energy remedies, in the end contributing to environmental sustainability and lowering carbon emissions.

3 Implementation of Digital Transformation on the Energy Sector

Digital transformation has emerged as a notable catalyst of change within the energy sector, providing diverse advantages such as heightened efficiency, cost reduction, and enhanced customer experience [26]. The integration of digital technologies into energy companies can lead to a more efficient, reliable, and cost-effective energy system for consumers [27]. Research has demonstrated that digital transformation within the energy sector can result in decreased production costs, energy savings, increased productivity, and augmented innovation potential [33]. Advancements in digital technologies combined with declining costs and pervasive connectivity have expedited the digital transformation of the energy sector, especially in domains like the electricity grid [39].

Digital transformation has become a significant driver of change in the energy sector, offering various benefits such as enhanced efficiency, cost reduction, and improved customer experience [26]. The integration of digital technologies in energy companies can lead to a more efficient, reliable, and cost-effective energy system for consumers [27]. Studies have shown that digital transformation in the energy sector can result in reduced production costs, energy savings, increased productivity, and enhanced innovation potential [33]. The advancement of digital technologies, coupled with decreasing costs and widespread connectivity, has accelerated the digital transformation of the energy sector, particularly in areas like the electricity network [39].

Challenges in the energy sector related to digital strategy implementation include issues with strategy communication, evolving business models, and the need for managers to take on new roles as digital leaders within energy organizations [40]. Business model transformation is a key aspect of the ongoing changes in the German energy sector, with barriers and drivers influencing the smart and sustainable transformation process [41]. The transition toward sustainable energy in various countries necessitates parallel processes, one of which involves the digitalization of the energy sector [42]. Innovative approaches to shaping digitalization in the energy sector are crucial for successful business innovation projects. Recommendations for tackling the digital transformation challenge in the energy sector include systematic business innovation processes and collaboration with industry partners [43]. Identifying the key drivers of digitalization in the energy sector is essential for ensuring improvements in sustainability, stability, and economic performance within the industry [44]. The incorporation of digital technologies into energy equipment manufacturing has substantially transformed operations, resulting in enhanced quality control and operational efficiencies [45]. Within the oil industry, collaborations between companies, research institutions, and universities are growing to address challenges and capitalize on digital transformation technologies for drilling and production applications [46]. Furthermore, the adoption of digital tools and platforms in Nigeria's renewable energy sector signifies a transformative shift in human resource management practices, highlighting the impact of digital transformation on organizational strategies [47].

Digital transformation is reshaping the energy sector by driving efficiency, innovation, and sustainability. Adopting digital technologies and evolving business models are imperative for energy companies to thrive within the context of a landscape undergoing rapid change.

4 From AI Ethics to DT Ethics

The ethical implications of artificial intelligence have received considerable scrutiny in recent years. The discipline of AI ethics explores the relevant ethical theories, guidelines, policies, and principles with the goal of ensuring AI systems comport with ethical standards and behave ethically [48]. As AI technologies continue to progress and their applications to proliferate, concerns regarding the ethical consequences of AI have grown more prominent [49]. Issues like bias in decision-making, infringements of privacy, and possible harms from misuse of AI have underscored the importance of incorporating ethical viewpoints into AI development and implementation [50].

One key aspect of AI ethics is the development of frameworks and guidelines to ensure the ethical application of AI across various domains, akin to how biomedical ethics provides frameworks for ethical medical practice [51]. Incorporating ethical principles into AI development processes is crucial to evaluate different approaches for upholding ethical standards, especially in Responsible Living Labs (RLLs) [52]. Furthermore, the integration of ethical AI governance models can illuminate the practical and political implications of ethical AI, highlighting challenges in operationalizing AI ethics [53]. Educational initiatives focusing on AI ethics are also essential. For instance, AI ethics education can help individuals, including students, recognize ethical dilemmas that may arise with AI technology and provide guidelines for responding to these dilemmas [54]. Moreover, the effects of ethics education programs on AI developers can significantly influence the internalization of ethical guidelines and the practice of ethical principles during AI development [55].

The rapidly expanding field of AI ethics requires a multidisciplinary approach to addressing the ethical challenges posed by AI technologies. By developing global ethical standards, integrating ethics into AI development processes, and raising ethical awareness through education, stakeholders can work to ensure that AI technologies are developed and deployed in a responsible, ethical, and sustainable manner.

The shift from AI ethics to digital twin ethics reflects the evolving landscape of digital technologies and their societal impacts. While AI ethics focuses on the moral implications of artificial intelligence systems, digital twin ethics encompasses the unique challenges posed by creating virtual replicas of physical entities or processes:

Expanding Scope

DT ethics builds upon AI ethics but extends to cover issues specific to the creation, management, and use of digital twins. This includes data privacy, consent, and potential misuse of highly detailed virtual representations [56].

Data Integrity and Ownership

A key concern in DT ethics is ensuring the accuracy and appropriate ownership of the vast amounts of data used to create and update digital twins. This raises questions about data governance and the rights of individuals or entities represented by digital twins [57].

Simulation and Prediction Ethics

DT ethics must address ethical implications of using digital twins for simulation and prediction, especially when these predictions might influence real-world decision-making [58].

Integration of Physical and Digital Worlds

As digital twins blur lines between physical and digital realms, ethical frameworks must evolve to address implications of this integration on privacy, autonomy, and human agency [59].

Responsibility and Accountability

DT ethics must grapple with questions of responsibility and accountability when actions in the physical world are influenced by digital twin simulations or predictions [60].

5 Analysis of Case Study of Energy 4.0: Principles, Ethics, and Limitations

Energy 4.0, a concept derived from Industry 4.0, represents the application of advanced digital technologies to the energy sector. This case study examines the implementation of Energy 4.0 in a European region, highlighting its core principles, the ethical, and the limitations of its application.

Principles of Energy 4.0:

- **Interconnectivity**: Energy 4.0 relies on the interconnection of energy systems via the Internet of Things (IoT). In our case study, a network of smart sensors was deployed across the entire regional power grid, enabling real-time monitoring of energy production and consumption [8].
- **Information Transparency**: The collected data is made accessible and comprehensible through advanced user interfaces. Consumers and grid managers can visualize real-time energy flows and associated costs [61].

- **Technical Assistance**: Artificial intelligence systems have been implemented to optimize energy distribution and predict demand peaks, thereby reducing losses and improving the overall efficiency of the system [62].
- **Decentralized Decision-Making**: The system allows for decentralized energy management, with microgrids capable of making autonomous decisions to balance local supply and demand [63].

Ethical:

- **Privacy**: The massive collection of energy consumption data raises concerns about consumer privacy. Encryption and anonymization measures have been put in place, but questions remain about the potential use of this data [64].
- **Equity and accessibility**: The initial implementation of Energy 4.0 widened inequalities between well-connected urban areas and less-equipped rural areas. Efforts were needed to ensure equitable access to the benefits of this technology [12].
- **Cybersecurity**: Increased reliance on digital systems has increased the network's vulnerability to cyberattacks. Significant investment in IT security has proved necessary [65].
- **Autonomy and control**: Increasing automation of the energy system raises questions about the degree of control users should retain over their energy consumption [66].

Limitations:

- **Infrastructure costs**: Setting up the infrastructure needed for Energy 4.0 has required significant investment, posing financing and short-term profitability challenges [67].
- **Technical complexity**: Managing such a complex, interconnected energy system raised significant technical challenges, requiring a highly skilled workforce and constant upgrades [68]
- **System resilience**: although aimed at improving grid stability, the heavy reliance on digital technologies has sometimes made the system more vulnerable to technical failures [69].
- **Social acceptance**: The adoption of these new technologies has met with resistance from some consumers, concerned about the complexity of the system and the perceived loss of control [70].

6 Conclusion and Perspectives

As we conclude this chapter, it becomes evident that the convergence of the energy transition and digital transformation presents both significant opportunities and complex challenges. The integration of advanced digital technologies into the energy sector offers the potential to revolutionize how we produce, distribute, and consume energy. This transformation can lead to a more sustainable and efficient energy system, characterized by increased use of renewable energy, optimized grid management, and enhanced energy security.

However, the journey toward an Energy 4.0 paradigm is not without its obstacles. The ethical implications and societal impacts of digital technologies in the energy context require careful consideration. Issues such as privacy protection, cybersecurity, and equitable access to technology must be addressed to ensure that the benefits of this transformation are shared broadly and fairly. Moreover, the global nature of these challenges calls for international collaboration and the development of universal ethical standards that respect diverse cultural and socio-economic contexts.

Looking ahead, several key perspectives emerge for future exploration:

- **Technological Advancements**: Continued innovation in AI, IoT, and blockchain will likely yield new applications and efficiencies within the energy sector. Research and development should focus on overcoming current technical limitations and expanding the capabilities of these technologies.
- **Ethical Framework Development**: There is a pressing need to develop comprehensive ethical frameworks that can guide the implementation of digital technologies in the energy sector. This includes addressing issues of data privacy, transparency, accountability, and inclusivity.
- **Policy and Regulation**: Governments and regulatory bodies must establish clear guidelines and policies to govern the digital transformation of the energy sector. This includes creating incentives for renewable energy adoption, ensuring data protection, and promoting equitable access to technology.
- **Global Cooperation**: The challenges of the energy transition and digital transformation are global in nature. International cooperation will be essential to share knowledge, resources, and best practices and to develop harmonized standards that can facilitate the global adoption of sustainable energy solutions.
- **Human-Centered Design**: As we move forward, it is crucial to keep human needs and values at the center of technological innovation. This involves not only focusing on technological efficiency but also considering the social, cultural, and ethical implications of new technologies.

The intersection of digital transformation and the energy transition offers a unique opportunity to reshape the future of energy. By embracing a multidisciplinary approach and adhering to strong ethical principles, we can navigate the complexities of this transformation and work toward a future that is sustainable, efficient, equitable, and human-centered.

References

1. IEA (2021) World energy outlook 2021. International Energy Agency, Paris
2. IPCC (2022) Climate change 2022: impacts, adaptation and vulnerability. In: Contribution of working group II to the sixth assessment report of the intergovernmental panel on climate change

3. Smil V (2010) Energy transitions: history, requirements, prospects. ABC-CLIO
4. REN21 (2023) Renewables 2023 global status report. REN21 Secretariat, Paris
5. Gielen D, Boshell F, Saygin D, Bazilian MD, Wagner N, Gorini R (2019) The role of renewable energy in the global energy transformation. Energ Strat Rev 24:38–50
6. Schwab K (2016) The fourth industrial revolution. World Economic Forum
7. Sioshansi FP (ed) (2019) Consumer, prosumer, prosumager: how service innovations will disrupt the utility business model. Academic Press
8. Tuballa ML, Abundo ML (2016) A review of the development of Smart Grid technologies. Renew Sustain Energy Rev 59:710–725
9. Voyant C, Notton G, Kalogirou S, Nivet ML, Paoli C, Motte F, Fouilloy A (2017) Machine learning methods for solar radiation forecasting: a review. Renewable Energy 105:569–582
10. Andoni M, Robu V, Flynn D, Abram S, Geach D, Jenkins D et al (2019) Blockchain technology in the energy sector: a systematic review of challenges and opportunities. Renew Sustain Energy Rev 100:143–174
11. Floridi L, Cowls J, King TC, Taddeo M (2020) How to design AI for social good: seven essential factors. Sci Eng Ethics 26(3):1771–1796
12. Sovacool BK, Dworkin MH (2015) Energy justice: conceptual insights and practical applications. Appl Energy 142:435–444
13. Mylrea M, Gourisetti SNG (2017) Cybersecurity and optimization in smart "autonomous" buildings. In: IEEE international conference on autonomic computing (ICAC). IEEE, pp 151–156
14. Stonehouse G, Konina N (2020) Management challenges in the age of digital disruption.https://doi.org/10.2991/aebmr.k.200201.001
15. Warner K, Wäger M (2019) Building dynamic capabilities for digital transformation: an ongoing process of strategic renewal. Long Range Plan 52(3):326–349. https://doi.org/10.1016/j.lrp.2018.12.001
16. Zhao X (2023) Can digital transformation reduce corporate stock price crashes? PLoS ONE 18(12):e0295793. https://doi.org/10.1371/journal.pone.0295793
17. Wintarto R (2024) Digital organisational culture: capturing local banking in digital transformation. Int J Soc Sci Bus 7(4):1022–1029. https://doi.org/10.23887/ijssb.v7i4.72306
18. Uzule K, Budanceva J (2023) Digital barriers in digital transition and digital transformation: literature review. Econ Culture 20(1):125–143. https://doi.org/10.2478/jec-2023-0011
19. Sarfraz M, Ye Z, Drăgan F, Ivaşcu L, Artene A (2022) Digital transformation strategy and environmental performance: a case study. Int J Comput Commun Control 17(6). https://doi.org/10.15837/ijccc.2022.6.5029
20. Windawi A (2022) The diffusion of blockchain as a general purpose technology driving digital transformation. Technol Innovation Manage Rev 12(1/2). https://doi.org/10.22215/timreview/1478
21. Hagberg J, Sundström M, Egels-Zandén N (2016) The digitalization of retailing: an exploratory framework. Int J Retail Distrib Manage 44(7):694–712. https://doi.org/10.1108/ijrdm-09-2015-0140
22. Móricz P, Marciniak R, Baksa M (2022) Excellence and renewal. Digital transformation patterns in the Hungarian business services sector. Vezetéstudomány 53(5):32–44. https://doi.org/10.14267/veztud.2022.05.03
23. Rupeika-Apoga R, Petrovska K, Bule L (2022) Smes' digital transformation facilitated by covid-19. https://doi.org/10.20944/preprints202201.0340.v1
24. Orlikowski W, Scott S (2023) The digital undertow and institutional displacement: a sociomaterial approach. Organ Theory 4(2). https://doi.org/10.1177/26317877231180898
25. Kim J (2024) How does digital transformation improve supply chain performance: a manufacturer's perspective. Sustainability 16(7):3046. https://doi.org/10.3390/su16073046
26. Nazari Z, Musilek P (2023) Impact of digital transformation on the energy sector: a review. Algorithms 16(4):211. https://doi.org/10.3390/a16040211
27. Akberdina V, Osmonova A (2021) Digital transformation of energy sector companies. E3s Web Conf 250:06001. https://doi.org/10.1051/e3sconf/202125006001

28. Баринова В, Devyatova A, Lomov D (2021) The role of digitalization in the global energy transition. Int Organ Res J 16(4):125. https://doi.org/10.17323/1996-7845-2021-04-06
29. Gao J (2024) Digital transformation path design research of energy enterprise based on space analysis.https://doi.org/10.4108/eai.17-11-2023.2342725
30. Gao J (2024) Research on the evolution mechanism and enhancement strategy of dual synergy of energy innovation system based on multi-agent win-win situation.https://doi.org/10.4108/eai.17-11-2023.2342723
31. Gao Z (2024) Can digital transformation foster green innovation in a "dual-carbon" vision?——evidence from Chinese construction enterprises. https://doi.org/10.21203/rs.3.rs-3644257/v1
32. Wang L (2024) Digitization meets energy transition: shaping the future of environmental sustainability. Energies 17(4):767. https://doi.org/10.3390/en17040767
33. Al-lami A (2024) The digital transformation model for innovative marketing maturity in the oil companies. Rev Gestão Soc Ambiental 18(9):e06535. https://doi.org/10.24857/rgsa.v18n9-056
34. Zhang H (2022) Pathways to carbon neutrality in major exporting countries: the threshold effect of digital transition. Environ Sci Pollut Res 30(3):7522–7542. https://doi.org/10.1007/s11356-022-22592-x
35. Zhao S, Peng D, Wen H, Wu Y (2022) Nonlinear and spatial spillover effects of the digital economy on green total factor energy efficiency: evidence from 281 cities in china. Environ Sci Pollut Res 30(34):81896–81916. https://doi.org/10.1007/s11356-022-22694-6
36. Lin W (2024) The impact of digital economy development on the transformation of manufacturing industry under the background of low-carbon development. Adv Econ Manage Polit Sci 84(1):297–304. https://doi.org/10.54254/2754-1169/84/20240824
37. Wang Y (2023) The role of digital governance on carbon emission performance: evidence from the cities in yangtze river delta, china. Environ Res Commun 5(8):085013. https://doi.org/10.1088/2515-7620/acf2dc
38. Lyu Y, Zhang L, Wang D (2023) The impact of digital transformation on low-carbon development of manufacturing. Front Environ Sci 11.https://doi.org/10.3389/fenvs.2023.1134882
39. Lin J (2023) Dilemma and countermeasures for digital Paris agreement of renewable energy in China under the "dual carbon" goal. https://doi.org/10.20944/preprints202309.0263.v1
40. Trzaska R, Sulich A, Organa M, Niemczyk J, Jasiński B (2021) Digitalization business strategies in energy sector: solving problems with uncertainty under industry 4.0 conditions. Energies 14(23):7997. https://doi.org/10.3390/en14237997
41. Meyer D, Fauser J, Hertweck D (2021) Business model transformation in the German energy sector: key barriers and drivers of a smart and sustainable transformation process in practice. ISPRS Ann Photogramm Remote Sens Spatial Inf Sci VIII-4/W1-2021:73–80. https://doi.org/10.5194/isprs-annals-viii-4-w1-2021-73-2021
42. Loseva O, Karpova S, Rasteryaev K, Sokolova E, Makar S, Kharchilava K (2020) Sustainable energy in island states: comparative analysis of new trends in energy digitalization and the experience of the UK, Japan, Indonesia And Cyprus. Int J Energy Econ Policy 10(6):722–731. https://doi.org/10.32479/ijeep.10571
43. Römer B, Gemsjäger B, Lembo G, Fröhner W (2017) How to shape digitalisation in the energy sector—a new approach for systematic business innovation. Cired—Open Access Proc J 2017(1):2767–2771. https://doi.org/10.1049/oap-cired.2017.0147
44. Światowiec-Szczepańska J, Stępień B (2022) Drivers of digitalization in the energy sector—the managerial perspective from the catching up economy. Energies 15(4):1437. https://doi.org/10.3390/en15041437
45. Zong J (2024) The enhancement of quality control in energy equipment manufacturing with digital technology. Ind Eng Innovation Manage 7(1):13–20
46. D'Almeida A, Bergiante N, Ferreira G, Leta F, Lima C, Lima G (2022) Digital transformation: a review on artificial intelligence techniques in drilling and production applications. Int J Adv Manuf Technol 119(9–10):5553–5582. https://doi.org/10.1007/s00170-021-08631-w

47. James O (2023) A review of the impact of digital transformation on hr practices and strategies in the nigerian renewable energy sector. J Trd Wrld Econ 1(1):36–46. https://doi.org/10.26480/jhcdc.01.2023.36.46
48. Huang C, Zhang Z, Mao B, Yao X (2023) An overview of artificial intelligence ethics. IEEE Trans Artif Intell 4(4):799–819. https://doi.org/10.1109/tai.2022.3194503
49. Morley J, Elhalal A, Garcia F, Kinsey L, Mökander J, Floridi L (2021) Ethics as a service: a pragmatic operationalisation of AI ethics. SSRN Electron J. https://doi.org/10.2139/ssrn.3784238
50. Kazim E, Koshiyama A (2020) A high-level overview of AI ethics. SSRN Electron J. https://doi.org/10.2139/ssrn.3609292
51. Weidener L (2024) Proposing a principle-based approach for teaching AI ethics in medical education. JMIR Med Educ 10:e55368. https://doi.org/10.2196/55368
52. Habibipour A (2024) Responsible living labs: what can go wrong? J Inf Commun Ethics Soc 22(2):205–218. https://doi.org/10.1108/jices-11-2023-0137
53. Ayinla B (2024) Ethical AI in practice: balancing technological advancements with human values. Int J Sci Res Arch 11(1):1311–1326. https://doi.org/10.30574/ijsra.2024.11.1.0218
54. Lee J (2024) Development of a content framework of artificial intelligence integrated education considering ethical factors. Int J Adv Sci Eng Inf Technol 14(1):205–213. https://doi.org/10.18517/ijaseit.14.1.19558
55. Choi J (2024) The effects of an ethics education program on artificial intelligence among middle school students: analysis of perception and attitude changes. Appl Sci 14(4):1588. https://doi.org/10.3390/app14041588
56. Grieves M, Vickers J (2017) Digital twin: mitigating unpredictable, undesirable emergent behavior in complex systems. In: Transdisciplinary perspectives on complex systems: new findings and approaches, pp 85–113
57. Tao F, Zhang H, Liu A, Nee AYC (2019) Digital twin in industry: state-of-the-art. IEEE Trans Industr Inf 15(4):2405–2415
58. Fuller A, Fan Z, Day C, Barlow C (2020) Digital twin: enabling technologies, challenges and open research. IEEE Access 8:108952–108971
59. Pang TY, Perera D, Lim KYH, Lim JYQ (2021) Ethical considerations of digital twins: a systematic literature review. J Eng Tech Manage 62:101626
60. Jones D, Snider C, Nassehi A, Yon J, Hicks B (2020) Characterising the digital twin: a systematic literature review. CIRP J Manuf Sci Technol 29:36–52
61. Strasser T et al (2015) A review of architectures and concepts for intelligence in future electric energy systems. IEEE Trans Industr Electron 62(4):2424–2438
62. Ramchurn SD et al (2012) Putting the 'smarts' into the smart grid: a grand challenge for artificial intelligence. Commun ACM 55(4):86–97
63. Lasseter RH, Paigi P (2004) Microgrid: a conceptual solution. In: IEEE power electronics specialists conference, vol 6, pp 4285–4290
64. McDaniel P, McLaughlin S (2009) Security and privacy challenges in the smart grid. IEEE Secur Priv 7(3):75–77
65. Leszczyna R (2018) Cybersecurity and privacy in standards for smart grids—a comprehensive survey. Comput Stand Interfaces 56:62–73
66. Floridi L et al (2018) AI4People—an ethical framework for a good AI society: opportunities, risks, principles, and recommendations. Mind Mach 28(4):689–707
67. Jamasb T, Pollitt M (2008) Liberalisation and R&D in network industries: the case of the electricity industry. Res Policy 37(6–7):995–1008
68. Gharavi H, Ghafurian R (2011) Smart grid: the electric energy system of the future. Proc IEEE 99(6):917–921
69. Panteli M, Mancarella P (2015) The grid: stronger, bigger, smarter?: Presenting a conceptual framework of power system resilience. IEEE Power Energ Mag 13(3):58–66
70. Mah DNY et al (2012) Consumer perceptions of smart grid development: results of a Hong Kong survey and policy implications. Energy Policy 49:204–216

A Systematic Literature Review on Modernizing Land Acquisition Process for Smart Decision Making in Urban Planning

Ayoub Ouchlif, Hicham Hajji, and Kenza Aitelkadi

Abstract Morocco has adopted an ambitious energy strategy aimed at developing renewable energies, particularly solar and wind energy, and reducing its dependence on fossil fuels. However, the implementation of this strategy often requires the acquisition of land for setting up energy projects. In this context, the country is also committed to achieving the Sustainable Development Goals (SDGs), notably SDG 11, which aims to make cities and human settlements inclusive, safe, resilient, and sustainable. To meet this growing demand for land, various stakeholders in the energy sector resort to land acquisition, once the land identification process has been approved, through the Directorate of State Domains. To handle the increasing demand for land, various actors in the energy sector turn to land experts, working through the Directorate of State Domains once the land identification process is complete. These experts are crucial for any transaction involving state-owned land, such as sales, rentals, purchases, assignments, or concessions, as they determine the market or rental value of the properties. However, the current method used by the administrative commission, which involves gathering a wide range of data on the property and comparing it to similar properties that have recently sold in the same area, has shown some shortcomings in accurately determining market value. In this context, the intelligent modernization of this process is imperative. Artificial Intelligence (AI) offers a valuable opportunity to improve land acquisition process. This chapter aims to provide a comprehensive analysis of studies on the integration of AI in land acquisition. It emphasizes the importance for provincial administrative commissions to adopt AI's intelligent acquistions methods to manage the vast amounts of data generated. Overall, AI offers enormous potential to transform traditional real estate

A. Ouchlif (✉) · H. Hajji · K. Aitelkadi
RU19-Geospatial Technologies for Intelligent Decision Making, Hassan II Agronomic and Veterinary Institute, Rabat, Morocco
e-mail: a.ouchlif@iav.ac.ma

H. Hajji
e-mail: h.hajji@iav.ac.ma

K. Aitelkadi
e-mail: k.aitelkadi@gmail.com

F. Latrache et al. (eds.), *Digital Technology for an Innovative Energy Transition*, Green Energy and Technology, https://doi.org/10.1007/978-3-031-89657-6_2

into smart real estate in Morocco and to significantly shape future practices of land acquisition process.

Keywords Artificial intelligence · Land acquisition · Provincial administrative commission · Energy

1 Introduction

Good urban planning is increasingly crucial in the context of climate change to mitigate the dangers of natural disasters, Effective planning involves comprehensive measures such as the integration of green infrastructure and strategic land use policies that privilege disaster-resistant design, and adopting energy transition practices. With this in mind, Morocco has signed up to international conventions, to the New Urban Agenda 2030, and to all the international mechanisms relating to urban development and planning issues. In particular, the eleventh goal of the SDGs, which aims to build cities that are accessible to all, healthy, intelligent, safe, adaptable, and resilient to disasters, and to take greater account of energy in planning. It is also actively working to incorporate these instruments into its national development strategies. These include digital transformation, which will be the key to the future in the age of massive data (Big Data) and artificial intelligence.

The Digital revolution is now recognized worldwide as the fourth industrial revolution, creating major economic and social opportunities, but also bringing unprecedented changes. The Moroccan real estate sector is one of the most active in the economic fabric, making a significant contribution to GDP and job creation. It is the focus of this strategic vision of digital transformation.

In its 2015 annual report, the Moroccan Court of Accounts raised dysfunctions in the setting of market value by the administrative appraisal commission. However, the correct assessment of property value should be based on objective criteria that take into account material elements (surface area, contents, geographical situation, general condition, soil quality, nature of relief, infrastructure, etc.) and legal means (registered property or property in the process of being registered, available amenities, situation in relation to town planning documents, etc.). Although this commission already lacks a set of rules clarifying how it operates and how it makes decisions and votes, it adopts a traditional method that is often based on the prices revealed by no more than three real estate transactions carried out for certain properties adjacent to or bordering on the property being appraised.

An analysis of recent prices shows that they generally do not reflect the reality of the Moroccan real estate market. The value of appraised land is sometimes lower than the price of comparable properties. The ultimate goal is to harness artificial intelligence algorithms in the Moroccan real estate sector to reliably manage information on land transactions and Moroccan property price indices.

Given the large number of stakeholders involved in observing and monitoring the evolution of the real estate market, and the heterogeneity of available data, intelligent valuation is emerging as a reliable and optimal method for determining market value. Technological developments have ensured that the massive data used in real estate appraisals can be correctly evaluated. This has facilitated the mass processing of real estate transactions, and conversely, the processing of individual properties has become more reliable. Either, the factors influencing value are interrelated. Nevertheless, the process of mobilizing Moroccan public land has shown that geographical location is the only factor determining real estate value.

Over the last few decades, the use of artificial intelligence (AI) techniques has developed rapidly, thanks to the increase in computing power and the availability of massive accessible data. For the case of real estate appraisal of the state's private real estate, the integration of AI techniques seems to offer a novel and revolutionary experience in public administration, as they enable rapid analysis and intelligent, reliable decision-making.

This study aims to provide a comprehensive analysis of studies on the integration of AI in real estate valuation. It emphasizes the importance for provincial administrative expertise commissions to adopt AI's intelligent valuation methods to manage the vast amounts of data generated. Overall, AI offers enormous potential to transform traditional real estate into smart real estate in Morocco and significantly shape future practices of real estate expertise.

2 Sustainable Development Goals (SDGs) and Horizon 2030 in Morocco: Integration and Digital Transformation

2.1 Generalities of Sustainable Development Goals

January 1, 2016, marks the start of the implementation of the 17 Sustainable Development Goals (SDGs) of the 2030 Agenda for Sustainable Development. These goals were adopted at a historic United Nations summit in September 2015. They apply universally to all countries and aim to mobilize efforts to eradicate poverty in all its forms, reduce inequality, and combat climate change while ensuring that no one is left behind. With their 169 sub-goals, these objectives reflect the scope and ambition of the program (Fig. 1).

The Sustainable Development Goals are a call to action for all countries, encouraging them to promote prosperity while preserving the planet. They recognize that poverty eradication must be combined with economic growth strategies that address the needs of society, such as education, health, social protection, and job creation. All this must be achieved in the context of the fight against climate change, biodiversity loss, and environmental protection. The relationship between people and land resources has a direct and indirect impact on all these objectives.

The eleventh SDG "Make cities and human settlements inclusive, safe, resilient and sustainable" aims to make urban centers inclusive and sustainable. In this context,

Fig. 1 Sustainable Development Goals (SDGs)

real estate expertise is of paramount importance, particularly in contributing to intelligent urban planning, the construction of safe and healthy housing, and the efficient management of land resources.

The use of artificial intelligence in real estate appraisal can enhance and enrich the state's land reserve, facilitating the construction of sustainable and resilient cities. For example, using data analysis algorithms and advanced predictive models, AI can help assess real estate market trends, identify at-risk areas, and propose innovative solutions for property management.

What's more, accurate and transparent real estate appraisal, facilitated by AI, can help reduce inequalities in access to housing and promote social inclusion in cities. By fairly assessing the value of real estate assets, real estate appraisal can ensure that the state's land resources are used efficiently and equitably, in line with the principles of sustainable development.

The synthesis of the 11th SDG targets highlights several crucial aspects relevant to our research problem. Firstly, access to decent housing (11.1) and the need to ensure adequate and safe basic services at an affordable cost underline the importance of effective real estate expertise to value property and guarantee housing accessible to all. Secondly, the target on safe, accessible, and sustainable transport (11.2) highlights the role of AI in optimizing urban infrastructure and planning transport systems, taking into account the specific needs of different segments of the population. Thirdly, promoting sustainable urbanization (11.3) requires effective land data management and integrated urban planning, where AI can play a key role in data analysis and decision-making. In addition, heritage preservation (11.4) and disaster prevention (11.5) highlight the importance of accurate, real-time data for risk assessment and crisis management strategies. Finally, the construction of sustainable and resilient buildings (11.c) requires real estate expertise capable of integrating best practices in green building and disaster resilience, where AI can help optimize the design and materials used.

2.2 Sustainable Development Goals in Morocco and Perspectives of Horizon 2030

"[…] The choice of your theme, "Implementing the New Urban Agenda in the Arab Region", attests to our collective commitment to implementing the Urban Agenda as endorsed by the international community at the Habitat III Conference, hosted by Ecuador last October. It also reflects our shared determination to strengthen the Agenda 2030 to achieve the Sustainable Development Goals, in particular the eleventh, which calls for the construction of cities that are inclusive, healthy, intelligent, safe, adaptable and disaster resilient […] Recognizing the scale of today's challenges, Morocco adheres to international conventions, the New Urban Agenda and all international mechanisms relating to urban development and planning issues. It is also actively seeking to incorporate these instruments into its national development strategies." **Message from His Majesty King Mohammed VI to participants at the 2nd Arab Ministerial Forum on Housing and Urban Development, held in Rabat, December 21, 2017.**

The Kingdom of Morocco is firmly committed to the implementation of the New Urban Agenda (**NAU**), adopted by the international community in Quito in 2016. Although remarkable progress has been made in achieving the NAU goals, Morocco recognizes the importance of maintaining the mobilization of all stakeholders to improve and accelerate reforms, particularly with a view to achieving Sustainable Development Goal 11.

The drafting of the first quadrennial report provides an excellent opportunity to demonstrate this will and commitment. As a member of the UN-Habitat Executive Council since 2019, Morocco will strengthen its relations with member countries, in particular African countries, by sharing its national policies and experience in the fields of regional planning, urban development, housing, and urban policy. At the same time, Morocco will take inspiration from the experiences of other countries to ensure the well-being of all, in line with the ultimate goal of leaving no one behind.

The development of an integrated territorial approach to housing policy and programs requires a new instrument capable of supporting this approach and governing relations between the State and local authorities in the field of Housing. The Regional Housing Plan (RHP) offers a territorialized vision of the housing sector, highlighting its needs and priorities.

The RPH proposes concrete resources, particularly in terms of land and finance, to implement priority actions aimed at meeting housing needs. It will also serve as a basis for contractual agreements between the State and local authorities, enabling public housing initiatives and public–private partnerships (PPPs) to be programmed.

3 Digital Transformation: The Integration of AI in the Land Acquisition

The implementation of digital transformation strategies in Morocco, such as the "Digital Morocco 2030" program and the creation of specialized organizations like the Digital Development Agency (DDA) and the National Commission for the Supervision of Personal Data Protection (NCDP), provides a favorable context for the integration of artificial intelligence (AI) in the real estate valuation of the State's private domain.

These initiatives reflect a desire to modernize administrative services and improve the quality of interactions between citizens and administrations. What's more, they aim to increase the economy's productivity and competitiveness, while helping to reduce social and spatial inequalities, objectives in line with the inclusive and responsible digital transformation advocated by the Economic, Social and Environmental Council (ESEC).

However, despite these advances, several challenges remain, including the delay in implementing digital transformation policies in various sectors, the lack of broadband Internet infrastructure, the patchy legislative framework, and the absence of a national roadmap for artificial intelligence.

To overcome these challenges, the Economic, Social, and Environmental Council proposes various measures, such as adopting digital as the default means of public service delivery, putting in place a comprehensive regulatory framework for digitalization, strengthening cyber security and digital sovereignty, and developing a national digital ecosystem.

By integrating artificial intelligence as a national priority in this digital transformation process, Morocco can optimize production costs, meet the differentiated needs of its citizens, and reinforce its progress toward a competitive and inclusive digital economy.

Summary of malfunctions relating to the mobilization of the State's private domain for investment:

The Court of Accounts carried out an audit of the management of the Directorate of State Domains under the Ministry of Economy and Finance, focusing specifically on the use of the State's private domain for investment purposes. In the course of this mission, the Court of Accounts identified dysfunctions compromising the efficiency of State land mobilization. These include both legal and structural shortcomings, as well as deficiencies in land planning, programming, mobilization, and valorization.

In summary, the report highlights several shortcomings in the mobilization of the State's private domain for investment in Morocco:

- Low priority given to land in investment code reforms: Reforms have not sufficiently taken into account the crucial role of land in encouraging both domestic and foreign investment.
- No clear definition of the private domain and its legal regime: The lack of a precise legal definition of the State's private domain complicates its management and development, as well as guaranteeing the State's rights as owner.
- Heterogeneous land tenure structure and lack of completion of legal reorganization: The complexity of the land tenure structure, combined with the lack of completion of legal reorganization, hinders the effective mobilization of State-owned land.
- Lack of a land policy and a structure dedicated to observing land markets: The lack of a planned land policy and structures dedicated to observing land markets limits the government's ability to anticipate land needs.
- Lack of multi-year assessment and programming of land needs: There is a lack of coordination and programming between the ministerial departments concerned to assess future land needs.
- Multiplicity of players and transfer procedures: The transfer procedure is complex and slow due to the multitude of players involved, which undermines the efficiency and transparency of the process.
- Dithering over the adoption of new instruments for land development: It is suggested that new instruments such as public/private partnerships and long leases be explored to optimize land development.

Based on this, the Court of Accounts proposed recommendations and paths around the modernization of tools for mobilizing the private domain of the state for investment. Among these recommendations, we focus on those related to real estate valuation, which are:

1. Evaluation of land expenses: Establish a system for evaluating land expenses to highlight the state's incentive effort for investment.
2. Reform of the investment charter: Reform the investment charter to integrate the land component into investment incentive measures.
3. Observation and monitoring of land and real estate markets: Establish a dedicated framework for observing and monitoring land and real estate markets for better understanding.
4. Standards for rational allocation of state-owned land: Establish standards for rationally allocating state-owned land based on sectors of activity and project scope.
5. Information system dedicated to state-owned land management: Establish an information system dedicated to state-owned land management for better organization and monitoring.
6. Multi-year programming of state-owned land needs: Develop multi-year programming of state-owned land needs for more efficient management.
7. Involvement of the Ministry of Finance in the development of urban planning documents: Involve the Ministry of Finance in the development of urban planning documents for better coordination.

8. Exploration of new legal, technical, and financial instruments: Consider the establishment of new instruments for land reserve reconstitution, in line with a forward-looking approach.
9. Reform of procedures for the transfer of the private domain of the State: Reform transfer procedures to make them adapted to the needs and requirements of investors.
10. Reflection on new modes of transfer of state-owned land: Consider new modes of transfer of state-owned land to meet changing market needs.

Based on these recommendations, and as part of this thesis project section, which aims to develop a roadmap for the integration of artificial intelligence in the cession process, the following section will address the Provincial Administrative Expertise Commission (AEC) in Morocco.

Real estate acquisitions and assignments of interests in private state property are of paramount importance to meet the needs of public administrations in terms of land reserves necessary for the establishment of public facilities such as those dedicated to administration, health, education, etc. Therefore, the rationalization and management of these two processes are of utmost importance to overcome the shortcomings observed in the current procedures.

In this context, the head of government issued circular No. 02-21-cab of 12/02/2021 specifying the procedure to be followed for the management of acquisition and assignment operations carried out by the State Property Department following the provisions of Article 13 of Decree No. 02.07.995 of 23 Chaoual 1429 (23 October 2008) relating to the attributions and organization of the Ministry of Economy and Finance.

4 Critical Analysis of Land Acquisition Process Conducted by the Administrative Expertise Commission (AEC)

The audit mission carried out by the Court of Accounts highlighted that the presence of numerous stakeholders and complex procedures in the land cession or acquisition process considerably hampers the conduct of these operations. Indeed, the cession procedure is complex and slow due to the multitude of actors involved, which undermines the efficiency and transparency of the expertise process. Furthermore, the Prime Minister's circular No. 209 of June 26, 1976, regulating the administrative expertise commission, did not provide a precise definition of expertise and its legal scope. In addition, this commission, as we have already mentioned, does not have a regulation clarifying its mode of operation and the decision-making and voting procedures. Moreover, it did not recommend the measures to be taken in case of objections formulated by one of its members on the proposed price. To overcome and remedy these dysfunctions in this matter, the head of government published circular No. 02-21-cab of 12/02/2021 specifying the procedure to be followed for the management of acquisition and assignment operations. The latter is considered a roadmap

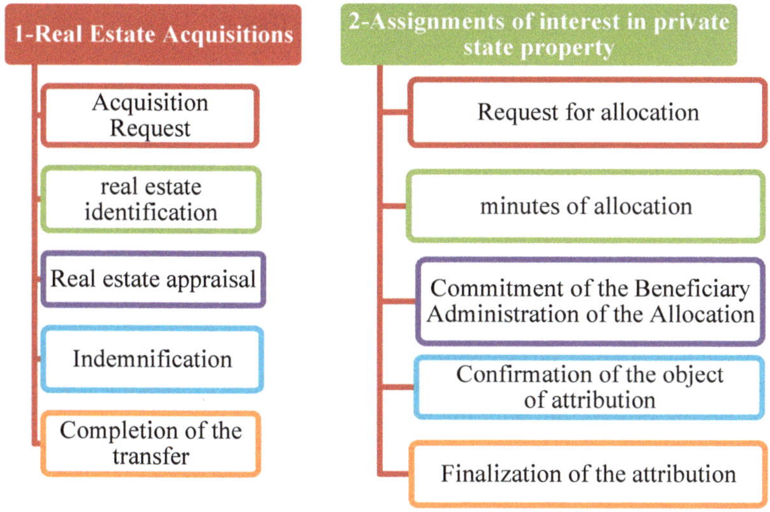

Fig. 2 Land acquisition and attribution process in Morocco

for properly conducting acquisition and transfer procedures by the administrative expertise commission.

Here is an organizational chart outlining the steps to follow for land acquisition and attribution related to the private domain of the state according to circular n° 02-21-cab (Fig. 2):

5 Real Estate Acquisitions

5.1 Submission of Acquisition Requests

As part of an annual program or action plan, administrations submit their acquisition requests to the Budget Directorate of the Ministry of Economy and Finance to obtain approval. Subsequently, the State Property Directorate carries out the acquisitions as stipulated in provisions 2, 3, and 4 of section three of article 3 of decree no. 2.07.995. This occurs after determining the measures related to financing the acquisition and mobilizing the land to avoid any land conflict and ensure effective management of the land subject to acquisition while respecting current legislation.

5.2 Identification of Real Estate Properties

The identification of real estate properties remains the responsibility of an administrative commission when the administration requesting the acquisition has not been able to determine the desired property. This commission, chaired by the governor or their representative, its consists of a representative of the administration requesting the acquisition of the property, a representative of the government authority responsible for urban planning and housing, the land registry service, the directorate of state domains, and a representative of the government authority responsible for agriculture if the property is located outside the urban perimeter. Moreover, the administrative commission meets every three months or as needed to examine the real estate properties proposed within the framework of the action plans of the concerned administrations. Before proceeding with the acquisition, the administration must conduct a technical study of the land. This study is entrusted to the state property directorate in the form of a detailed technical sheet, including land characteristics as well as a list of owners and beneficiaries, justified by a division plan that takes into account the area necessary for the project. A technical division file is also prepared in case of partial acquisition of registered land. Priority is given to lands with a legal and compliant legal situation in accordance with the nature of the project and current urban planning regulations, in order to proceed with an amicable acquisition.

Real estate transactions related to acquisition, conducted by the State Property Directorate, are subject to a real estate appraisal. This appraisal is entrusted to a provincial administrative commission responsible for determining the market value of the property in question.

The composition of the administrative appraisal commission includes:

- The Governor or their representative as president;
- The State Property Delegate or their representative;
- A representative of the tax administration;
- A representative of the administration that submitted the acquisition request;
- A representative of the urban planning authority;
- It is also possible for this commission to include a representative of the government authority responsible for:
- Agriculture, if the land is intended for agricultural use;
- Wakf, if the land is part of the habous domain.
- Additionally, a representative of the concerned ethnic community may also join the commission if the land in question belongs to them.

Upon receiving the request from the State Property Delegate, the Governor or their representative convenes the commission within a period of 15 days. The determination of the market value is based on comparative elements and subjective parameters such as legal situation, surface area, urban planning and architectural dispositions, type of use, proximity to services and amenities, accessibility, and the shape of the property. In general, the commission takes into account all available data collected from other public or private administrations.

In this context, the State Property Directorate transmits to the tax service all data related to the property subject to appraisal (material and legal characteristics). This service then establishes an appropriate list of comparative elements, taking into account the characteristics of the surface area, zoning, geographical location, date, and price of real estate transactions.

If the State Property Directorate finds that the market value set by the administrative appraisal commission does not reflect the real value of the property, it may request an additional meeting for re-evaluation.

The notification of the compensation proposal to the owner is done through legal channels. In case of acceptance of the proposed market value by the owner, the promise of sale is formalized, accompanied by a decision issued by the minister in charge of finances in accordance with paragraph 82 of royal decree no.330.66. Subsequently, the contract is signed in compliance with the provisions of Article 4 of the Code of Real Rights, and the management of the compensation is transferred to the services of the General Treasury of the Kingdom for approval, payment, or consignment of compensations.

In case of refusal of the market value set, land charges on the property, or nonjustification of ownership, expropriation procedures for public utility reasons are initiated in accordance with the current legal provisions. The approval of the General Treasury of the Kingdom is subject to the presentation of the property ownership document.

To sum up, this section presents the process of real estate appraisal in the context of acquisition operations carried out by the State Property Directorate in Morocco. After an in-depth analysis of this process, several shortcomings have been identified:

- Subjectivity of evaluation criteria: Although some objective criteria are considered, such as legal situation and surface area, other elements, such as use and proximity to services, may be interpreted subjectively, which could lead to subjective evaluations.
- Lack of transparency: The process of selecting comparison posts and weighting subjective criteria is not clearly defined, which could compromise the transparency of the evaluation process.
- Absence of specific evaluation methods: The document does not refer to the use of specific real estate evaluation methods, such as the comparison method or the income method, which raises doubts about the rigor of the evaluation.
- Quality control and revision: It is not specified whether quality control or revision mechanisms are in place to ensure the accuracy and reliability of the evaluations carried out by the administrative commission.

In conclusion, although the described process seems well organized and involves collaboration between various government entities, improvements are necessary to ensure transparency, objectivity, and rigor in real estate appraisals.

6 Previous Studies on Real Estate Evaluation by AI and Discussion

Achieving the objectives outlined in this chapter depends on a systematic literature review. According to Jaillet et al. [1], a literature review is a targeted, in-depth, and critical study of the main existing works carried out on a particular topic. It allows mastery of the research domain and the acquisition of primary knowledge on the research work carried out in this domain. This enabled us to collect relevant publications for this research. To do this, we conducted research in journals dealing with land, real estate, and artificial intelligence themes. These reviews are listed in scientific databases such as Google Scholar, Springer, Sciencesdirect, Francis & Taylor, Cairn, Mdpi, and IGI Global. We also targeted journals specialized in the following themes: Journal of Real Estate Research (JRER); Journal of Property Research; Journal of Artificial Intelligence in Real Estate (JAIRE); Journal of Housing Research (JHR); International Journal of Geographical Information Science (IJGIS); Computers, Environment and Urban Systems; Land Use Policy; Journal of Artificial Intelligence Research (JAIR).

This research strategy led us to compile a database of 35 references. Table 1 represents the publications retained after analysis and classification by approach adopted (Table 2).

The inclusion criteria set for the final synthesis are:

- Royal speeches addressing digital transformation;
- Inclusion of targets from the 11th Sustainable Development Goal (SDG);
- Ministerial guidelines, opinions, notes, and circulars related to the real estate sector and access to information;
- Publications dealing with the use of AI, Machine Learning (ML), Deep Learning (DL), Big Data in real estate price prediction processes;
- Publications addressing the exploitation of geospatial data coupled with AI to predict real estate prices.

These queries facilitated the identification of the current integration of AI algorithms in the real estate appraisal process, as well as the type of Big Data resulting from this process.

6.1 *The Advantages of AI in Real Estate Valuation*

Artificial Intelligence (AI) is revolutionizing the real estate industry, particularly in the realm of property valuation. By leveraging advanced algorithms and data processing capabilities, AI offers several specific advantages that enhance the accuracy, efficiency, and scalability of real estate valuation.

Table 1 Target of the 11th SDG

Target	Description
Housing and basic services	11.1: By 2030, ensure access for all, to adequate, safe, and affordable housing and basic services, and clean up slums
Public transport systems	11.2: By 2030, ensure access for all, to safe, accessible, and sustainable transport systems at an affordable cost, by improving road safety, in particular by developing public transport, with particular attention to the needs of people in vulnerable situations, women, children, the disabled, and the elderly
Sustainable urbanization	11.3: By 2030, strengthen sustainable urbanization for all and the capacity for participatory, integrated, and sustainable planning and management of human settlements in all countries
Cultural and natural heritage	11.4: Strengthen efforts to protect and preserve the world's cultural and natural heritage
Resilience to disasters	11.5: By 2030, significantly reduce the number of people killed and the number of people affected by disasters, including water-related disasters, and significantly reduce the amount of economic losses directly caused by such disasters as a proportion of global gross domestic product, with a focus on protecting the poor and vulnerable
Urban air quality and waste management	11.6: By 2030, reduce the negative environmental impact of cities per capita, including by paying particular attention to air quality and waste management, especially municipal waste management
Urban green and public spaces	11.7: By 2030, ensure access for all, especially women and children, the elderly, and people with disabilities, to safe green spaces and public areas
Urban planning	11.a: Promote positive economic, social, and environmental links between urban, peri-urban, and rural areas by strengthening national and regional development planning
Disaster risk management policies	11.b: By 2020, significantly increase the number of cities and settlements that adopt and implement integrated policies and action plans for: inclusive living, resource efficiency, climate change adaptation and mitigation, and disaster resilience, and develop and implement, in line with the Sendai Framework for Disaster Risk Reduction (2015–2030), comprehensive disaster risk management at all levels
Sustainable and resilient buildings	11.c: Help the least developed countries, including through financial and technical assistance, to build sustainable and resilient buildings using local materials

6.1.1 Improved Accuracy

One of the most significant advantages of AI in real estate valuation is its ability to improve the accuracy of price predictions. Advanced algorithms, such as artificial neural networks (ANNs) and deep learning techniques, can capture complex, nonlinear relationships in real estate data that traditional linear models often miss. This capability leads to more accurate and reliable valuations.

Table 2 Summary of previous studies on AI in real estate valuation

Publication	Authors	Approach	Date	Key words	Summary
The application of artificial intelligence in real estate valuation: a systematic review	Nasrin Mahmoud Alsawan Muhammad Turki Alshurideh	AI DL GIS	[2]	Artificial intelligence Real estate Property Valuation Appraisal GIS Satellite imaging	The article offers a comprehensive review of studies published between 2020 and 2022, exploring how artificial intelligence is being used in real estate evaluation. It shows that machine learning models, like Artificial Neural Networks (ANN), can significantly improve the accuracy and reduce errors in property valuations. However, these models need to be tailored to specific research goals and data types to be most effective. Additionally, the article discusses how combining Geographic Information Systems (GIS) and satellite imagery with AI models can further enhance real estate evaluation practices
Artificial intelligence algorithms to predict Italian real estate market prices	Luca Rampini Fulvio Re Cecconi	AI ML	[3]	Artificial intelligence; house market price; real estate; machine learning; deep learning	This study employs three machine-learning techniques to predict housing prices in two Italian cities. The findings indicate that the artificial neural network performs the best, although all models struggle with the most expensive cases due to insufficient data. This approach provides a practical method for predicting housing prices, but further research could enhance its accuracy

(continued)

Table 2 (continued)

Publication	Authors	Approach	Date	Key words	Summary
Identifying real estate opportunities using machine learning	Alejandro Baldominos Iván Blanco Antonio José Moreno Rubén Iturrarte Óscar Bernárdez Carlos Afonso	AI ML	[4]	Real estate; appraisal; investment; machine learning; artificial intelligence	This article introduces a machine learning application designed to spot real-time bargains in the real estate market, specifically homes listed well below their market value. The app caters to investors interested in the real estate market, with a focus on properties in Madrid's Salamanca district listed on a major Spanish housing sales and rentals website. The article details how the application was set up as a regression problem and explores the selection and comparison of several machine-learning algorithms, including regression trees, k-nearest neighbors, support vector machines, and neural networks
Research on urban land price assessment based on artificial neural network model	Cai shousong gu xiaomin wang xiaoguang Chen ying	AI ANN	[5]	Artificial intelligence; smart city; land price assessment	To tackle challenges like limited data on residential land prices and imbalanced class distribution, this study introduces a land price feature extraction method using transfer learning. Various land price evaluation models are employed to assess accuracy and identify suitable models and methods. Experimental results demonstrate that the proposed transfer learning-based algorithm outperforms traditional methods such as principal component analysis and linear normalization, with average accuracy improvements of 10.9% and 4.73%, respectively. This algorithm effectively meets the precision needs of real-world residential land price evaluation

(continued)

Table 2 (continued)

Publication	Authors	Approach	Date	Key words	Summary
Artificial intelligence in price estimation of real estate	Vykhodtsev, N. A	AI, ML	[6]	IA; ML; Modèle Prédictif; immobilier	The study examines information systems used in real estate sales to identify those that incorporate artificial intelligence. It includes a correlation analysis of real estate features and compares AI algorithms using coefficients of determination, errors, and deviations. Experiments are conducted to fine-tune the algorithms' hyperparameters and identify features that positively impact model fitting
Application of artificial intelligence for predicting real estate prices: the case of Saudi Arabia	Elham Alzain Ali Saleh Alshebami Theyazn H. H. Aldhyani Saleh Nagi AlsUbari	ANN	[7]	Artificial intelligence; investment; prediction house time series; prediction model	This study presents the use of Artificial Neural Networks (ANN) to accurately predict real estate prices in Saudi Arabia. The results demonstrate a strong correlation between experimental and predicted values, highlighting the high performance of the developed model. This approach can inform government decisions and real estate investments, addressing a critical need in the market
A survey on application of artificial intelligence in real estate industry	Woubishet Zewdu Taffese	ANN	[8]	Real estate valuation; artificial intelligence; artificial neural networks; expert systems; hybrid systems	This study explores the use of artificial intelligence (AI) in the real estate industry, with a focus on artificial neural networks (ANN) and expert systems (ES) for property valuation. It compares the predictive performance of ANN with that of multiple regression analysis (MRA) models and discusses the shared features of ANN and ES. Additionally, it investigates the application of hybrid systems for property valuation, which leverage the strengths of different intelligent systems

(continued)

Table 2 (continued)

Publication	Authors	Approach	Date	Key words	Summary
Review on the application of artificial neural networks in real estate valuation	Nkolika J. Peter Hilary I. Okagbue Emmanuela C. M. Obasi Adedotun O. Akinola	ANN	[9]	Appraisal; artificial neural networks; price prediction; real estate valuation; risk assessment	This paper aims to tackle the complexity of real estate valuation by highlighting the advantages of artificial neural networks, an innovative model for property valuation. The literature review shows that neural network models outperform traditional methods. This work opens up new perspectives and enhances the understanding of property valuation principles
Real estate valuation models performance in price prediction	Adela DEACONU Anuța BUIGA Helga TOTHĂZAN	ANN	[10]	Real estate; valuation reports review; artificial neural networking; price prediction	Using data from 900 apartments in Cluj-Napoca, Romania, our study compares the predictive performance of Artificial Neural Networks (ANN) and Generalized Linear Models (GLM) for price prediction. The results confirm the superiority of ANN in predicting sales prices with greater stability, aligning with previous studies. Additionally, the study addresses criticisms about result transparency by demonstrating ANN's ability to highlight the significance of different real estate attributes using appropriate statistical indicators. These findings have implications for various real estate valuation purposes, including the revision of valuation reports

(continued)

Table 2 (continued)

Publication	Authors	Approach	Date	Key words	Summary
Prediction on housing price based on deep learning	Li Yu Chenlu Jiao Hongrun Xin Yan Wang Kaiyang Wang	DL	[11]	Deep learning; Convolutional neural network; LSTM; housing prediction	This study proposes the use of deep learning to develop housing price prediction models using two approaches: one centered on real estate features, comparing CNN and LSTM models to logistic regression, and another focused on time series, comparing an LSTM model to ARMA. Through a comprehensive study of housing prices in Beijing, we created an optimal model for evaluating and predicting prices in the real estate industry
Real estate properties assessment using deep neural network	A. Shinde N. Dange N. Patane S. Gholap V. Beera	DL	[12]	Deep neural network; layered knowledge graph; structurally connected; deep learning approach	The study employs Deep Neural Networks (DNN) to effectively assess real estate values by creating a layered knowledge graph. This structured DNN approach requires less training data and learns from the most recent data, enhancing its accuracy and efficiency
Deep learning approach in predicting property and real estate indices	Seng Hansun, Alethea Suryadibrata, Donn Rithalna Sandi	DL	[13]	Bi-LSTM networks; deep learning; LQ45; property and real estate; stock price prediction	In this study, they employed bidirectional LSTM networks to predict the closing prices of six real estate stocks, achieving reasonable results with accuracies of up to 90% for some stocks

(continued)

Table 2 (continued)

Publication	Authors	Approach	Date	Key words	Summary
Investigations on deep learning techniques for property assessment	Junchi Bin Zheng Liu Eric Li Jian Liu Dimitry Sediako, Nicolas Miguel Peleato	DL	[14]	DL; ANN; property assessment; urban data	The thesis explores the application of deep learning techniques to tackle three specific challenges in real estate valuation: peer dependence, topography, and the integration of urban data. Three studies are conducted, each proposing an innovative method to address these challenges. The first study introduces a peer-dependent valuation model using a recurrent neural network (RNN). The second study proposes a multimodal fusion method based on attention mechanisms to estimate real estate values by integrating topographic data. Finally, the third study presents a deep learning framework for merging multi-source urban data in real estate valuation. The experimental results of each study demonstrate the effectiveness of the proposed approaches
Housing prices prediction with deep learning: an application for the real estate market in Taiwan	Choujun Zhan Zeqiong Wu Yonglin Liu Zefeng Xie Wangling Chen	DL	[15]	Deep learning; housing prices; prediction; time series	The study focuses on real estate price prediction in Taiwan between 2013 and 2018, noting that price fluctuations are influenced by numerous factors and exhibit nonlinear relationships that classical methods struggle to capture. To address this, the study proposes models based on deep learning, specifically Backpropagation Neural Networks (BPNN) and Convolutional Neural Networks (CNN). The results indicate that CNN, when incorporating housing features, provides the most accurate predictions. This approach not only enhances the understanding of the real estate market but also facilitates effective interventions

(continued)

Table 2 (continued)

Publication	Authors	Approach	Date	Key words	Summary
Deep learning with XGBOOST for real estate appraisal	Yun Zhao Girija Chetty Dat Tran	DL	[16]	Deep neural network; real estate; image assessment; visual content analysis; EXtreme gradient; boosting	This paper explores the use of deep learning combined with Extreme Gradient Boosting (XGBoost) to predict real estate prices. By analyzing historical sales data and online house images, the study found that this combined approach improves the performance of house price predictions
Deep learning et agriculture	Chaire AgroTIC	DL	2018	Deep learning; agriculture	Deep Learning represents a promising innovation in the agricultural field, but its adoption still requires efforts to explain its results and build the necessary trust. Challenges related to regulation and supporting farmers need to be addressed to ensure the successful integration of these technologies
A literature survey on housing price prediction	Sushant Suresh Yalgudkar N. V. Dharwadkar	DL	[17]	Housing price; prediction; OLS; gradient boosting; deep learning	This study aims to predict real-time real estate prices in Pune using data from websites such as 99acres.com and magicbricks.com. It employs regression techniques, including Multiple Linear Regression (OLS), Lasso, and XGBoost, to create predictive models and compares their accuracy to identify the best approach
A machine learning approach to price indices applications	Felipe D. Calainho Alex M. van de Minne Marc K. Francke	ML	[18]	Commercial real estate; price indices; machine learning	This study analyzes data from 29,998 commercial real estate transactions in New York between 2000 and 2019. The results show that machine learning (ML) models have higher prediction accuracy than linear models. However, ML algorithms are more dependent on the data used for calibration, producing less stable results with small samples and potentially exhibiting estimation bias

(continued)

Table 2 (continued)

Publication	Authors	Approach	Date	Key words	Summary
Data analysis on the influencing factors of the real estate price	Huang-Mei He Yi Chen Jia-Ying Xiao Xue-Qing Chen Zne-Jung Lee	ML	[19]	Data analysis; real estate price; decision tree; random forest	The article proposes a data analysis method to identify factors influencing real estate prices. The method begins with data cleaning and conversion. It then discretizes real estate prices using ranges based on the mean and standard deviation and establishes decision tree and random forest models for six different scenarios. When the dataset is divided into training data (70%) and test data (30%), the highest test accuracy is achieved. By analyzing the importance of each input variable, the main factors influencing real estate prices are identified as cost, interior decoration, location, and condition. The results suggest that the real estate industry and buyers should focus on these factors when adjusting or purchasing properties
Housing price prediction using machine learning algorithms in COVID-19 times	Raul-Tomas Mora-Garcia, Maria-Francisca Cespedes-Lopez V. Raul Perez-Sanchez	ML	[20]	ML; mass appraisal; real market; partial dependence plots; COVID-19	This study demonstrates that machine learning algorithms outperform traditional linear models in analyzing complex real estate market data, thanks to their ability to adapt to nonlinearities. Models based on bagging tend to overfit, while those based on boosting show better performance. This research is one of the first to use machine learning and microdata to study the impact of the COVID-19 pandemic on housing prices in Spain

(continued)

Table 2 (continued)

Publication	Authors	Approach	Date	Key words	Summary
Housing price prediction via improved machine learning techniques	Quang Truong Minh Nguyen Hy Dang Bo Mei	ML	[11]	Housing price; prediction; hybrid regression; ML; stacked generalization	This study aims to enhance the accuracy of housing price predictions by leveraging advanced machine learning techniques. The authors propose a hybrid regression model that combines multiple machine learning algorithms through stacked generalization. This approach is designed to capture the complex relationships and nonlinearities inherent in real estate data, thereby improving predictive performance
Housing prices: testing machine learning methods	Isabelle Chardon Francisco Javier Martínez Concha Alexandre Bergel	ML	[21]	Machine learning; housing prices; importance of variables	This article applies artificial neural networks, support vector machines, and random forests to predict real estate values in Santiago, Chile. While the current literature on using these algorithms for land value estimation is limited, they appear well-suited for modeling the complexity of the problem. After comparing the three algorithms, the study concludes that the random forest model is the most effective for predicting values and evaluating the relative importance of different variables. This algorithm is applied to the entire dataset and then differentiated by income to assess the model's performance

(continued)

Table 2 (continued)

Publication	Authors	Approach	Date	Key words	Summary
Machine learning building price prediction with green building determinant	Thuraiya Mohd Syafiqah Jamil Suraya Masrom	ML	[22]	Algorithm; green building; ML; phyton; prediction; skirt-learn	This article presents an empirical study aimed at predicting construction prices based on green building features and other common determinants. The experiments utilized five common machine learning algorithms—linear regression, decision tree, random forest, Ridge, and Lasso—and tested them on a real dataset covering the district of Kuala Lumpur, Malaysia. The results demonstrated that the random forest algorithm outperformed the other four on the tested dataset and that the green building determinant had a significant positive impact on the model's performance
Machine learning for spatial analyses in urban areas: a scoping review	Ylenia Casali Nazli Yonca Aydin Tina Comes	ML	[23]	ML; urban areas; review; spatial analyses; geospatial data	This article presents a review of studies that use machine learning with geospatial data to analyze urban areas. It highlights the most commonly studied topics, data sources, machine learning methods, and parameter selection approaches, as well as the challenges faced. This analysis identifies knowledge gaps in machine learning methods for spatial data science, providing guidance for future research

(continued)

Table 2 (continued)

Publication	Authors	Approach	Date	Key words	Summary
Machine learning for property price prediction and price valuation: a systematic literature review	Nur Shahirah Ja'afar Junainah Mohamad Suriatini Ismail	ML	[24]	Machine learning; real estate; property price prediction, valuation	This study examines the use of machine learning techniques for predicting and evaluating real estate prices. It aims to identify the best solutions for predicting property price indices to help decision-makers assess the overall economic situation. The results indicate an increasing adoption of these methods, with the random forest algorithm emerging as the most successful due to its better adaptability to available data
Machine-learning-based prediction of land prices in Seoul, South Korea	Ping-Feng Pai Wen-Chang Wang	ML	[25]	Real estate prices; machine learning; predict	Both studies employ four machine learning models to predict real estate prices: Least Squares Support Vector Regression (LSSVR), Classification and Regression Trees (CART), General Regression Neural Network (GRNN), and Backpropagation Neural Network (BPNN). The numerical results indicate that LSSVR outperforms the other three models in terms of prediction accuracy, delivering superior prediction results compared to previous studies. Therefore, the machine learning-based model, particularly LSSVR, is a substantial and feasible method for predicting real estate prices, offering relatively competitive and satisfactory results

(continued)

Table 2 (continued)

Publication	Authors	Approach	Date	Key words	Summary
Housing price prediction using machine learning algorithms: The Case of Melbourne City, Australia	The Danh Phan	ML	[26]	House price prediction; RT; NN; SVM; stepwise; PCA	The literature explores the use of machine learning techniques to analyze real estate transactions in Australia, revealing large price disparities between the most expensive and affordable suburbs of Melbourne. The experiments demonstrate that the combination of Stepwise regression and Support Vector Machine (SVM) provides a competitive approach based on the mean squared error measure
Predicting property prices with machine learning algorithms	Winky K. O. Ho Bo-Sin Tang Siu Wai Wong	ML	[1]	ML; algorithms; SVM; RF; GBM; property valuation	This study compares three machine learning algorithms for evaluating real estate prices in Hong Kong: Support Vector Machine (SVM), Random Forest (RF), and Gradient Boosting Machine (GBM). RF and GBM outperform SVM in terms of predictive power, demonstrating superior performance in Mean Squared Error (MSE), Root Mean Squared Error (RMSE), and Mean Absolute Percentage Error (MAPE). Despite this, SVM remains useful for adjusting data and providing accurate predictions within tight deadlines. In conclusion, machine learning offers a promising alternative for real estate price prediction

(continued)

Table 2 (continued)

Publication	Authors	Approach	Date	Key words	Summary
Real estate market price prediction framework based on public data sources with case study from Croatia	Leo Mrsic(B) Hrvoje Jerkovic Mislav Balkovic	ML	[27]	Real estate price; prediction; machine learning; real estate prices	This study employs machine learning algorithms to develop a model for predicting apartment prices in Zagreb, Croatia. The analysis is based on real estate ads from the largest Croatian online service, njuskalo.hr, collected between April and May, totaling around 8000–9000 ads each time. Using data from 7416 apartments in Zagreb, the study developed a price prediction model utilizing algorithms such as Random Forest, GradientBoosting, AdaBoost, and XGBoost. The XGBoost model stands out for its superior accuracy in predicting real estate prices
Real estate pricing models using machine learning schemes	Ingrid Rosales Carlos Hernández	ML	[28]	Machine learning scheme; artificial neural network; real estate market; cross validation; regression analysis	This study compares three real estate price prediction models using artificial neural networks. The models were developed and evaluated on a set of 228 properties using error measures such as Mean Absolute Percentage Error (MAPE), Mean Absolute Error (MAE), and Root Mean Squared Error (RMSE). The results show that all models achieved a MAPE between 14 and 16%, demonstrating the effectiveness of machine learning schemes for predicting real estate market prices

(continued)

Table 2 (continued)

Publication	Authors	Approach	Date	Key words	Summary
The use of machine learning in real estate research	Lennon H. T. Choy Winky K. O. Ho	ML	[29]	Extra trees; k–nearest neighbors; random forest; property prices	This study demonstrates how machine learning can provide more accurate price predictions in the real estate sector. Using nearly 25,000 transaction records, the study compares the effectiveness of three machine learning algorithms with a traditional hedonic pricing model. The results indicate that machine learning algorithms significantly outperform classical statistical techniques in terms of accuracy and explanatory power. However, the growing adoption of machine learning raises ethical and social concerns that require special attention
Using machine learning algorithms for predicting real estate values in tourism centers	Tansu Alkan Yeşim DOKUZ Alper ECEMIŞ Aslı BOZDAG Süleyman Savaş DURDURAN	ML GIS	[30]	Real estate valuation; machine learning; algorithm; k-nearest neighbors; random forest; support vector machines; GIS	This study employs artificial intelligence techniques, specifically machine learning, to objectively assess real estate properties in the Alanya district of Turkey, an area significantly influenced by tourism. Real estate value criteria are determined, and data are collected from real estate websites. The property values are predicted using various machine learning algorithms. The results demonstrate that Support Vector Machines (SVM) provide the most accurate predictions, highlighting the applicability of machine learning techniques to real estate valuation

(continued)

Table 2 (continued)

Publication	Authors	Approach	Date	Key words	Summary
Machine Learning Applications to Land and Structure Valuation	Michael Mayer Steven C. Bourassa Martin Hoesli Donato Scognamiglio	ML DL	[31]	Land and structure valuation; machine learning; structured additive regression; gradient boosting; deep learning; interpretability; transparency; hedonic modeling	In this example, machine learning techniques enhance predictive accuracy in hedonic real estate valuation but may lack interpretability. Structured additive regression models (STAR) offer a solution by combining the accuracy of machine learning techniques with improved interpretability. We illustrate how STAR models can be used for supervised dimension reduction and to transparently describe the effects of covariables. This approach is applied to the valuation of residential land and structures, yielding promising results in terms of both interpretability and predictive performance
Systematic review of deep learning and machine learning for building energy	Sina Ardabili Leila Abdolalizadeh Csaba Mako Bernat Torok Amir Mosavi	ML DL	[32]	Deep learning; machine learning; building energy; energy demand; energy consumption; smart grid; Internet of things; data science	This study examines the performance of machine learning (ML) and deep learning (DL) techniques in building energy systems (BE). Hybrid and ensemble models prove to be the most robust for predicting energy demand, while DL and Support Vector Machine (SVM) methods stand out for predicting energy consumption. Models based on linear regression are found to be less robust
Estimating the price of real estate properties with the help of online images	Dasari Ashok Chaganti Lavanya	RNN	[33]	Visual content analysis; real estate; deep neural networks; property value; estimation	Real estate investment evaluation is complex for buyers, but new opportunities are emerging with the exploitation of mobile data and online images. Real estate agents play a crucial role in providing easy access to property information. This study examines the predictive power of online housing forecasts using computer vision algorithms

AI models can also be customized and adjusted based on specific research objectives, purposes, and data types. This adaptability ensures that the models are fine-tuned to the unique characteristics of different real estate markets, further enhancing their accuracy. For instance, studies have shown that algorithms like Random Forest, XGBoost, and deep neural networks (DNNs) often outperform traditional methods in terms of accuracy and explanatory power.

6.1.2 Enhanced Efficiency

AI models can automate the valuation process, significantly reducing the time and resources required for real estate appraisal. This automation allows for faster and more efficient processing of large volumes of data. AI techniques can also efficiently handle data cleaning, conversion, and discretization. For example, the method proposed by He et al. [19] demonstrates how AI can preprocess data to identify key factors influencing real estate prices, streamlining the valuation process.

Moreover, AI models can learn from the most recent data, ensuring that valuations are up-to-date and reflect current market conditions. This capability is particularly useful in dynamic real estate markets where prices can fluctuate rapidly.

6.1.3 Ability to Handle Large Datasets

AI techniques, particularly deep learning models, are designed to handle large, complex datasets. They can process vast amounts of data efficiently, making them ideal for real estate valuation where large datasets are common. AI models can also integrate various data types, including geospatial data, urban data, and even satellite imagery. This integration enhances the comprehensiveness of the valuation process, leading to more accurate and reliable predictions.

For instance, the deep neural network approach by Shinde et al. [12] demonstrates how AI can effectively evaluate real estate values using structured knowledge graphs, requiring less training data and learning from the most recent data. This ability to learn from big data is a significant advantage in the real estate industry, where the availability of large datasets is increasingly common.

In summary, AI offers several specific advantages for real estate valuation, including improved accuracy through advanced algorithms, enhanced efficiency through automation and data preprocessing, and the ability to handle large datasets through scalability and integration of diverse data sources. These advantages make AI a powerful tool for real estate professionals, enabling them to make more informed and timely decisions.

However, it is essential to address challenges such as the need for large datasets, potential estimation bias, interpretability issues, and ethical considerations to fully leverage the benefits of AI in real estate valuation. Future research should aim to improve AI models while ensuring they meet the needs of real estate professionals and comply with ethical and legal standards.

6.2 Expediting Land Acquisition for Renewable Energy Projects

Morocco's transition to renewable energy is a vital part of its sustainable development strategy. One of the biggest hurdles in this transition is the efficient and timely acquisition of land suitable for renewable energy projects. Improved valuation methods, particularly those powered by AI, can significantly speed up this process.

- **Improved Accuracy and Reliability:**

  Precise Valuations: AI-driven valuation models can provide more accurate and reliable property valuations. By analyzing large datasets and capturing complex relationships in the data, AI models can generate precise estimates of land values. This accuracy helps in negotiating fair prices with landowners, reducing the time spent on price negotiations and expediting the acquisition process.
  Identification of Key Factors: AI can identify the main factors influencing land prices, such as location, topography, access to infrastructure, and environmental conditions. This detailed analysis helps in selecting the most suitable and cost-effective sites for renewable energy projects, ensuring that the land acquisition process is efficient and targeted.

- **Enhanced Efficiency:**

  Automated Valuations: AI models can automate the valuation process, significantly reducing the time required for appraisals. This automation allows for faster processing of large volumes of data, enabling quicker decision-making and expediting the land acquisition process.
  Real-Time Updates: AI models can learn from the most recent data, ensuring that valuations are up-to-date and reflect current market conditions. This capability is particularly useful in dynamic real estate markets where prices can fluctuate rapidly. Real-time updates help in making timely decisions and securing land before prices escalate.

- **Handling Large Datasets:**

  Scalability: AI techniques, particularly deep learning models, are designed to handle large, complex datasets. They can process vast amounts of data efficiently, making them ideal for real estate valuation where large datasets are common. This scalability ensures that valuations are comprehensive and reliable, even when dealing with extensive land areas.
  Integration of Diverse Data Sources: AI models can integrate various data types, including geospatial data, urban data, and even satellite imagery. This integration enhances the comprehensiveness of the valuation process, leading to more accurate and reliable predictions. For renewable energy projects, this means that AI can consider a wide range of factors, such as solar irradiation, wind patterns, and environmental impact, to identify the most suitable sites.

- **Supporting Morocco's Energy Transition**

  Optimized Land Use: By providing accurate and efficient valuations, AI can help optimize land use for renewable energy projects. This optimization ensures that the most suitable and cost-effective sites are selected, maximizing the potential for renewable energy generation and supporting Morocco's energy transition goals.
  Reduced Time and Costs: The expedited land acquisition process enabled by AI-driven valuation methods reduces the time and costs associated with project development. This efficiency allows for faster deployment of renewable energy infrastructure, accelerating Morocco's transition to clean energy.
  Enhanced Decision-Making: AI models provide detailed and comprehensive analyses of land values and suitability for renewable energy projects. This information supports informed decision-making by policymakers, developers, and investors, ensuring that resources are allocated effectively and that projects are aligned with Morocco's energy transition strategy.

6.2.1 Potential Long-Term Benefits of Modernizing Real Estate Valuation for Morocco

Modernizing real estate valuation through the adoption of advanced technologies, such as AI, can bring about significant long-term benefits for the broader economy and society in Morocco. These benefits extend beyond the real estate sector, influencing various aspects of the country's development and sustainability (Table 3):

6.3 Discussion

After conducting and examining a systematic literature review of various AI techniques for real estate evaluation and prediction, several general remarks can be made. Regarding artificial neural networks, the literature highlights their high performance, which depends on the techniques used for calibration. However, it is observed that these networks produce less stable results when applied to small samples and may introduce estimation bias. Additionally, the complexity of ANN models can limit their interpretability and adoption by real estate professionals. In general, the studies conducted do not propose generalizable solutions. Each study seems to focus on real estate evaluation in a single scenario, which limits the identification of the problem's characteristics and the proposal of generalizable methods.

During the analysis of previous studies, it appeared that 16 studies used various machine learning algorithms, such as decision trees, random forests, and support vector machines, for real estate evaluation. The results vary depending on the algorithm and context, but it is generally observed that machine learning (ML) models are more accurate than traditional methods. However, Shinde et al. [12] used a deep neural network (DNN) to effectively evaluate real estate values by creating a layered

Table 3 Long-term benefits

Category	Benefits
Economic growth and investment	– Attracts foreign investment through reliable and transparent valuations – Stimulates domestic investment by providing clear and reliable data
Urban development and planning	– Optimizes land use by identifying suitable and cost-effective sites – Supports sustainable urban planning through comprehensive data integration
Energy transition and sustainability	– Expedites land acquisition for renewable energy projects – Promotes green buildings and energy efficiency
Social and community benefits	– Supports affordable housing initiatives – Fosters community development and economic revitalization
Enhanced decision-making and governance	– Supports informed policy decisions through detailed market analyses – Enhances transparency and accountability in the real estate sector
Technological innovation and skills development	– Stimulates technological innovation and new business opportunities – Drives skills development and educational programs in technology and data science

knowledge graph. This structured DNN requires less training data and learns from the most recent data, an approach validated by Bin [14]. These researchers examined the use of deep learning techniques to address three specific challenges in real estate valuation: peer dependence, topography, and integration of urban data. Three studies were conducted, each proposing an innovative method to address these challenges. The experimental results of each study demonstrated the effectiveness of the proposed approaches. However, Nasrin Mahmoud Alsawan et al. [2] highlighted the sensitivity of machine learning models to research objectives, purpose, and data types, indicating the need for adjustments and customization. They also discussed the integration of Geographic Information Systems (GIS) and satellite imagery with AI models.

According to Calainho et al. [18], machine learning (ML) models demonstrate higher prediction accuracy than linear models. However, these machine-learning algorithms rely more heavily on the data used for calibration; they produce less stable results when applied to small samples and may exhibit estimation bias. To address this challenge, He et al. [19] proposed a data analysis method to identify factors influencing real estate prices. This method starts by cleaning and converting the data used. Then, it discretizes real estate prices using ranges based on the mean and standard deviation and establishes decision tree and random forest models for six different situations. When the dataset is divided into training data (70%) and test data (30%), it achieves the highest test accuracy. By observing the importance of each input variable, the main factors influencing real estate prices are identified as cost, interior decoration, location, and condition3. These results suggest that the real estate industry and buyers should pay attention to these factors to adjust or purchase real estate properties.

Other studies, such as Mora-Garcia et al. [20], used microdata and machine learning for the first time to study the impact of the COVID-19 pandemic on real estate prices in Spain. They demonstrated that machine-learning algorithms outperform traditional linear models in analyzing complex real estate market data due to their adaptation to nonlinearities. Models based on bagging show problems with overfitting, while those based on boosting exhibit better performance.

The challenge of exploiting spatial data is evident given the multiplicity of data sources. This challenge, as well as the most commonly used machine learning methods and parameter selection approaches, was highlighted by Casali et al. [23], who presented a review of the scope of studies on machine learning using geospatial data to analyze urban areas.

Machine learning offers a promising alternative for real estate price prediction. Mrsic(B) et al. [27] delved into the parameter selection of their study where they developed a model for predicting apartment prices in Zagreb, Croatia. Based on algorithms such as Random Forest (RF), Gradient Boosting AdaBoost, and XGBoost, their model analyzed real estate ads from the largest Croatian online service, njuskalo.hr, between April and May, collecting around 8000–9000 ads each time. Based on data from 7416 apartments in Zagreb, the XGBoost model stood out for its superior accuracy in predicting real estate prices. Thuraiya [22] confirmed in their empirical study aiming to predict construction prices based on green buildings and other common determinants. They used five common machine-learning algorithms, namely linear regression, decision tree, random forest, Ridge, and Lasso, tested on a real dataset covering the district of Kuala Lumpur, Malaysia. The results showed that the Random Forest (RF) algorithm outperformed the other four on the tested dataset and that the green building determinant contributed to promising effects on the model.

In the same vein, Ja'afar et al. [24], in order to help decision-makers better assess the overall economic situation, examined the use of machine learning techniques for predicting and evaluating real estate prices. They identified the best solutions for predicting price indices, with the results showing an increase in the use of this method, with Random Forest as the most successful algorithm due to its better adaptability to available data.

The study conducted by Choy et al. [29] highlights the ability of machine learning to provide more accurate price predictions in the real estate sector. By collecting 25,000 transaction records, they compared the effectiveness of three machine-learning algorithms with a traditional hedonic pricing model. The results clearly indicate that machine-learning algorithms significantly outperform classical statistical techniques in terms of accuracy and explanatory power. However, the growing adoption of machine learning raises ethical and social concerns that require special attention.

To guide future research, this study proposes some guidelines. It is essential to combine different methods and increase the amount of data while seeking to simplify techniques without compromising their performance. The goal is to implement solutions that are generalizable to different real estate markets.

In conclusion, the exploration of artificial intelligence techniques for real estate evaluation reveals significant potential to improve the accuracy of price predictions. Machine learning models, such as artificial neural networks and deep learning, show promising performance, although some challenges remain, notably regarding the need for big data and the stability of results. Despite these challenges, the growing adoption of AI in the real estate sector raises ethical and social concerns, highlighting the need for responsible and ethical use of these technologies. For the future, it is essential to continue research aimed at improving AI models while ensuring that they meet the needs of real estate professionals and comply with ethical and legal standards.

7 Conclusions and Perspectives

This study set out to examine the real estate evaluation process in Morocco, particularly focusing on the acquisition operations by the Directorate of State Domains. Through our analysis, we identified several critical issues and potential solutions. The following key findings encapsulate the core results of our research:

- Our study highlighted several shortcomings in Morocco's real estate evaluation process, particularly within the Directorate of State Domains. We found that evaluation criteria are often subjective, transparency is lacking, and there are no consistent methods or quality control measures in place.
- These issues are holding back progress toward Sustainable Development Goal 11, which aims to make our cities more inclusive, safe, resilient, and sustainable.
- On a positive note, Artificial Intelligence (AI) has great potential to revolutionize real estate valuation in Morocco. Techniques like machine learning, including artificial neural networks and deep learning have already shown they can improve price forecasting.
- However, to fully harness AI's potential, we need to overcome challenges related to data availability and ensuring the stability of results.
- Embracing AI in real estate could also help speed up energy transition projects, creating new opportunities for better decision-making and progress.
- This modernization could lead to better control over land prices and support both environmental sustainability and economic resilience.
- Nevertheless, as we move forward with these technologies, it is crucial to address the ethical considerations to ensure we use AI responsibly and sustainably.

Looking ahead, our findings point to several important directions for further research and practical action. One key area is creating strong frameworks that make real estate evaluations more transparent and consistent while incorporating advanced

AI techniques. As AI is growing and changing, it is crucial to develop policies and guidelines that tackle ethical issues, especially around data privacy, bias in AI algorithms, and making sure the benefits of technology are shared fairly.

Our work also highlights the need for a solid data infrastructure to support AI in real estate. This will require teamwork between government agencies, private companies, and universities to ensure that the data is accurate, accessible, and secure.

However, it is important to remember that AI is not a magic fix. Future efforts should aim to combine traditional methods with AI-driven approaches to create a more flexible and resilient evaluation system. By doing so, Morocco can not only bring its real estate sector into the modern era but also set an example in using technology to support sustainable urban development and economic growth.

In conclusion, while there are still challenges ahead, integrating AI into real estate evaluation offers a promising opportunity to transform the sector, promote sustainability, and strengthen economic resilience in Morocco. Moving forward, it is essential to strike a balance between embracing technological innovation and ensuring ethical responsibility to fully realize these benefits.

References

1. Ho WKO, Tang B-S, Wong SW (2021) Predicting property prices with machine learning algorithms. J Prop Res 38(1):48–70. https://doi.org/10.1080/09599916.2020.1832558
2. Alsawan NM, Alshurideh MT (2022) The application of artificial intelligence in real estate valuation: a systematic review. In: Hassanien AE, Snášel V, Tang M, Sung TW, Chang KC (eds) Proceedings of the 8th international conference on advanced intelligent systems and informatics 2022. AISI 2022. Lecture notes on data engineering and communications technologies, vol 152. Springer, Cham. https://doi.org/10.1007/978-3-031-20601-6_11
3. Rampini L, Re Cecconi F (2022) Artificial intelligence algorithms to predict Italian real estate market prices. J Property Investment Finance 40(6):588–611. https://doi.org/10.1108/JPIF-08-2021-0073
4. Baldominos A, Blanco I, Moreno AJ, Iturrarte R, Bernárdez Ó, Afonso C (2018) Identifying real estate opportunities using machine learning. Appl Sci 8(11):2321. https://doi.org/10.3390/app8112321
5. Shousong C, Xiaomin G, Xiaoguang W, Ying C (2019) Notice of violation of IEEE publication principles: research on urban land price assessment based on artificial neural network model. IEEE Access 7:180738–180748. https://doi.org/10.1109/ACCESS.2019.2958978
6. National Research Tomsk Polytechnic University (NR TPU), Vykhodtsev NA (2021) Artificial intelligence in price estimation of real estate. Proc TUSUR Univ 24(1):68–72. https://doi.org/10.21293/1818-0442-2021-24-1-68-72
7. Alzain E, Alshebami AS, Aldhyani THH, Alsubari SN (2022) Application of artificial intelligence for predicting real estate prices: the case of Saudi Arabia. Electronics 11(21):3448. https://doi.org/10.3390/electronics11213448
8. Taffese WZ (2006) A survey on application of artificial intelligence in real estate industry. In: 3rd International conference on artificial intelligence in engineering and technology, Kota Kinabalu, Malaysia
9. Peter NJ, Okagbue HI, Obasi EC, Akinola AO (2020) Review on the application of artificial neural networks in real estate valuation. Int J 9(3):5–11
10. Deaconu A, Buiga A, Tothăzan H (2022) Real estate valuation models performance in price prediction. Int J Strateg Prop Manag 26(2):86–105. https://doi.org/10.3846/ijspm.2022.15962

11. Yu L, Jiao C, Xin H, Wang Y, Wang K (2018) Prediction on housing price based on deep learning. World Acad Sci Eng Technol Int J Comput Electr Autom Control Inf Eng 12:90–99
12. Shinde A, Dange N, Patane N, Gholap S, Beera V (2019) Real estate properties assessment using deep neural network. Zenodo. https://doi.org/10.5281/zenodo.3234639
13. Hansun S (2022) Deep learning approach in predicting property and real estate indices. Int J Adv Stud Comput Sci Appl 14(1):61–71. https://doi.org/10.15849/IJASCA.220328.05
14. Bin J (2019) Investigations on deep learning techniques for property assessment. Electronic Theses and Dissertations (ETDs) 2008+. T, University of British Columbia. https://doi.org/10.14288/1.0376072
15. Zhan C, Wu Z, Liu Y, Xie Z, Chen W (2020) Housing prices prediction with deep learning: an application for the real estate market in Taiwan. In: IEEE 18th international conference on industrial informatics (INDIN). IEEE, pp 719–724. https://doi.org/10.1109/INDIN45582.2020.9442244
16. Zhao Y, Chetty G, Tran D (2019) Deep learning with XGBoost for real estate appraisal. In: IEEE symposium series on computational intelligence (SSCI), pp 1396–1401. https://doi.org/10.1109/SSCI44817.2019.9002790
17. Yalgudkar SS, Dharwadkar NV (2022) A literature survey on housing price prediction. J Comput Sci Comput Math 12(3):41–45. https://doi.org/10.20967/jcscm.2022.03.002
18. Calainho FD, Van De Minne AM, Francke MK (2022) A machine learning approach to price indices: applications in commercial real estate. J Real Estate Financ Econ. https://doi.org/10.1007/s11146-022-09893-1
19. He H-M, Chen Y, Xiao J-Y, Chen X-Q, Lee Z-J (2021) Data analysis on the influencing factors of the real estate price. Artif Intell Evol 52–66. https://doi.org/10.37256/aie.222021966
20. Mora-Garcia R-T, Cespedes-Lopez M-F, Perez-Sanchez VR (2022) Housing price prediction using machine learning algorithms in COVID-19 times. Land 11(11):2100. https://doi.org/10.3390/land11112100
21. Chardon I (n.d.) Housing prices: testing machine-learning methods
22. Mohd T, Jamil S, Masrom S (2020) Machine learning building price prediction with green building determinant. Int J Artif Intell 9(3):379. https://doi.org/10.11591/ijai.v9.i3.pp379-386
23. Casali Y, Aydin NY, Comes T (2022) Machine learning for spatial analyses in urban areas: a scoping review. Sustain Cities Soc 85:104050. https://doi.org/10.1016/j.scs.2022.104050
24. Ja'afar NS, Mohamad J, Ismail S (2021) Machine learning for property price prediction and price valuation: a systematic literature review. Property Manage 19. https://doi.org/10.21837/pm.v19i17.1018
25. Kim J, Won J, Kim H, Heo J (2021) Machine-learning-based prediction of land prices in Seoul, South Korea. Sustainability 13(23):13088. https://doi.org/10.3390/su132313088
26. Phan TD (2018) Housing price prediction using machine learning algorithms: The case of Melbourne City, Australia. In: International conference on machine learning and data engineering (iCMLDE). IEEE, pp 35–42. https://doi.org/10.1109/iCMLDE.2018.00017
27. Mrsic L, Jerkovic H, Balkovic M (2020) Real estate market price prediction framework based on public data sources with case study from Croatia. In: Sitek P, Pietranik M, Krótkiewicz M, Srinilta C (eds) Intelligent information and database systems. Springer, pp 13–24. https://doi.org/10.1007/978-981-15-3380-8_2
28. Rosales I, Hernández C (2021) Real estate pricing models using machine learning schemes. In Proceedings of the international conference on industrial engineering and operations management. IEOM Society International, pp 2250–2258. https://doi.org/10.46254/SA02.20210706
29. Choy LHT, Ho WKO (2023) The use of machine learning in real estate research. Land 12(4):740. https://doi.org/10.3390/land12040740
30. Alkan T, Dokuz Y, Ecemiş A, Bozdag A, Durduran SS (2022) Using machine learning algorithms for predicting real estate values in tourism centers. https://doi.org/10.21203/rs.3.rs-1757533/v1
31. Mayer M, Bourassa SC, Hoesli M, Scognamiglio D (2022) Machine learning applications to land and structure valuation. J Risk Financ Manage 15(5):193. https://doi.org/10.3390/jrfm15050193

32. Ardabili SF, Abdilalizadeh L, Mako C, Torok B, Mosavi A (2022) Systematic review of deep learning and machine learning for building energy. https://doi.org/10.31219/osf.io/fxtmz
33. Ashok D (2022) Estimating the price of real estate properties with the help of online images. ECS Trans 107(1):16933–16941. https://doi.org/10.1149/10701.16933ecst
34. Court of Auditors (2015) Annual report of the court of auditors. https://www.courdescomptes.ma/wp-content/uploads/2023/01/domaine-prive.pdf. 10 Mar 2024
35. Economic, Social and Environmental Council (2021) Towards a responsible and inclusive digital transformation
36. Government of Morocco (2021) Circular No. 02-21-cab of 30 Joumada II 1442 (12 February 2021) on Real Estate Acquisition and Assignments of Interests in Private Property of the State
37. Jaillet A, Mabilon-Bonfils B (2021) Chapitre 1. Qu'est-ce qu'une revue de littérature? Je réussis mon mémoire de Master MEEF. 1er degré: professeur des écoles. Vuibert, pp 42
38. Ministry of National Territorial Planning, Urban Planning, Housing and City Policy (2022) 1st ed. https://www.urbanagendaplatform.org/sites/default/files/2022-03/Kingdom%20of%20Morocco%20NUA%20Report%2018%20March%202022.pdf. (15 Mar 2024)
39. Nkolika J (2020) Review on the application of artificial neural networks in real estate valuation. Int J Adv Trends Comput Sci Eng 9(3):2918–2925. https://doi.org/10.30534/ijatcse/2020/66932020
40. Truong Q, Nguyen M, Dang H, Mei B (2020) Housing price prediction via improved machine learning techniques. Procedia Comput Sci 174:433–442. https://doi.org/10.1016/j.procs.2020.06.111
41. Vykhodtsev NA (2021) Using artificial intelligence to assess the value of real estate property. Rep TUSUR 24(1):68–72

Optimization Financing Innovations Models for Next-Generation Sustainable Clean Energy Technologies

Karri Pallavi and Sonali Srivastava

Abstract India's unique development of sustainable clean energy technologies is a way to enhance its energy security, improve the living standards for millions, and also to play a leadership role internationally and create jobs domestically. Clean technologies can meet energy needs in a cleaner, more efficient, and environmentally friendly way than traditional approaches. For several decades, scientific discoveries and technological innovations in renewable energy, energy efficiency, and energy storage and conversion technologies have advanced the prospects for next-generation clean technologies to help address our climate and energy security challenges. This chapter tends to understand the energy systems that are financed and their financing models. It also analyzes the role of technological advancements and associated risks within them and concludes by stating the future scope of such renewable clean energy technology models and about the current dynamics of the finance and technology markets. The investment gap in sustainable infrastructure is a result of political, financial, and institutional challenges. Overall, and in contrast to the underlying technology they are supporting, financing models for renewable energy technologies have changed less over the past two or three decades than they have in other sectors. Across other areas of technology and business, the emergence of disruptions caused by the falling costs of renewables and other forms of clean energy generation ensures that new models are developing at a rapid pace. For energy in general and clean energy technologies in the integration of blockchain technology is associated with improved transparency, streamlined processes to increased project bankability.

Keywords Renewable energy (RE) · Environmental sustainability · Clean energy · Optimized financing models · Risk and artificial intelligence

K. Pallavi (✉)
VIT Bhopal University, Sehore, Madhya Pradesh, India
e-mail: pallavi.k2022@vitbhopal.ac.in

S. Srivastava
Jaypee Business School, Jaypee Institute of Information Technology, Noida, India

F. Latrache et al. (eds.), *Digital Technology for an Innovative Energy Transition*, Green Energy and Technology, https://doi.org/10.1007/978-3-031-89657-6_3

1 Introduction

Green energy or clean energy is booming in present times as governments of various countries as well as people are striving hard to attain sustainable development. The industrial growth and development have led to depletion of fossil fuels, deforestation, increase in carbon emissions that ultimately lead to increase in global warming thus raising sea levels above normal. These alarming scenarios have drifted the economies to shift towards adopting renewable energy systems that promote environmental betterment and promoting sustainable development and economic growth too. But these renewable energy systems sometimes prove to be riskier investments from stakeholders' point of view as investing in high-end and costly renewable energy infrastructure may or may not attain expected returns. Thus, investors who are risk-averse do not intend to invest in these. But government needs to promote such renewable energy enhanced projects that encourage investors and also guarantee them a threshold return percent to provide them safety net from these riskier investments. This can also be achieved by governments choosing, developing and promoting optimized clean energy models. A cleaner future depends on the switch towards renewable energy (RE) replacing traditional fossil fuels [1], and clever solutions are required to get past the challenging issues that come up throughout this shift, like separating from outdated and ineffective ways of achieving energy needs [2]. The Renewable Energy Transition (RET) is referred to as the shift from energy sources based on fossil fuels and towards renewable energy sources [3]. A substantial sum of money will be needed to accomplish a successful RET [4]. Technology improvements have led to a significant fall in the total costs involved in RE generation in recent years [5], but there has not been any significant rise in investment patterns. Because of the quantity of money necessary and policy changes, investors are unwilling to take on investment risk, which makes financing the RET perhaps one of the largest issues of the twenty-first century.

The problem of improving rural residents' quality of living can also be resolved by moving towards renewable energy [6]. Even though it can be difficult to access some isolated locations, building grid connectivity with nearby nations is a practical solution that can promote peace in the region. Other advantages of this kind of project include less reliance on a single nation for energy demands, the development of isolated rural areas through the provision of electricity, and—above all—the promotion of international friendship.

This chapter tends to understand the energy systems that are financed and their financing models. It also analyzes the role of technological advancements and associated risks within them. The chapter concludes by stating the future scope of such renewable clean energy technology models.

1.1 Literature Review

India is uniquely positioned on its development pathway to bypass the long historical periods made necessary by the industrial revolution in the West and the process of industrial catch-up. This is the result of a variety of factors related to demographic and economic basics [7, 8]. It presents a huge societal challenge for the energy sector in order to have a sufficient technological leapfrog to sustainable energy systems as soon as possible, to avoid the legacy issues of historical investments that have developed most of the world throughout economic growth periods [9]. The enormous challenges set by the Indian case on both fronts provide a crucial case study to balance the priorities of sustainable energy development with the political and socio-economic issues at stake in the governance of modernizing economies. The case is a particularly daunting one because energy is essential to growth, and traditionally poverty alleviation had relied on its utilization [10, 11]. The relentless advance of climate change has posed these objectives in opposition, with carbon-based energies accounting for a very large share of anthropogenic CO_2. Understandably, many developing countries, including India, have resisted setting targets for emissions reductions as they saw this as a tool for maintaining the colonial arena in which developed countries could industrialize more rapidly and more cheaply while shifting the related externalities to the detriment of the inhabitants of the governments' territories. However, with affordable deployment of renewable and sustainable energies, the technological frontier has moved on. Due to the growing and significant scale of its energy system, India has been pursuing technological innovation with the central priority of reducing unit costs of renewable energy production [12–14].

Public policies with supportive human activity need at locations are the conditions under which family's neighborhood, community, and locality receive public values from activities in metropolitan zones that generate durable and sustainable increased value sustainability [15, 16]. Urban scarcity implies local geographically specific adjustments and public bargaining outcomes within the metropolitan area. Urban infill scarcity regimes suggest that through urban land and capital market exchanges, the location of housing in neighborhood, community, and locality can increase the generation of normative desirable qualities and diverse activities. For instance, albeit the capital gain from the changeovers is capitalized on rental and land value prices at the equity of public good allocations and services provided by social community, local public goods favorable towards individuals' quality of life are only a localized problem [17–19]. In the present context, public policies address issues of social welfare concerned with generating options for resolving institutional, framework, and policy-regulatory governance for making cities more attuned to economic and environmental sustainability [15, 20]. Public interest is evident through calls for perspective, analysis, and design in environmental economic activities and should promote all the significant sustainable tasks because environmental sustainability suggests normatively desirable economic performance. Many additional local public good internalities make local governing of local economic activities and opportunities important resources contributing to urban sustainability. Structures and facilities of urban environmental conditions at local public government levels offer proximate reach to residential housing sites. Those opportunities available to individuals at locations provided through housing or social community values and extended imitation, positively supporting local public policy for environmental protection [21].

The World Economic Forum's report on the global governance gaps in advancing technologies has noted the tardy advancement and innovation of transformative sustainable technologies as an important driver of the global sustainability crisis [22, 23]. Despite the readily available but under-actualized potential for sustainability, accelerating the development, dissemination, and innovation of mature energy technologies to address climate change is a key challenge [9, 24]. In the context of India's technological capabilities and pressing energy needs, sustainable clean energy innovation assumes added importance. India has large unmet energy needs, one of the highest levels of energy poverty, substantial opportunities for industrial, high living cost, and emission reductions, and challenges providing access to clean and affordable energy to its citizens and industries [25, 26].

India's unique development of sustainable clean energy technologies is a way to enhance its energy security, improve the living standards for millions, and also to play a leadership role internationally and create jobs domestically. In this context, the relevance of financing innovative and emerging technologies at the pre-commercial level, where the ability to draw upon future cash flows for investments is limited, cannot be overemphasized due to the long-time horizons for investments coupled with the high uncertainty linked to successfully commercializing these technologies [27, 28]. Green finance, when applied to the problems of renewable energy and currently available clean energy technologies that could be expensive when compared to the existing conventional energy generation technologies, will have implications on innovation and commercialization policies [29, 30]. The same effects would be realized from an increase in the effectiveness of financing (lower transaction costs and higher speed) and the decrease in risk. Institutional support and financial mechanisms can primarily aid in technology transfer between the developed and the developing world. These then lead the new technologies to make a quick entry into the country [30, 31]. India aspires to establish an economy that is strong, equitable, and built on the pillars of sustainability. It is also indispensable to the realization of the climate goals that can mitigate emissions [32, 33].

Policy measures aimed at deploying renewable power generation technologies have been evolving, facilitating technological learning-by-doing that is typically stimulated by a well-designed policy mix. India's vulnerability to disruptions in the supply of critical hydrocarbon resources leads to a set of markets with under-priced imported inputs and under-exploited local resources [34]. The ultimate objective of most energy problems is to improve social welfare, specifically by increasing living standards. Modern societies depend upon large amounts of energy for their functioning, but the production and consumption of energy generate a number of environmental problems as well [35, 36]. The biggest challenge in the sustainable and renewable energy sector is not the development of cost-effective and efficient technologies, but the commercialization, adaptation, adjustment, and widespread implementation of these technologies. Carbon Funds (CFs) are depositories of money for investing in developing countries for clean energy transaction reduction projects. In some countries, there is a varying design of particular requirements that must be met by prospective ventures to be eligible for CF transactions. Sulyok [37] the large-scale investment needed for sustainable energy systems, global financial models need innovative and flexible financial instruments and mechanisms. Clean energy business models and private and public financial techniques are proposed for some next-generation sustainable clean energy technologies [15, 38].

1.2 The Importance of Optimizing Financing Models to Accelerate the Adoption of Sustainable Technologies

For the last 20 years, a number of important themes have occupied the concerns of the sustainable technology community: the increasing global deployment of renewable-based energy solutions and their competitiveness level with fossil-fuel-based technologies at the retail level, the potential decline of lithium-ion battery costs combined with potential commercialization of long-term stationary energy storage chemistries contributing to cost-effective grid-connected renewable energy solutions, the modest but slow growing worldwide usage of passenger electric vehicles, and the reduction of their purchased cost [39, 40]. Simultaneously, there is a growing problem that is increasing faster than any of the technology trends in the preceding list: growing awareness that a transition away from dirty, polluting diesel and gasoline fuels for both the transportation of goods and people has not occurred and, without change, will not occur. Trends and data suggest that less rather than more passenger mobility will be enabled with no demand on natural resources [41–43].

Humankind is at a critical inflection point in which the burden of our human industrialization on the Earth's natural and ecological systems must be kept within their sustainable limits [44]. The era of rapid acceleration of many technologies, including energy and transportation, has come with both significant positive and negative environmental consequences [45, 46] (Adebayo and Kirikkaleli). On the other hand, and considering these financial markets, new opportunities and applications for emerging areas such as the environment, cognitive and autonomous cars, information technology, blockchain, artificial intelligence, and more specifically, industry 4.0, which we usually designate "smart technologies", have emerged [47–49].

The financial markets, which emerged as important promoters of the wealth creation process of the planet, increasingly growing further away from the concept of fundamental axioms on the economy's role [50, 51]. Instead, excessive financial leverage, speculative behaviors, and asset bubbles have increased the financial risks and led the economy to, more than once, a situation of near collapse with serious macroeconomic consequences [52, 53]. Indeed, resource mobilization is an essential means of implementation to foster countries' synergies to achieve their Sustainable Development Goals [54]. The significance that bringing more focus and attention to it, imbuing it with a powerful collective intention, and leveraging fiscal and financial instruments to incentivize the development of green financial services are components of the finance, technology, and capacity-building framework that is required [55]. In recent years, many technological solutions with significant potential to help societies become cleaner, more resilient, and more efficient have become available. There are thousands of state-of-the-art solutions within the fields of renewable energy, energy efficiency, waste management, water efficiency and supply enhancement, transmission and distribution, and sustainable transportation [56]. The level of interest in sustainable finance is increasing. This becomes apparent in its clarity and variety in reports, ratings, and investment, policies serving a variety of purposes on mounting constraints for private risk-taking; and distinctions that are drawn among "green," "ESG," "social," and related criteria [57, 58]. Strategies for fostering demand in the building energy efficiency sector need to couple with business models and financial products that match growth-oriented development, deployment, and reimbursement [59]. Engaging private-sector finance, at scale, in the implementation of sustainable building technologies, strengthening self-sustainable markets in these industries will help to drive the green recovery at a higher pace [60].

1.3 Innovative Financing

Innovative financing is an emerging consideration that could facilitate the breakthrough in energy technology. So far, less attention has been given to financing engineering and low-carbon energy R&D policies and technology commercialization activities [31, 61]. Whether it is on clean energy and climate technology market development or product design, limited options and maturity are the existing solutions [62, 63]. There seems to be an optimized proper financing structure that combines explicit and implicit support in an efficient order. Small and medium-sized next-generation innovative clean energy technology companies would be facing unique capital challenges than traditional energy generation and distribution companies throughout various stages of their life cycle [64, 65].

Globally, the average temperature is rising while sea levels are getting higher. As energy is crucial for addressing the full dimensions of the climate challenge, it is vital to implement a new, innovative, ambitious, and clear global framework [66, 67]. However, due to its high cost and poor performance, the energy system's transition to a zero-emissions generating system is considered a very challenging task [68]. The financial crisis, high private R&D risk, and climate policy uncertainty slow down the market's pace. To address these problems, existing financial and regulatory instruments are not enough, and for which models of innovative financing are needed.

The energy sector simply uses corporate finance for a particular purpose: to monetize the long-dated, contractual cash flows from its generation and supply activities [69, 70]. Project finance, useful to isolate the project in a special purpose company and to reduce project default risk, is another peculiarity of the energy sector rather than of other industrial sectors. Due to its specifics, such as the presence of long-term liabilities to be backed with similar maturities assets, it should be less subjected to the adverse effects of boom-and-bust cycles [38, 71]. However, in practice, the energy sector is more exposed to business cycles than other industrial sectors due to its capital-intensive character, the geopolitical instability that reduces the confidence of financiers, corporatist behaviors by producers and consumers, and the increasing technology obsolescence that may reduce the capability to earn the income needed to honor loan repayments and that increases investors' perception of technological risks [72–74].

Innovative financing refers to mechanisms for raising funds and for the way these funds are used by a variety of public and private sector stakeholders (Shames & Scherr) [75, 76]. Generally, innovative financing fills a gap in the supply of or the demand for funding for a set of activities. Despite the energy sector's traditional reliance on capital-intensive assets, innovative energy finance has outperformed other sectors in the finance industry. In so doing, it has maintained its traditional role in providing essential goods and services [77–79]. It is increasingly recognized that overcoming market barriers to commercialization will require breakthrough technologies—new ways of capturing, converting, storing, and transmitting energy more efficiently and in a manner less damaging to human health, the environment, and the economy [80, 81].

Clean technologies can meet energy needs in a cleaner, more efficient, and environmentally friendly way than traditional approaches. For several decades, scientific discoveries and technological innovations in renewable energy, energy efficiency, and energy storage and conversion technologies have advanced the prospects for next-generation clean technologies to help address our climate and energy security challenges [82–84]. Despite recent progress, the transition to cleaner energy is hindered by the enormous scale of global capital stock and continuing dependence on high levels of fossil-based energy consumption [85, 86]. The growing energy consumption and its associated environmental impacts, including greenhouse gas emissions that drive global climate change, emphasize the importance of clean energy future [87, 88]. Furthermore, concerns that global supplies of petroleum are or might become limited or that oil product prices could be subject to future instability because of geopolitical factors underscore the interest in developing alternative energy technologies that minimize the impact of a petroleum-constrained economy [89]. Although energy generation and distribution based on coal, natural gas, and electricity transmission and distribution infrastructure can make significant contributions from improved technologies, achieving a clean energy economy will be difficult without additional technology options [90, 91]. The scale of the clean energy challenge is even greater when factoring in global economic development. The path to a cleaner energy future will therefore continue to require the systematic employment of a range of administrative, legislative, and regulatory measures to ensure that the right market signals get generated.

Next-generation clean energy technologies need to be scaled up significantly in order to make a meaningful contribution to the reduction of greenhouse gas emissions. The scale of the challenge and the urgency of addressing it cannot be overstated [92, 93]. Each innovative technology requires rigorous screening and performance evaluation against mature technologies, along with accurate forecasting of future trends in a traditional and changing financial environment. Data-driven technologies, such as blockchain and artificial intelligence, have shown considerable promise and have potentials to facilitate the scale-up and spread of emerging clean energy solutions. Several questions need answering to predict the future viability of new pavements that carve ways for clean energy technologies amidst ever-increasing barriers [77, 94–96].

1.4 Financing Landscape for Clean Energy Technologies

The societies all over the economies have been striving hard to achieve sustainable environment. Sustainable environment majorly talks about the sustainable energy systems. People usually demand energy services such as heating, cooking, lighting, mobility, etc. Comprising a combination of energy careers and technologies [97]. Usually, such environmentally driven services require higher capital investments to finance high-end technologies and lower energy costs. Additionally, energy services obtained from renewable and traditional sources like wind and solar energy require negligible (almost nil) energy inputs.

Practically speaking about the energy system, there are broadly three areas in which long-term financing is very important, that is, (1) Investing in renewable energy technologies, for example, huge solar, photovoltaic and wind systems, (2) Investments in electricity grid infrastructure, district heating, electric vehicle (EV) charging stations and fuel cell vehicle (FCV) hydrogen refuelling stations, storage, etc. [98], (3) Investment in cutting demand for energy and thus promoting energy efficiency, for example, buildings insulated with thermal insulations also known as "Energy Performance Contracting". Primarily, the markets are brimming with short-term and long-term investments both, within which the short-term investments required by start-ups can utilize electricity available from photovoltaic and wind systems, in decentralized battery storage systems, etc. but within the context of long-term, the investment needs appropriate financing strategies since the investments are large and energy requirements are huge too.

While analyzing the financing strategies for such investments, two major viewpoints need to be considered. Firstly, the cost of capital (includes the interest rate and weighted average cost of capital-WACC), secondly the underlying risk for the investor. Cost of capital are usually determined by the interest rate on debt financing and on the expected rate of return to the investor. As return is considered so comes the risk associated with it. When the investor's anticipated rate of return deviates from the actual acquired rate of return it is known to be the associated risk [99] (Table 1).

Some of the examples of less risky investments are expanding or installing energy-saving lighting technologies in residential buildings, and installing thermal retrofit in a residential apartment. These are known to be low-risk investments as they are well established and have long life span of operation and are cost-effective too. Examples of some of the high risk involved investments are, installation of charging and refuelling stations for hydrogen fuel cells or electric battery-operated vehicles.

Table 1 Depicting important financing energy systems

Energy systems	Authors	Cost of capital	Investment risk
Investment in renewable energy technologies, for example, huge solar, photovoltaic and wind systems	[98]	3–6%	Low
Investments in electricity grid infrastructure, district heating, electric vehicle (EV) charging stations and fuel cell vehicle (FCV) hydrogen refueling stations, storage, etc.	[98]	4–15%	High
Investment in cutting demand for energy and thus promoting energy efficiency, for example, buildings insulated with thermal insulations also known as "Energy Performance Contracting"	[98]	5–25%	Low

Source Developed by author

These are considered to be high risk associated as they involve factors like life time, size of the technology design, service demand, market prices, or the existence of a compensation system. The more the risk more shall be the Return on Investment (ROI) factor and vice versa.

Previous studies have advocated the need for financing these energy-efficient and clean energy investments. It is been understood that not only energy efficient investments play an important role in the society at large but also, inculcating green finance for sustainable development activities and energy security implements economic development and environmental sustainability [100]. Increasing energy consumption levels and associated carbon emissions have alarmed the increasing carbon footprints which have led to adopting energy-efficient and sustainable activities by governments of economies [101].

1.5 Objective and Rationale of the Study

This chapter intends to understand the financing patterns of clean energy sector and interrelate the risk and return associated with it. The chapter encloses various optimization models that analyzes the uncertainty and risk involved with the technological advancements in implementing the RE infrastructural projects. It analyzes that how low-income countries who are well worse in technological know-hows and cannot afford to import high-end RE infrastructure can implement optimized models on their own to successfully implement RE designs. It also enlightens the various policy and regulatory guidelines by various countries and how lessons can be taken from them so that countries with low technical expertise can implement RE projects. The rest of the chapter entails the future of RE-optimized models and conclusion stating the positive impact of these optimized models on RE infrastructural projects.

2 Modelling Technological Uncertainty and Risk

Uncertainty in technological know-hows emerges through factors such as availability of resources, costs, as well as changing market conditions like relative pricing [102]. Basically, there are two major ways in which technology uncertainty can be incorporated into energy systems modeling: (1) Portfolio Theory (2) Scenario Analysis.

2.1 *Scenario Analysis*

Scenario Analysis includes assumptions with respect to energy or models of climatic-related policy to explore the unexplored contingency scenarios. Some examples of such contingencies includes, high or low energy demands, increase and fall in carbon prices, etc. It is also learned that scenario analysis with sensitivity analysis can be a way to identify key uncertainty parameters which can be modelled stochastically. A large-scale scenario study [103] specifically answered the question of how GHG mitigation technologies changes under various important uncertainties such as energy demand, resource constraints, technology availability and costs, and finally the scope of GHG emission limits. All these uncertainties were bifurcated into 22 scenarios. The scenarios were categorized using the bottom-line programming model MESSAGE, which is particularly suitable for this type of analysis because it presents the energy technologies in detail or measuring the effect of uncertainty in scenarios, the metric used is cumulative future GHG emission reductions were implemented using different scenario specifications, incorporating the idea of reduction measures in the form of "wedges" [104]. Riahi et al. [103] suggested a new way to evaluate various mitigation strategies by introducing a "robustness" metric stated as the spread of values between the lowest and highest points in all scenarios analyzed. Thus, scenario analysis helps identify technology innovation intricacies and so help decision makers in deciding technology portfolios. It helps analyze magnitude of anticipated uncertainties. The only drawback with scenario approach is that it does not take into account full impact of what uncertainty scenarios have on decisions related to technology adoption. In parlance, decisions usually depend upon risk preferences of decision makers whether they are ready to take more risk for attaining more return on his/her investment and vice versa. An investor can decide to bet all his/her investment options on low-cost clean energy technology where anticipated technology promise might or might not flourish creating a considerable risk zone. Thus, this combination of diversify risk with anticipated technology and assessing investment portfolios can be useful in hedging risks. This combination of resulting investment portfolios depends on risk attitudes (premium of risk taken) of the investor and technological innovation uncertainty.

2.2 *Portfolio Theory*

The portfolio theory addresses the issues of decision-making in investment with associated risks [105, 106]. Portfolio theory applied to energy sector helps mitigate the impact of risk taken and return obtained by fixing the risk at a specified level [107]. This theory basically marks an edge above as it inculcates risk-averse decision makers mindset. Risk aversion results in increased levels of acceptance.

Presently, more expensive technological alternatives like contemporary biomass and renewable sources, as well as carbon utilizing capture and storage (CCS) as a strategy to mitigate uncertainty in innovation. Therefore, the model recommends increased investments in advanced technology in the short to medium term despite the higher costs at present technologies facing risk aversion.

3 Integration of Effective Optimization Models in Clean Energy Finance

As we have seen that financial constraints and risk associated with risk have high impact on firm's decisions that compel them to invest in renewable or green and sustainable energy. Although, various mathematical models have been derived that exemplify the effects of factors associated with green energy investments. For example, using Analytic Network Process (ANP), it is identified that government mandates and policies are most impacting factors that drive investment in renewable energy [108]. Sim and Kim [109] developed a model using systems dynamics modeling technique that signifies that 'waste converted to energy' technology is least risk investment. Using a Cournot duopoly model, a model developed by Yang et al. [110] showed that green credit prices and carbon taxes can promote investment in green and clean technologies. An econometric model developed by Stucki [111] showed that low costs of energy investments hinder green investments. Another multivariate econometric model developed by Bauwens [112] analysing community-based investment decisions on renewable energy which reveal strong influence of social networks. Other than these descriptive models, another range of practical models also do exist. For example, Multi-Criterion Decision Analysis (MCDA) techniques have been used to analyze investment options based on environmental, social, and economic factors [113]. A renowned model developed combines goal programming, fuzzy TOPSIS, and analytic hierarchy process (AHP) developing a hybrid model to understand renewable energy and its significance. A graphical performance analysis model was developed by Bandyopadhyay et al. [114] for supplementing funds to promote and foster conservation of energy projects in industrial plants. A study amalgamated the concept of financial risk while analysing the impact of investments in energy-efficient projects in context of paper, pulp, and cement industries. Later,

Mixed Integer Linear Programming (MILP) model was developed by Roychaudhuri and Bandyopadhyay [115] which is based on source-sink concept for renewable energy investments.

4 Case Studies of Analyzing the Optimized Models for Clean Energy

In today's world of diminishing energy sources and increasing energy consumption and demands, it is important to keep self-dependence over energy production in order to sustain energy security. Such demand can be fulfilled with the help of renewable energy (RE) [116]. The non-uniform distribution of energy resources over some areas over the other gives them an advantage that they can dig and fetch the energy resources underground but due to harsh environmental conditions it costs them high for extraction and also increases the level of greenhouse gases emissions produced during the process of extraction. Some of the major factors that play essential role in the conservation and sustenance of renewable energy (RE) are the public funding of the R&D of RE [117, 118], policy guidelines for RE [119] which aim to reduce GHG emissions [120].

Apart from all these, energy storage technologies play significant role in providing one-stop solution for issues associated with RE [121]. As suggested by Gallo et al. [122], inability to meet capital and regulatory requirements hinders the optimum utilization of energy storage technologies. Since the cost of establishing and usage of RE storage technologies is high, one can think of community investment to distribute the incurring cost burden of it. This is possible by involving community citizens directly in energy storage systems and also encourages people to invest in clean energy technologies without bearing much risk [123].

As per the previous studies, the China Government has been addressing this challenge of coping with the climate change bring in environmentally friendly action plan in the area of wind energy sector. China has so far invested much time and money in investments on R&D to understand its energy requirements and with the help of various government-owned agencies like China National Renewable Centre, Energy Research Institute of National Development and Reform Commission, Chinese Academy of Engineering and Chinese Academy of Sciences have proposed the development targets for wind, solar, and other non-fossil energies by 2020, 2030, and 2050 prospectively. Specifically, for wind energy generation targets have been set for 2020, 2030, and 2050 as total electricity generation would reach 5.0%, 8.5%, and 15.9%, respectively. Previous literature studies the need for China to invest in energy-efficient modern and low-carbon technologies and investing in sophisticated technologies to utilizing more wind energy for fulfiling the gap of energy requirements (Fig. 1).

The above model shows optimum balance between funding sources and innovation projects based on ROI (Return on investments) and TRL (technology readiness level)

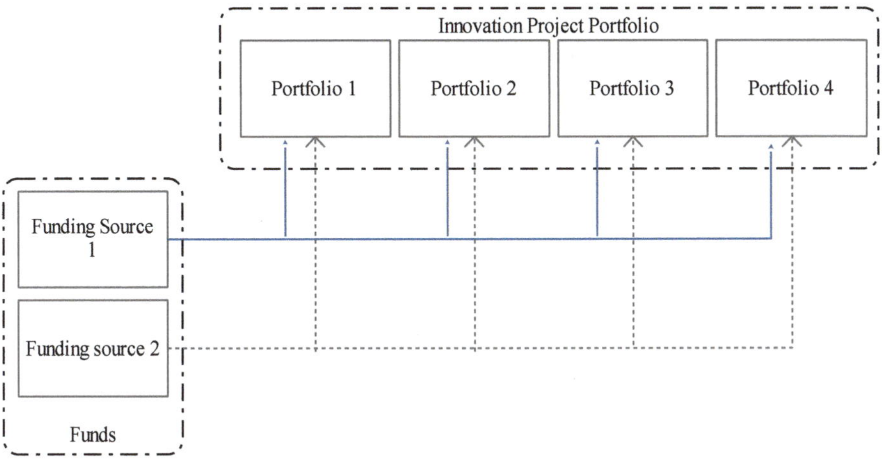

Fig. 1 Showing model for optimum balance between funding sources and innovation projects

considerations. The model was developed as a bi-objective MILP model based on source sink allocation [124].

On the other hand, China has witnessed the increasing solar photovoltaic (PV) industry market and thus, in order to promote it further with the very aim of sustainable development of China's economy, it has introduced the policy of 'Solar Rooftop Program' introduced in 2009 and 2010 which provided considerable subsidies either to solar PV project employers or to solar PV product manufacturers up to 17–20 Yuan/watt power based on the type of building material to 13–15 Yuan/Watt power for installation for PV type building project. Another solar project by Chinese government "Golden-sun project" gives subsidy to project employers up to 70% of total investment within the remote areas and up to 50% of total investment in non-remote areas.

4.1 Incentives and Strategies for Promoting RE Investments

Incentives relying on the government R&D incentives can only encourage to invest in RE projects [125, 126]. For countries who cannot afford to import high-end RE infrastructural equipment from other countries as it creates a burden on their pockets, they should focus on self-investing and developing in R&D which can help them shift towards RE. The incentives can be in the form of:

- Tax and fiscal incentives: The government can think of tax rebates and fiscal subsidies when RE is imported in countries where RE is in less developed stage, so that it encourages RE to be imported more. One such example is China which implemented such incentives to promotes those technologies which are under development stage in China. Other way around, fossil fuel generation can be taxed heavily

by governments so that it creates a competitive space among fossil fuel generating power companies which can strive to keep their taxes at minimum [127]. Another way to promote RE infrastructure and usage is that provide cheaper loans to construct and establish RE plants for residential and domestic use [128, 129].
- RE market incentives: Investors entry barriers to the budding investors and obtaining government permissions and subsidies can be challenging which can weaken the confidence of investors. This issue can be addressed by encouraging investors more rate of return and easing entry and exit restrictions, providing standard testing and certifications can encourage RE development.
- Grid network and Tariff incentives: The contract made with government bodies while agreement for RE is signed promises that the government shall purchase the extra electricity generated and a suitable tariff shall be paid to the investor. But post installation of costlier grid connection, this does not hold true. Either the electricity generated goes waste or the suitable price is not paid for purchasing the produced excess electricity. To prevent this, such studies have advocated that the payment for purchasing the excess electricity could be made in the form of digital currency [130].

Below Table 2 provides an overview of the main incentive policies used within the five main incentive types mentioned:

Few strategies which can help develop RE are:

- Government-owned institutions: Government can endorse projects by ensuring volatility, risk involved, and all related factors with the project so that it gives the investors the confidence that the project is investable as it is endorsed by government thus safe to invest in. Also, the government can think of taxing more to those who produce more carbon emissions and less to those who produce less. Countries like Sweden, Switzerland, and Norway have already tried implementing this and have been successful in doing so [131]. Governments should also think of globalizing the RE grids so that the gaps in production and supply of RE be met from other countries producing in excess with that of country producing in deficit. The needs of RE infrastructure not necessarily be met by government but can also be contributed by loans from microfinance institutions and other financial institutions.
- Strategies developed by corporates: Indian Oil Corporate companies like Petrobras, Chevron, BP, Eni, etc. and tech companies like Google, Apple need to come forward to take some steps towards energy conservation and transition investing in them or thinking of reducing carbon emissions. Corporates should think of energy storage technologies such as battery storage units to preserve the excess energy produced.
- Individual and community level: Cooperatives and communities should strive to create RE non-profit entities at local level that promote and invest in sustainable development. Such communities should encourage individuals to invest more in RE investment projects keeping "one member one vote" policy giving equal right to all stakeholders. This will promote RE investment and RE development activities at grass root level [132].

Table 2 Showing an overview of all the incentives based on market, policies, subsidies, tax, R&D with context of different countries

Studies	Country basis	Tax incentive		R&D incentive		Market-based incentives				Grid connection incentives		Policy and regulatory incentives		
		(a)	(b)	(c)	(d)	(e)	(f)	(g)	(h)	(i)	(j)	(k)	(l)	(m)
Zhao et al.	China	✔	✔	✔	✔			✔	✔	✔	✔			
Melović and Ćirović	Montenegro		✔											
Zhi et al. [128, 129]	China		✔	✔						✔				
Aquila et al.	Brazil					✔	✔			✔	✔			
Abdmouleh et al.	USA, India,Brazil, Australia, and the EU	✔	✔	✔						✔	✔	✔	✔	
Eichhammer et al.	Germany		✔					✔	✔	✔				
Štreimikiene et al.	Baltic States													✔
Malik et al.	GCC Region		✔	✔						✔				
Li et al. [3]	Australia									✔				✔
Mihaylov et al. [130]	Belgium									✔	✔			

Source Compiled by author
Where: (a): Tax aid, (b): Loan subsidy, (c): R&D support, (d): Development of R&D centres, (e): Auctions, (f): Quotas, (g): Market restrictions, (h): Development agenda, (i): Feed in tariff (FIT), (j): Net Metering (NM), (k): Green certificates, (l): Power purchase legislations, (m):Defined targets

4.2 Policy and Regulatory Considerations for Optimized Clean Energy Finance

Industrial policies are drifting more towards adopting clean and green policy initiatives [133]. to foster environmental sustainability and low-carbon markets. Parallelly, the cost competition in emerging RE technologies leads to keeping domestic electricity prices under control [134]. Countries utilize RE deployment by merging it with local community incentives and policies. Studies have evidenced the impact of low-carbon green energy policy with technological advancements can foster green industries at local community level [135]. Energy sector is often bound to policy restrictions as the impact of negative externalities, monopolies, and dependencies on technological advancements and competition is more in energy markets [136]. Studies from [137] have developed policy frameworks that interlinks hierarchical micro-level to macro-level policy design and implementation. Basically, in such frameworks, the policy design starts at macro-level and the agenda for implementation reaches the micro-level. A few policies design example in various countries such as Mexican policy development for renewable energy, they implemented the policy of auctioning of energy generation with the help of private sector-based auction policy which gives them the lower cost of energy buy and utilization and a competition among the purchasing parties. Similar is the case of South Africa who brought in a policy framework of Integrated Resource Plan (IRP) that set the goals and objectives of how long-term energy requirements be met, offering affordable prices for electricity supply within the economy and creating local job opportunities. These two examples signify that energy policy design has to be implemented at micro levels. It also signifies that besides having stringent and robust policy framework, market size and stability along with proper value chain networks within the energy sector play an important role in better implementation of policy designs.

4.3 The Future of Optimization Models in Clean Energy Finance

Implementing green energy and clean energy initiatives requires proper policy design frameworks and its implementation at macro as well as micro-level. Adopting proper green initiatives and related clean energy projects by government endorsing them can lead to confidence boost in investors that may guarantee them with low risk and moderate return patterns at low costs. This also promotes implementation of technological advancements in RE infrastructure promoting overall boost to the energy sector. As models have also depicted the understanding of balance between ROI and equity investments by investors for better implementing RE projects. But these optimized models still lack in a few areas such as financing with respect to green bonds, green infrastructural bonds, financing on natural calamities prevention remains un-investigated. Also, finance from corporate level is not prominent yet the community level financing has been well explored. The sole responsibility of reducing carbon footprints and attaining sustainable RE development not only lies with the public undertakings and government but corporates if step into these initiatives can contribute much. The impact of domestic government-based initiatives for energy conservation against that of legal policy system are different. The community as well as corporates respond better to government-endorsed energy conservation RE policy initiatives as they are made aware about their responsibilities and they act responsibly towards it. But the legal system-imposed policy guidelines worldwide can be understood by them as a burden and they might not wilfully act upon it. Thus, local communities at micro-level and hierarchical macro-level RE policy guidelines be planned well and proper awareness be brought in thus better implementing policy guidelines. Future research works can develop models which introduce digital payments and digital currencies in the RE project financing. Optimized models can also consider yield on RE investment portfolio and profit margins along with debt and equity factors for RE projects.

4.4 Future Implications

The private sector has recognized that sustainable finance has the potential to be structured in a variety of forums and financial products and is committed to deploying cost or price estimates as benchmarks for its own carbon pricing externalities [138, 139]. The investment gap in sustainable infrastructure is a result of political, financial, and institutional challenges. A robust market for financing has not yet been developed, making it difficult for small-scale, experimental ideas in clean energy, sometimes referred to as demonstration projects, to receive the financing and support they need. The public and private costs of federal compliance programs that are designed to maximize the deployment of renewable energy for climate benefits have been determined using social cost of carbon estimates that deviate substantially from the approach that has been used in parallel programs designed to maximize local co-benefits of these programs. Democratized and decentralized energy architectures are often seen as hallmarks of future financing systems for generating technologies, which can take investors closer to the point of generation. Overall, and in contrast to the underlying technology they are supporting, financing models for renewable energy technologies have changed less over the past two or three decades than they have in other sectors. Across other areas of technology and business, the emergence of disruptions caused by the falling costs of renewables and other forms of clean energy generation ensures that new models are developing at a rapid pace.

Emerging technologies can impact the financing of clean energy technologies in a variety of ways. Some of these are technologies themselves, such as electric vehicles. Other enabling technologies include blockchain, artificial intelligence, and the use of big data. Big data allows for the identification of new potential sites for renewable and clean technology projects, removes uncertainty about project quality, and can reduce the time needed for monitoring purposes. For energy in general and clean energy technologies in particular, the integration of blockchain technology is associated with improved transparency, streamlined processes, increased resistance to fraud, and additional bankable merchant solutions enabling increased project bankability. Artificial intelligence can be used to lower project execution costs by predicting environmental changes and engaging in predictive maintenance once projects are in operation.

5 Conclusion

Green energy or clean energy is booming in present times as governments of various countries as well as people are striving hard to attain sustainable development. The industrial growth and development have led to depletion of fossil fuels, deforestation, increase in carbon emissions that ultimately lead to increase in global warming thus raising sea levels above normal. These alarming scenarios have drifted the economies to shift towards adopting renewable energy systems that promote environmental betterment and promoting sustainable development and economic growth too. But these renewable energy systems sometimes prove to be riskier investments from stakeholders' point of view as investing in high-end and costly renewable energy infrastructure may or may not attain expected returns. Thus, investors who are risk-averse do not intend to invest in these. But the government needs to promote such renewable energy-enhanced projects that encourage investors and also guarantee them a threshold return percent to provide them safety net from these riskier investments. This can also be achieved by governments choosing, developing, and promoting optimized clean energy models. As models have also depicted the understanding of balance between ROI and equity investments by investors for better implementing RE projects. But these optimized models still lack in a few areas such as financing with respect to green bonds, green infrastructural bonds, financing on natural calamities prevention remains un-investigated. Also, finance from corporate level is not prominent yet the community level financing has been well explored. The sole responsibility of reducing carbon footprints and attaining sustainable RE development not only lies with the public undertakings and government but corporates if step into these initiatives can contribute much. The impact of domestic government-based initiatives for energy conservation against that of legal policy system are different. The community as well as corporates respond better to government-endorsed energy conservation RE policy initiatives as they are made aware about their responsibilities and they act responsibly towards it. But the legal system-imposed policy guidelines worldwide can be understood by them as a burden and they might not wilfully act upon it. Thus, local communities at micro-level and hierarchical macro-level RE policy guidelines be planned well and proper awareness be brought in thus better implementing policy guidelines. Future research works can develop models which introduce digital payments and digital currencies in the

RE project financing. Optimized models can also consider yield on RE investment portfolio and profit margins along with debt and equity factors for RE projects.

6 Competing Interests

The authors declare to have no competing interests.

7 Ethics Approval

The authors confer that they have adhered to the accepted ethical standards.

References

1. Imteyaz B, Lawal DU, Tahir F, Rehman S (2021) Prospects of large-scale photovoltaic-based power plants in the Kingdom of Saudi Arabia. Eng Rep 3(9):e12398
2. Alasseri R, Rao TJ, Sreekanth KJ (2020) Institution of incentive-based demand response programs and prospective policy assessments for a subsidized electricity market. Renew Sustain Energy Rev 117:109490
3. Li HX, Edwards DJ, Hosseini MR, Costin GP (2020) A review on renewable energy transition in Australia: an updated depiction. J Clean Prod 242:118475
4. Hall S, Foxon TJ, Bolton R (2017) Investing in low-carbon transitions: energy finance as an adaptive market. Climate policy 17(3):280–298
5. Dhabi A (2020) Irena. Renewable energy statistics
6. Sen S, Ganguly S (2017) Opportunities, barriers and issues with renewable energy development—a discussion. Renew Sustain Energy Rev 69:1170–1181
7. Hodgson GM (2021) Financial institutions and the British Industrial revolution: did financial underdevelopment hold back growth? J Inst Econ
8. Williams G, Mahadevia D, Schindler S, Chattaraj S (2023) Megaprojects, mirages and miracles: territorializing the Delhi-Mumbai Industrial Corridor (DMIC) and state restructuring in contemporary India. Territory Politics Governance 11(3):456–477
9. Ahmad T, Zhang D, Huang C, Zhang H, Dai N, Song Y, Chen H (2021) Artificial intelligence in sustainable energy industry: status quo, challenges and opportunities. J Cleaner Prod 289
10. Dong K, Taghizadeh-Hesary F, Zhao J (2022) How inclusive financial development eradicates energy poverty in China? The role of technological innovation. Energy Econ
11. Murshed M, Ozturk I (2023) Rethinking energy poverty reduction through improving electricity accessibility: a regional analysis on selected African nations. Energy
12. Ahmed R, Liu G, Yousaf B, Abbas Q, Ullah H, Ali MU (2020) Recent advances in carbon-based renewable adsorbent for selective carbon dioxide capture and separation—a review. J Cleaner Prod 242
13. Sarfraz M, Naseem S, Mohsin M, Bhutta MS (2022) Recent analytical tools to mitigate carbon-based pollution: new insights by using wavelet coherence for a sustainable environment. Environ Res
14. Shewchuk SR, Mukherjee A, Dalai AK (2021) Selective carbon-based adsorbents for carbon dioxide capture from mixed gas streams and catalytic hydrogenation of CO_2 into renewable energy source: a review. Chem Eng Sci

15. Lu Y, Khan ZA, Alvarez-Alvarado MS, Zhang Y, Huang Z, Imran M (2020) A critical review of sustainable energy policies for the promotion of renewable energy sources. Sustainability 12(12):5078
16. Pata UK, Caglar AE, Kartal MT, Depren SK (2023) Evaluation of the role of clean energy technologies, human capital, urbanization, and income on the environmental quality in the United States. J Cleaner Prod
17. Barman D (2023) Land redevelopment, real estate and capital in urban place-making: a case study of Siliguri, India. Environ Urbanization ASIA
18. Bonde-Hansen M (2021) The dynamics of rent gap formation in Copenhagen: an empirical look into international investments in the rental market
19. Mou YG, Ma M (2022) Measuring the role of consuming natural resource domination and social media for green economic recovery. Resour Policy
20. Bekun FV (2022) Mitigating emissions in India: accounting for the role of real income, renewable energy consumption and investment in energy. Int J Energy Econ Policy
21. Zeemering ES (2021) Sustainability management, strategy and reform in local government. In: Sustainable public management
22. Mohammad WA, Nazih Diab Y, Elomri A, Triki C (2023, April) Innovative solutions in last mile delivery: concepts, practices, challenges, and future directions. Supply Chain Forum Int J 24(2):151–169
23. Tardy T (2021) NATO 2030: new technologies, new conflicts, new partnerships
24. Brabec CJ, Distler A, Du X, Egelhaaf HJ, Hauch J, Heumueller T, Li N (2020) Material strategies to accelerate OPV technology toward a GW technology. Adv Energy Mater 10(43)
25. Lan J, Khan SU, Sadiq M, Chien F, Baloch ZA (2022) Evaluating energy poverty and its effects using multi-dimensional based DEA-like mathematical composite indicator approach: findings from Asia. Energy Policy
26. Rafi M, Naseef M, Prasad S (2021) Multidimensional energy poverty and human capital development: empirical evidence from India. Energy Econ
27. Delina R, Gróf M, Dráb R (2021) Understanding the determinants and specifics of pre-commercial procurement. J Theor Appl Electron Commer Res 16(2):80–100
28. Dimitri N (2020) Budget allocation design in the EU pre-commercial procurement for innovation. J Public Procurement
29. Li CZ, Umair M (2023) Does green finance development goals affects renewable energy in China. Renewable Energy
30. Zhang L, Saydaliev HB, Ma X (2022) Does green finance investment and technological innovation improve renewable energy efficiency and sustainable development goals. Renewable Energy
31. Qadir SA, Al-Motairi H, Tahir F, Al-Fagih L (2021) Incentives and strategies for financing the renewable energy transition: a review. Energy Rep
32. Govindan K, Shankar KM, Kannan D (2020) Achieving sustainable development goals through identifying and analyzing barriers to industrial sharing economy: a framework development. Int J Prod Econ 227
33. Udemba EN, Güngör H, Bekun FV, Kirikkaleli D (2021) Economic performance of India amidst high CO_2 emissions. Sustain Prod Consumption 27:52–60
34. Hayashi D (2020) Green innovation in China and India: technological learning, technology transfer, and public policies in the wind power and Solar Photovoltaic Industries
35. Al-Shetwi AQ (2022) Sustainable development of renewable energy integrated power sector: trends, environmental impacts, and recent challenges. Sci Total Environ
36. Kumar M (2020) Social, economic, and environmental impacts of renewable energy resources. In: Wind solar hybrid renewable energy system
37. Sulyok A (2022) Green aspects of the government securities purchase programme and a snapshot of Hungarian green government bonds. Monetary Policy Sustain
38. Bertoldi P, Economidou M, Palermo V, Boza-Kiss B, Todeschi V (2021) How to finance energy renovation of residential buildings: review of current and emerging financing instruments in the EU. Wiley Interdisc Rev: Energy Environ 10(1):e384

39. Mahmood A, Qasim M, Khan MF (2021) Renewable energy policy around the globe. Eng Appl Sci Lett
40. Oyekale J, Petrollese M, Tola V, Cau, G. (2020). Impacts of renewable energy resources on effectiveness of grid-integrated systems: succinct review of current challenges and potential solution strategies. Energies
41. Gürsan C, de Gooyert V (2021) The systemic impact of a transition fuel: does natural gas help or hinder the energy transition? Renew Sustain Energy Rev
42. Neves SA, Marques AC (2021) The substitution of fossil fuels in the US transportation energy mix: are emissions decoupling from economic growth? Res Transp Econ
43. Zeng J, Bao R, McFarland M (2022) Clean energy substitution: the effect of transitioning from coal to gas on air pollution. Energy Econ
44. Upreti G (2023) Climate change and its threat to humanity in the anthropocene. In: Ecosociocentrism: the earth first paradigm for sustainable living. Springer Nature Switzerland, Cham, pp 137–162
45. Adedoyin FF, Bekun FV, Driha OM, Balsalobre-Lorente D (2020) The effects of air transportation, energy, ICT and FDI on economic growth in the industry 4.0 era: evidence from the United States. Technol Forecast Soc Chang 160
46. Haldar A, Sethi N (2022) Environmental effects of information and communication technology-exploring the roles of renewable energy, innovation, trade and financial development. Renew Sustain Energy Rev
47. Bag S, Yadav G, Dhamija P, Kataria KK (2021) Key resources for industry 4.0 adoption and its effect on sustainable production and circular economy: an empirical study. J Cleaner Prod
48. Bai C, Dallasega P, Orzes G, Sarkis J (2020) Industry 4.0 technologies assessment: a sustainability perspective. Int J Prod Econ 229
49. Kumar R, Singh RK, Dwivedi YK (2020) Application of industry 4.0 technologies in SMEs for ethical and sustainable operations: analysis of challenges. J Cleaner Prod
50. Bahoo S, Alon I, Paltrinieri A (2020) Sovereign wealth funds: past, present and future. Int Rev Financ Anal
51. Goldstein I (2023) Information in financial markets and its real effects. Rev Finan
52. Issa S, Gevorkyan AV (2022) Optimal corporate leverage and speculative cycles: an empirical estimation. Struct Change Econ Dyn
53. Simsek A (2021) The macroeconomics of financial speculation. Ann Rev Econ
54. Annan-Aggrey E, Arku G, Atuoye K, Kyeremeh E (2022) Mobilizing 'communities of practice'for local development and accleration of the sustainable development goals. Local Econ 37(3):219–229
55. Shen M, Ma N, Chen Q (2024) Has green finance policy promoted ecologically sustainable development under the constraints of government environmental attention? J Cleaner Prod
56. Huang H, Kong W, Zhou S, Zheng Z, Guo S (2021) A survey of state-of-the-art on blockchains: theories, modelings, and tools. ACM Comput Surv (CSUR) 54(2):1–42
57. Kumar S, Sharma D, Rao S, Lim WM, Mangla SK (2022) Past, present, and future of sustainable finance: insights from big data analytics through machine learning of scholarly research. Ann Oper Res 1–44
58. Wang X, Wang Q (2021) Research on the impact of green finance on the upgrading of China's regional industrial structure from the perspective of sustainable development. Resour Policy
59. Bürer MJ, de Lapparent M, Capezzali M, Carpita M (2022) Governance drivers and barriers for business model transformation in the energy sector. In: Swiss energy governance: political, economic and legal challenges and opportunities in the energy transition, pp 195–243
60. Debrah C, Chan APC, Darko A (2022) Green finance gap in green buildings: a scoping review and future research needs. Build Environ
61. Habiba U, Xinbang C, Anwar A (2022) Do green technology innovations, financial development, and renewable energy use help to curb carbon emissions? Renewable Energy
62. Jayachandran M, Gatla RK, Rao KP, Rao GS, Mohammed S, Milyani AH et al (2022) Challenges in achieving sustainable development goal 7: affordable and clean energy in light of nascent technologies. Sustain Energy Technol Assessments 53

63. Nevzorova T, Karakaya E (2020) Explaining the drivers of technological innovation systems: the case of biogas technologies in mature markets. J Cleaner Prod
64. Jaiswal KK, Chowdhury CR, Yadav D, Verma R, Dutta S, Jaiswal KS, Karuppasamy KSK (2022) Renewable and sustainable clean energy development and impact on social, economic, and environmental health. Energy Nexus 7:100118
65. Usman FO, Ani EC, Ebirim W, Montero DJP, Olu-lawal KA, Ninduwezuor-Ehiobu N (2024) Integrating renewable energy solutions in the manufacturing industry: challenges and opportunities: a review. Eng Sci Technol J 5(3):674–703
66. Chan S, Boran I, Van Asselt H, Ellinger P, Garcia M, Hale T et al (2021) Climate ambition and sustainable development for a new decade: a catalytic framework. Global Policy 12(3):245–259
67. Elia G, Margherita A, Petti C (2020) Building responses to sustainable development challenges: a multistakeholder collaboration framework and application to climate change. Bus Strateg Environ 29(6):2465–2478
68. Bistline JET (2021) Roadmaps to net-zero emissions systems: emerging insights and modeling challenges. Joule
69. McCracken-Hewson MD, McCracken-Hewson DM (2023) Brookfield and MidOcean application for merger authorisation for proposed acquisition of origin energy
70. Szepanski A (2022) The financial system and the state. Financial capital in the 21st century: a new theory of speculative capital. Springer International Publishing, Cham, pp 219–267
71. Polzin F, Sanders M (2020) How to finance the transition to low-carbon energy in Europe? Energy Policy
72. Epstein B, Shapiro AF (2021) Increasing domestic financial participation: implications for business cycles and labor markets. Rev Econ Dyn
73. Luciani G (2022) Macroeconomics of the energy transition. The Palgrave handbook of international energy economics. Springer International Publishing, Cham, pp 495–509
74. Yu W, Jin X (2024) Economic policy uncertainty and firm performance: evidence from the energy-intensive industry in China. Energy Environ
75. Stefani U, Schiavone F, Laperche B, Burger-Helmchen T (2020) New tools and practices for financing novelty: a research agenda. Eur J Innov Manag 23(2):314–328
76. Tirumala RD, Tiwari P (2022) Innovative financing mechanism for blue economy projects. Mar Policy
77. Liu H, Khan AR, Aslam S, Rasheed AK, Mohsin M (2022) Financial impact of energy efficiency and energy policies aimed at power sector reforms: mediating role of financing in the power sector. Environ Sci Pollut Res 1–14
78. Qamruzzaman M, Kler R (2023) Do clean energy and financial innovation induce SME performance?: Clarifying the nexus between financial innovation, technological innovation, clean energy, environmental degradation, and SMEs performance in Bangladesh. Int J Energy Econ Policy 13(3):313–324
79. Razzaq A, Sharif A, Ozturk I, Skare M (2023) Asymmetric influence of digital finance, and renewable energy technology innovation on green growth in China. Renewable Energy
80. Hu G, Chen C, Lu HT, Wu Y, Liu C, Tao L et al (2020) A review of technical advances, barriers, and solutions in the power to hydrogen (P2H) roadmap. Engineering 6(12):1364–1380
81. Shah KJ, Pan SY, Lee I, Kim H, You Z, Zheng JM, Chiang PC (2021) Green transportation for sustainability: review of current barriers, strategies, and innovative technologies. J Cleaner Prod 326
82. Butt OM, Zulqarnain M, Butt TM (2021) Recent advancement in smart grid technology: future prospects in the electrical power network. Ain Shams Eng J
83. Pareek A, Dom R, Gupta J, Chandran J, Adepu V, Borse PH (2020) Insights into renewable hydrogen energy: recent advances and prospects. Mater Sci Energy Technol 3:319–327
84. Su Y, Fan Q (2022) Renewable energy technology innovation, industrial structure upgrading and green development from the perspective of China's provinces. Technol Forecast Soc Change

85. Akpan J, Olanrewaju O (2023) Sustainable energy development: history and recent advances. Energies
86. Ansari D, Holz F (2020) Between stranded assets and green transformation: fossil-fuel-producing developing countries towards 2055. World Dev
87. Nielsen KS, Nicholas KA, Creutzig F, Dietz T, Stern PC (2021) The role of high-socioeconomic-status people in locking in or rapidly reducing energy-driven greenhouse gas emissions. Nat Energy
88. Tang KHD (2022) Climate change policies of the four largest global emitters of greenhouse gases: their similarities, differences and way forward. J Energy Res Rev
89. Jefferson M (2020) A crude future? COVID-19s challenges for oil demand, supply and prices. Energy Res Soc Sci
90. Husin H, Zaki M (2021) A critical review of the integration of renewable energy sources with various technologies. Prot Control Mod Power Syst 6(1):1–18
91. Sinsel SR, Riemke RL, Hoffmann VH (2020) Challenges and solution technologies for the integration of variable renewable energy sources—a review. Renewable Energy
92. Kabeyi MJB, Olanrewaju OA (2022) Sustainable energy transition for renewable and low carbon grid electricity generation and supply. Front Energy Res
93. Lima MA, Mendes LFR, Mothé GA, Linhares FG, De Castro MPP, Da Silva MG, Sthel MS (2020) Renewable energy in reducing greenhouse gas emissions: reaching the goals of the Paris agreement in Brazil. Environ Dev 33
94. Sun H, Hua W, You M (2023) Blockchain and artificial intelligence technologies for smart energy systems
95. Suvarna M, Yap KS, Yang W, Li J, Ng YT, Wang X (2021) Cyber–physical production systems for data-driven, decentralized, and secure manufacturing—a perspective. Engineering
96. Teng SY, Touš M, Leong WD, How BS, Lam HL, Máša V. (2021) Recent advances on industrial data-driven energy savings: digital twins and infrastructures. Renew Sustain Energy Rev 135
97. Walker IO, Wirl F (1993) Irreversible price-induced efficiency improvements: theory and empirical application to road transportation. Energy J 183–205
98. Ajanovic A, Haas R (2018) Economic prospects and policy framework for hydrogen as fuel in the transport sector. Energy Policy 123:280–288
99. Corporate Finance Institute (2021) Risk—definition, types, adjusment and measurement. Corporate Finance Institute. Available from: https://corporatefinanceinstitute.com/resources/knowledge/finance/risk/
100. Sachs JD, Schmidt-Traub G, Mazzucato M, Messner D, Nakicenovic N, Rockström J (2019) Six transformations to achieve the sustainable development goals. Nat Sustain 2(9):805–814
101. Gallaher MP, Link AN, O'Connor A (2012) Public investments in energy technology. Edward Elgar Publishing
102. Nakicenovic N, Alcamo J, Davis G, Vries BD, Fenhann J, Gaffin S et al (2000) Special report on emissions scenarios
103. Riahi K, Grübler A, Nakicenovic N (2007) Scenarios of long-term socio-economic and environmental development under climate stabilization. Technol Forecast Soc Chang 74(7):887–935
104. Pacala S, Socolow R (2004) Stabilization wedges: solving the climate problem for the next 50 years with current technologies. Science 305(5686):968–972
105. Markowitz HM (1952) Portfolio selection. J Financ 7:77–91
106. Markowitz HM (1959) Portfolio selection: efficient diversification of investments. Wiley, New York
107. Bazilian M, Roques F (2008) Analytic methods for energy diversity and security. Applications of mean variance portfolio theory. A tribute to Shimon Awerbuch
108. Ozorhon B, Batmaz A, Caglayan S (2018) Generating a framework to facilitate decision making in renewable energy investments. Renew Sustain Energy Rev 95:217–226
109. Sim J, Kim CS (2019) The value of renewable energy research and development investments with default consideration. Renewable Energy 143:530–539

110. Yang D-X, Chen Z-Y, Yang Y-C, Nie P-Y (2019) Green financial policies and capital flows. Phys Stat Mech Appl 522:135–146
111. Stucki T (2019) Which firms benefit from investments in green energy technologies?—The effect of energy costs. Res Policy 48(3):546–555
112. Bauwens T (2019) Analyzing the determinants of the size of investments by community renewable energy members: findings and policy implications from Flanders. Energy Policy 129:841–852
113. Kumar A, Sah B, Singh AR, Deng Y, He X, Kumar P, Bansal RC (2017) A review of multi criteria decision making (MCDM) towards sustainable renewable energy development. Renew Sustain Energy Rev 69:596–609
114. Bandyopadhyay S, Foo DC, Tan RR (2016) Feeling the pinch? Chem Eng Prog 112(11):46–49
115. Roychaudhuri PS, Bandyopadhyay S (2018) Financial pinch analysis: minimum opportunity cost targeting algorithm. J Environ Manage 212:88–98
116. Hamed TA, Bressler L (2019) Energy security in Israel and Jordan: the role of renewable energy sources. Renewable Energy 135:378–389
117. Bointner R, Pezzutto S, Grilli G, Sparber W (2016) Financing innovations for the renewable energy transition XE "energy transition" in Europe. Energies 9(12):990
118. de Negri JF, Pezzutto S, Gantioler S, Moser D, Sparber W (2020) A comprehensive analysis of public and private funding for photovoltaics research and development in the European Union, Norway, and Turkey. Energies 13.https://doi.org/10.3390/en13112743
119. Schiffer HW, Trüby J (2018) A review of the German energy transition: taking stock, looking ahead, and drawing conclusions for the Middle East and North Africa. Energy Transit 2:1–14
120. Al Arrouqi RA, Ellabban O, Rasheed MB, Al-Fagih L (2019) An assessment of different electricity tariffs on residential photovoltaic system profitability: Australian case study. In: 2nd International conference on smart grid and renewable energy. SGRE, IEEE, pp 1–6. https://doi.org/10.1109/SGRE46976.2019.9021078
121. Yao L, Yang B, Cui H, Zhuang J, Ye J, Xue J (2016) Challenges and progresses of energy storage technology and its application in power systems. J Mod Power Syst Clean Energy 4:519–528. https://doi.org/10.1007/s40565-016-0248-x
122. Gallo AB, Simões-Moreira JR, Costa HKMM, Santos MM, Moutinho dos Santos E (2016) Energy storage in the energy transition context: a technology review. Renew Sustain Energy Rev 65:800–822. https://doi.org/10.1016/j.rser.2016.07.028
123. Parra D, Swierczynski M, Stroe DI, Norman SA, Abdon A, Worlitschek J, O'Doherty T, Rodrigues L, Gillott M, Zhang X, Bauer C, Patel MK (2017) An interdisciplinary review of energy storage for communities: challenges and perspectives. Renew Sustain Energy Rev 79:730–749. https://doi.org/10.1016/j.rser.2017.05.003
124. Tan RR, Aviso KB, Ng DKS (2019) Optimization models for financing innovations in green energy technologies. Renew Sustain Energy Rev 113:109258
125. He DX (2016) Coping with climate change and China's wind energy sustainable development. Adv Clim Chang Res 7(1–2):3–9. https://doi.org/10.1016/j.accre.2016.06.003
126. He D-X (2016) Coping with climate change and China's wind energy sustainable development. Adv Clim Change Res 7(1–2):3–9. https://doi.org/10.1016/j.accre.2016.06.003
127. Lipp J (2007) Lessons for effective renewable electricity policy from Denmark, Germany and the United Kingdom. Energy Policy 35:5481–5495. https://doi.org/10.1016/j.enpol.2007.05.015
128. Zhi Q, Sun H, Li Y, Xu Y, Su J (2014) China's solar photovoltaic policy: an analysis based on policy instruments. Appl Energy 129:308–319
129. Zhi Q, Sun H, Li Y, Xu Y, Su J (2014) China's solar photovoltaic policy: an analysis based on policy instruments. Appl Energy 129:308–319. https://doi.org/10.1016/j.apenergy.2014.05.014
130. Mihaylov M, Rădulescu R, Razo-Zapata I, Jurado S, Arco L, Avellana N, Nowé A (2019) Comparing stakeholder incentives across state-of-the-art renewable support mechanisms. Renew Energy 131:689–699. https://doi.org/10.1016/j.renene.2018.07.069

131. Funke F, Mattauch L (2018) Why is carbon pricing in some countries more successful than in others?—Our World in Data [WWW Document]. https://ourworldindata.org/carbon-pricing-popular
132. Yildiz Ö, Rommel J, Debor S, Holstenkamp L, Mey F, Müller JR, Radtke J, Rognli J (2015) Renewable energy cooperatives as gatekeepers or facilitators? Recent developments in Germany and a multidisciplinary research agenda. Energy Res Soc Sci 6:59–73. https://doi.org/10.1016/j.erss.2014.12.001
133. Bowen A, Fankhauser S (2011) The green growth narrative: paradigm shift or just spin? Glob Environ Chang 21(4):1157–1159
134. Matsuo T, Schmidt TS (2017) Hybridizing low-carbon technology deployment policy and fossil fuel subsidy reform: a climate finance perspective. Environ Res Lett 12(1):014002
135. Baker L, Sovacool BK (2017) The political economy of technological capabilities and global production networks in South Africa's wind and solar photovoltaic (PV) industries. Polit Geogr 60:1–12
136. Gillingham K, Sweeney J (2010) Market failure and the structure of externalities. In: Harnessing renewable energy in electric power systems. Routledge, pp 69–91
137. Howlett M, Cashore B (2009) The dependent variable problem in the study of policy change: understanding policy change as a methodological problem. J Comp Policy Anal 11(1):33–34
138. Macchiavello E, Siri M (2022) Sustainable finance and fintech: can technology contribute to achieving environmental goals? A preliminary assessment of 'green fintech' and 'sustainable digital finance.' Eur Company Financ Law Rev 19(1):128–174
139. Ziolo M, Bak I, Cheba K (2021) The role of sustainable finance in achieving sustainable development goals: does it work? Technol Econ Dev Econ 27(1):45–70

Promoting the Energy Transition Throughout Dealing with the Climate Change Issue

Hamza El Kezazy and Yassine Hilmi

Abstract Climate change, driven largely by the burning of fossil fuels like coal, oil, and natural gas, is causing significant environmental and societal impacts worldwide. Transitioning to renewable energy sources such as solar, wind, and hydroelectric power is crucial in mitigating these effects by reducing greenhouse gas emissions. The energy transition encompasses this shift, aiming for a sustainable, low-carbon future. This chapter discusses the energy transition as an essential response to the challenges of climate change, emphasizing the reduction of greenhouse gas emissions and the enhancement of sustainability. The growing impact of climate change, primarily driven by the combustion of fossil fuels, has necessitated a shift toward renewable energy sources and innovative energy solutions. This shift is critical to mitigating environmental impacts and fostering a more sustainable future. The chapter explores technological innovations, international policies like the Paris Agreement, and national strategies from countries such as Germany, China, and the USA, which promote increased use of renewable energies. Practical examples illustrate the progress and challenges encountered. The chapter also highlights the importance of digital technologies and AI in efficient energy management and environmental monitoring. Challenges such as resistance from fossil fuel industries, the need for financing, and social justice are discussed. In conclusion, coordinated action among governments, the private sector, and communities is essential for a successful energy transition.

Keywords Energy transition · Climate change · Technological innovations · Energy management · Environmental monitoring

H. E. Kezazy (✉)
Economie, Management et Finance des Organisations (EEMFO), Université Sultane Moulay Slimane (USMS), Beni Mellal, Morocco
e-mail: hamza.elkezazy@usms.ac.ma

Y. Hilmi
Laboratory of Studies and Research in Economic Sciences and Management National School of Commerce and Management, Chouaib Doukkali University, El Jadida, Morocco
e-mail: hilmi.y@ucd.ac.ma

F. Latrache et al. (eds.), *Digital Technology for an Innovative Energy Transition*, Green Energy and Technology, https://doi.org/10.1007/978-3-031-89657-6_4

1 Introduction

Climate change, driven by the accumulation of greenhouse gases from fossil fuel combustion, poses significant environmental and socio-economic challenges globally. Transitioning from fossil energy, which includes coal, oil, and natural gas, to renewable energy sources such as solar, wind, and hydroelectric power is essential for reducing carbon emissions. The energy transition involves not only the adoption of cleaner technologies but also the restructuring of economic and social systems to support sustainable and resilient energy practices.

Climate change is an undeniable reality of the twenty-first century, characterized by rising global temperatures, shifting precipitation patterns, and an increased frequency of extreme weather events [1]. The primary cause of climate change is the increased concentration of greenhouse gases in the atmosphere, mainly due to the burning of fossil fuels for energy production, deforestation, and industrial activities. These changes lead to severe environmental, economic, and social impacts, including sea level rise, loss of biodiversity, and adverse health effects.

Fossil energy refers to energy sources derived from the remains of ancient plants and animals, primarily coal, oil, and natural gas. These sources are finite and contribute significantly to greenhouse gas emissions when burned for energy.

The energy transition involves shifting from fossil fuels to renewable and less carbon-intensive energy sources. This transition is crucial for reducing greenhouse gas emissions and mitigating the effects of climate change. Renewable energy sources, such as solar, wind, and hydroelectric power, are sustainable alternatives to fossil fuels. They are inexhaustible and have a minimal environmental footprint compared to traditional energy sources.

Renewable energy is defined as energy generated from natural processes that are continuously replenished. This includes sunlight, wind, rain, tides, waves, and geothermal heat. Unlike fossil fuels, renewable energy sources do not deplete over time and have a lower impact on the environment.

Sustainability refers to meeting the needs of the present without compromising the ability of future generations to meet their own needs. It encompasses environmental, economic, and social dimensions, ensuring that development can continue indefinitely without depleting natural resources or causing ecological damage.

Faced with the challenges of climate change, the energy transition emerges as a crucial strategy. According to Masson-Delmotte et al. [2], this transition is vital for limiting the increase in global temperatures and mitigating environmental impacts. The energy transition is central to global efforts to meet the objectives of the Paris Agreement on climate, which aims to keep the global temperature rise well below 2 °C [3]. Creutzig et al. [4] in Environmental Research Letters emphasize that without a rapid and extensive energy transition, these climate goals will remain unattainable, exacerbating global environmental and economic risks.

Efforts to promote energy transition and combat climate change have been significant. Initiatives such as COP28, the 28th Conference of the Parties to the United Nations Framework Convention on Climate Change (UNFCCC), aim to strengthen global commitments to reduce greenhouse gas emissions and enhance climate resilience. COP28 focuses on accelerating the deployment of renewable energy technologies, increasing energy efficiency, and fostering international cooperation to address climate change.

The theoretical framework used to analyze the energy transition includes various analytical frameworks and key definitions. This framework helps in understanding the dynamics of energy systems, the role of policy and technological innovation, and the socio-economic implications of transitioning to a sustainable energy future.

This chapter aims to explore the promotion of energy transition in the global fight against climate change by examining technological advances, policy frameworks, and economic challenges associated with this transition, providing a detailed analysis of successful strategies and obstacles encountered. The growing urgency to address climate change with sustainable and effective solutions has heightened interest in this topic. The energy transition is not only about environmental protection but also about redefining the economic and social foundations of societies worldwide. This chapter considers how different regions and sectors can effectively contribute to the energy transition while considering local economic and social realities, and how policies and technologies can be aligned to achieve a just and effective energy transformation. The analysis is based on a review of academic literature, reports from international organizations, and sector-specific case studies, using both quantitative and qualitative data to assess the effectiveness of various approaches to energy transition.

To understand and promote energy transition strategies, several models must be analyzed. These models provide perspectives on the dynamics of change within energy systems, drawing on theories from economics, sociology, and environmental studies. The Socio-Technical Transition Model (Multi-Level Perspective, MLP), developed by Geels [5], is widely used to study transitions in socio-technical systems, such as energy systems. This model describes three interconnected levels: niches (where radical innovations originate), regimes (dominant structures or status quos), and landscapes (external factors not influenced by individual actions). Transitions are seen as resulting from the alignment of these levels, leading to substantial changes in how energy is produced and consumed [5, 6]. The Technological Innovation Systems Theory, introduced by Carlsson and Stankiewicz [7], explains how technological innovations emerge, diffuse, and are supported or inhibited by configurations of institutional and organizational factors. This framework is particularly relevant for examining how new energy technologies such as solar and wind can be encouraged by public policies, corporate strategies, and market dynamics [8]. Rogers [9] Diffusion of Innovations Theory provides a model for understanding how, why, and at what rate new ideas and technologies spread across cultures. This theory is crucial for understanding the factors that affect the adoption of renewable technologies, including the characteristics of innovations, communication among participants, time, and the social context in which diffusion takes place. Finally, Environmental Economics and Transaction Costs Theory offer insights into the costs associated with different energy systems and the economic barriers to energy transition. These costs include not only direct financial expenses but also transaction costs associated with modifying energy infrastructures and implementing new technologies [10, 11].

The chapter explores the energy transition in the digital age, starting with an analysis of theoretical foundations, including various analytical frameworks and key definitions, and then addressing global policies and strategies influenced by international agreements like the Paris Agreement, and examining specific initiatives from countries such as Germany, China, and the USA. The focus shifts to green technologies and innovations, particularly advances in solar, wind, and solid-state battery technologies, with practical cases illustrating successes and challenges encountered. Economic and social implications are analyzed to assess the impacts of these transitions, and the chapter concludes with future perspectives, highlighting the importance of digital technologies and AI in promoting a sustainable energy future.

2 Power Plant Based on Fossil Energy

Fossil energy is derived from the decomposition of ancient organic matter, such as dead plants and animals, which have been transformed over millions of years under heat and pressure in the Earth's crust. The main forms of fossil energy include coal, oil, and natural gas. These energy sources are non-renewable because they form over geological timescales, far exceeding the periods of human consumption. When burned to produce energy, these fuels release carbon dioxide (CO_2) and other greenhouse gases, significantly contributing to climate change [1]. Consequently, the energy transition refers to a structural change in the global energy system, characterized by a significant replacement of fossil fuel-based energy sources with renewable and less carbon-intensive sources. According to Smil [12], the energy transition is not just a technical replacement but also an adjustment of the economic and social infrastructures that support energy production and consumption. Then renewable energy is defined as playing a crucial role in this transition, as it includes energy sources generated from natural processes that are continually replenished, such as solar, wind, hydroelectric, biomass, and geothermal energy. According to REN21 [13], these technologies are central to reducing greenhouse gas emissions and meeting the goals of the Paris Agreement on climate. As result, sustainability in the energy context involves managing resources to ensure long-term ecological, economic, and social stability. The 1987 Brundtland Report defines sustainable development as "development that meets the needs of the present without compromising the ability of future generations to meet their own." In the energy sector, this translates to investing in technologies and practices that minimize environmental impact while promoting economic and social equity [14].

Table 1 showcases various power plants based on fossil energy, highlighting their capacities, technologies used, and operational status. Notably, the Taichung Power Plant in Taiwan and the Belchatow Power Plant in Poland have the highest capacities, at 5500 MW and 5472 MW, respectively. The majority of these plants utilize coal as their primary energy source, with only a few, like Taweelah A2 in the United Arab Emirates and W.A. Parish in the USA, incorporating natural gas technology.

Table 1 Fossil fuel power plants

Power plant	Country	Capacity (MW)	Technology used	Current status
Jorf Lasfar	Morocco	2056	Coal	Operational
Mohammedia	Morocco	600	Coal	Operational
Safi	Morocco	1386	Coal	Operational
Taweelah A2	United Arab Emirates	710	Natural gas	Operational
RWE power plant	Germany	2200	Coal	Operational
Taichung power plant	Taiwan	5500	Coal	Operational
Suralaya power plant	Indonesia	4025	Coal	Operational
Belchatow power plant	Poland	5472	Coal	Operational
Neurath power plant	Germany	4400	Coal	Operational
W.A. Parish	USA	3653	Coal and natural gas	Operational

3 Challenges and Opportunities of Promoting Energy Transition

3.1 Challenges of the Energy Transition

3.1.1 High Initial Costs

The energy transition requires considerable investments in infrastructure. The construction of wind and solar farms, as well as the development of smart grids and energy storage systems, involves very high initial costs. These investments are particularly difficult to bear for developing countries, which often lack the necessary financial resources. According to the International Energy Agency [15], global investments in clean energy must triple in the coming decades to meet international climate goals.

3.1.2 Dependence on Fossil Fuels

Most global economies are heavily dependent on fossil fuels for their energy production. This dependence creates economic and social inertia that makes the transition difficult. The coal, oil, and natural gas industries provide significant jobs and revenues in many regions. The transition to renewable energy could cause significant economic and social disruptions, with job losses in traditional sectors [16].

3.1.3 Immature Technologies

Some energy storage and renewable energy production technologies are not yet mature enough to be deployed on a large scale reliably and economically. For example, energy storage batteries, necessary to compensate for the intermittency of renewable energy sources such as solar and wind, are still expensive, and have limitations in terms of capacity and lifespan [17].

3.1.4 Management and Regulation Issues

The energy transition also requires an appropriate regulatory framework and favorable policies. However, establishing these regulations can be complex and face resistance from various interest groups, notably the fossil fuel lobbies. The International Energy Agency [15] emphasizes the importance of international cooperation to harmonize regulations and facilitate the energy transition on a global scale.

3.2 Opportunities of the Energy Transition

3.2.1 Economic Opportunities

The energy transition offers considerable economic opportunities. Firstly, it is a driver of green job creation. The renewable energy sector is booming and is expected to generate millions of new jobs in the construction, operation, and maintenance of renewable infrastructure. According to IRENA [16], the number of jobs in the renewable energy sector could reach 42 million by 2050.

Furthermore, the energy transition stimulates technological innovation. The development of new technologies, such as energy storage systems, smart grids, and carbon capture and storage technologies, paves the way for new industries and markets. These innovations can also improve energy efficiency and reduce costs in the long term [17].

3.2.2 Environmental Opportunities

The environmental benefits of the energy transition are equally important. Moving to renewable energy significantly reduces greenhouse gas emissions, thus contributing to the fight against climate change. The Intergovernmental Panel on Climate Change [18] highlights that limiting global warming to 1.5 °C requires rapid and deep reductions in CO_2 emissions.

Moreover, renewable energy sources, being inexhaustible, reduce the pressure on limited natural resources and promote a more sustainable use of them. This is particularly crucial in the current context of fossil resource depletion and environmental degradation [15].

4 Global Policies and Strategies About Energy Transition and Climate Change

4.1 Global Policies and Strategies About Energy Transition and Climate Change

4.1.1 International Agreements and Global Initiatives

The global community has recognized the urgent need to address climate change, leading to the adoption of several landmark agreements and initiatives. The Paris Agreement, reached in 2015, stands out as a historic accord adopted by 196 parties to the United Nations Framework Convention on Climate Change (UNFCCC). Its primary aim is to limit global warming to well below 2 °C above pre-industrial levels, while pursuing efforts to limit the temperature increase to 1.5 °C. To achieve these goals, countries are required to submit Nationally Determined Contributions (NDCs) outlining their efforts to reduce emissions and enhance resilience, with a commitment to review and strengthen these commitments every five years to increase their ambition.

Looking ahead, the 28th Conference of the Parties to the UNFCCC (COP28) represents a crucial opportunity to strengthen global commitments to reduce greenhouse gas emissions and improve climate resilience. COP28 is expected to focus on accelerating the deployment of renewable energy technologies, increasing energy efficiency across sectors, and promoting international cooperation to combat climate change. By fostering collaboration and knowledge sharing among nations, COP28 aims to drive collective action toward achieving the objectives of the Paris Agreement.

One of the key initiatives launched in conjunction with the Paris Agreement is the Mission Innovation Initiative, announced at COP21 in 2015. This initiative brings together 24 countries and the European Commission, all committed to doubling their respective clean energy research and development budgets over five years. By accelerating clean energy innovation, Mission Innovation aims to make clean energy more affordable, attractive, and accessible, thereby driving the transition toward a sustainable, low-carbon future.

4.1.2 National and Regional Policies

Energy transition has become a global priority in response to the increasing challenges of climate change and energy security. Several countries have committed to policies and strategies to promote renewable energies and reduce their dependence on fossil fuels. Among these countries, Morocco stands out for its commitment to an ambitious energy transition.

Morocco has adopted a strategic approach to energy, defined by its National Energy Strategy of 2009, which aims to increase the share of renewable energies in its energy mix to 52% by 2030. This strategy has led to the realization of emblematic projects such as the Noor Ouarzazate and Noor Laayoune solar parks, as well as wind farms like Tarfaya. Additionally, the Renewable Energy Law of 2010 has stimulated private investments in the renewable energy sector by allowing the production, export, and sale of electricity produced from renewable sources.

In parallel, Morocco has also integrated measures for the sustainable management of natural resources into its Green Morocco Plan of 2008, primarily focused on agriculture. These measures contribute indirectly to the reduction of greenhouse gas emissions and the promotion of a greener economy.

Similarly, Germany has adopted a similar approach with its Energiewende policy, which aims to gradually reduce its dependence on nuclear and fossil fuels while increasing the use of renewable energies and improving energy efficiency. This policy is supported by the Renewable Energy Sources Act (EEG), which encourages the production of electricity from renewable sources through guaranteed feed-in tariffs and market premiums.

China, as the world's largest emitter of greenhouse gases, has implemented an ambitious five-year energy plan aimed at increasing its renewable energy capacity and reducing the carbon intensity of its economy. Additionally, China launched its National Emissions Trading System (ETS) in 2021, the world's largest, aiming to reduce emissions by setting a cap for major industries and allowing the trading of emission permits.

In the USA, the Biden administration has reaffirmed its commitment to climate action by launching the Clean Energy Revolution and Environmental Justice Plan, which aims to achieve 100% clean electricity by 2035 and net-zero emissions by 2050. Additionally, the USA rejoined the Paris Agreement in 2021, committing to reduce its greenhouse gas emissions and support international efforts against climate change.

4.1.3 Recent Innovations Driving the Energy Transition

The global energy landscape is undergoing a profound transformation, driven by the urgent need to address climate change and transition to sustainable energy sources. Recent innovations in renewable energy technologies, energy storage solutions, grid modernization, and policy frameworks are playing a crucial role in accelerating this transition.

One of the key innovations in the renewable energy sector is the development of floating solar panels. By installing solar panels on bodies of water, this technology not only increases energy production but also maximizes land use efficiency, making it particularly attractive in densely populated areas. Additionally, advancements in biogas production from organic waste are helping to reduce landfill emissions and provide a renewable source of energy [19].

In the wind energy sector, advanced wind turbine designs and materials are significantly improving efficiency and reducing costs. These innovations, coupled with the increasing scale of wind farms, are making wind energy more competitive with traditional fossil fuel sources [20].

Another significant development is the emergence of solar paint, a special paint that can generate electricity from sunlight. This innovation offers a low-cost way to integrate solar power into buildings, potentially revolutionizing the way we think about solar energy deployment in urban areas [21].

In the realm of energy storage, advancements in battery technology are enabling the storage of excess renewable energy for use during periods of low production. Similarly, developments in hydrogen storage are paving the way for the use of hydrogen as a clean energy carrier, particularly in transportation [22].

Grid modernization is also a critical aspect of the energy transition. Smart grids, for example, are intelligent electricity distribution networks that optimize energy use, reduce waste, and integrate renewable energy sources. These grids enable greater flexibility and reliability in energy supply, essential for the integration of variable renewable energy sources like solar and wind power [23].

On the policy and financial front, innovations such as green bonds and carbon pricing mechanisms are providing new avenues for financing renewable energy projects and incentivizing emission reductions. Renewable energy auctions, another innovative policy tool, are ensuring cost-effective deployment of renewable energy projects through competitive bidding processes [24].

5 Green Technologies and Recent Innovations

5.1 Green Technologies

Green technologies play a central role in the energy transition by providing renewable and efficient solutions to replace fossil fuels. These primarily include solar, wind, and hydroelectric energy, each characterized by specific innovations that enhance their efficiency and integration into energy networks.

Solar energy, harnessed through photovoltaic panels and solar thermal power plants, is one of the most promising renewable energy sources. According to Razykov et al. [25], technological advancements in semiconductor materials, such as monocrystalline silicon and perovskites, have significantly increased the efficiency of solar panels. Additionally, innovations such as solar trackers that follow the sun's trajectory to maximize exposure, and integrated energy storage systems, allow for more flexible and extensive use of solar energy.

Wind energy is captured using wind turbines that convert the kinetic energy of the wind into electricity. With significant developments in turbine design, including the increase in blade size and improvements in control systems, wind farms have become more efficient and less costly. According to Burton et al. [26], innovations in composite materials for blades and aerodynamic improvements have also reduced noise and increased the durability of turbines.

Hydroelectric energy, produced by converting the potential energy of water into electricity, is a stable and mature source of renewables. Paish [27] notes that innovations in the design of hydraulic turbines and water management systems now allow for energy production more adaptable to variations in energy demand. Small hydroelectric plants, in particular, represent an increasingly viable option for remote communities and off-grid applications.

5.2 Recent Innovations

Technological innovation in the energy sector plays a key role in transitioning to cleaner and more sustainable energy sources. Recent advances offer promising prospects for improving efficiency, reducing costs, and increasing the acceptance of renewable technologies. Here are some of the most impactful and recent innovations in the field.

- Solid-State Batteries:

Solid-state batteries represent a major advance in energy storage technology, offering superior capacities in terms of energy density and safety compared to traditional lithium-ion batteries. According to Janek and Zeier (2016), these batteries use solid electrolytes instead of liquids, making them less likely to catch fire and more efficient at extreme temperatures. This technology could revolutionize the electric vehicle market and facilitate the integration of intermittent renewable energies such as solar and wind.

- Perovskite Solar Cells:

Perovskite-based solar cells are a promising innovation in photovoltaics due to their low cost and potentially high efficiency. Park [28] reports that these materials offer a light-to-electricity conversion efficiency that could surpass that of silicon-based solar panels, while being less expensive to produce. Their flexibility and lightness open up new applications in integrating solar energy into buildings and portable devices.

- Floating Wind Turbines

Floating wind turbines allow for the exploitation of wind energy in deep-sea environments, where winds are stronger and more consistent. According to Sclavounos et al. (2018), this technology uses floating platforms anchored to the seabed, enabling installation in deep waters where traditional wind turbines are not viable. This could significantly increase the potential for wind energy production, especially in regions previously untapped.

- Green Hydrogen

Hydrogen produced by electrolysis of water using electricity from renewable sources offers a promising solution for long-term energy storage and as a substitute for fossil fuels in transportation and industry. According to Bertuccioli et al. [29], green hydrogen could play a crucial role in decarbonizing sectors that are difficult to electrify, such as steelmaking and maritime transport.

5.3 Upcoming Challenges

As the world progresses toward an energy transition, several significant challenges must be addressed to ensure the effective and equitable deployment of renewable energies and associated technologies. This section identifies and analyzes potential obstacles that could hinder future progress in this crucial field.

One major challenge is the resistance from fossil fuel industries, which have well-established economic interests and global value chains. According to McGlade and Ekins (2015), the transition to renewable energies threatens existing investments and infrastructures of fossil fuels, leading to strong political and economic opposition. This resistance can slow the adoption of policies favorable to renewable energies and negatively influence public opinion.

Another challenge is the intermittency of renewable sources such as solar and wind energy, which poses a significant technical challenge for managing energy networks. Renewable energy production is not always in phase with demand peaks, requiring advanced energy storage solutions and smart grids to maintain network stability. Acker et al. (2021) emphasize that the development of economically viable energy storage technologies is crucial for overcoming this challenge.

Financing the energy transition is also a major hurdle, as it requires enormous investments for research and development, as well as for the large-scale deployment of renewable technologies. Jones and Warner (2016) point out that access to capital for renewable energy projects can be limited, especially in developing countries where financial risks are higher. The creation of innovative financing mechanisms and the engagement of international financial institutions are essential for overcoming these financial barriers.

6 Practical Study in Promoting Energy Transition by Addressing Climate Change

The global energy landscape is evolving rapidly, driven by the urgent need to combat climate change and transition to sustainable energy sources. Recent innovations in renewable energy technologies have played a pivotal role in accelerating this transition. This study focuses on practical examples of renewable energy projects around the world, analyzing their current status, challenges faced, and future trends in energy transition to reduce the impact of climate change.

This chapter aims to explore the promotion of energy transition in the global fight against climate change by examining technological advances, policy frameworks, and economic challenges associated with this transition, providing a detailed analysis of successful strategies and obstacles encountered. The growing urgency to address climate change with sustainable and effective solutions has heightened interest in this topic. The energy transition is not only about environmental protection but also about redefining the economic and social foundations of societies worldwide. This chapter considers how different regions and sectors can effectively contribute to the energy transition while considering local economic and social realities, and how policies and technologies can be aligned to achieve a just and effective energy transformation. The analysis is based on a review of academic literature, reports from international organizations, and sector-specific case studies, using both quantitative and qualitative data to assess the effectiveness of various approaches to energy transition.

The present work explores the energy transition in the digital age, starting with an analysis of theoretical foundations, including various analytical frameworks and key definitions, and then addressing global policies and strategies influenced by international agreements like the Paris Agreement, and examining specific initiatives from countries such as Germany, China, and the USA. The focus shifts to green technologies and innovations, particularly advances in solar, wind, and solid-state battery technologies, with practical cases illustrating successes and challenges encountered. Specific examples, such as the Noor Ouarzazate and Noor Midelt concentrated solar plants in Morocco, the Kamuthi Solar Plant in India, and the Gansu Wind Farm in China, highlight both the challenges and opportunities presented by the transition to renewable energy sources. Economic and social implications are analyzed to assess the impacts of these transitions, and the chapter concludes with future perspectives, highlighting the importance of digital technologies and AI in promoting a sustainable energy future (Tables 2 and 3).

- Analysis and Challenges Faced

In Morocco, the Noor Ouarzazate and Noor Laayoune solar plants stand as symbols of the country's dedication to renewable energy. Despite this commitment, challenges such as grid integration and energy storage need to be addressed to facilitate further expansion of renewable energy sources.

Table 2 Power plants using renewable energy

Power plant	Country	Capacity (MW)	Technology used	Current status
Noor Ouarzazate	Morocco	510	Concentrated solar	Operational
Noor Midelt	Morocco	800	Concentrated solar	Under construction
Noor Laayoune	Morocco	80	Photovoltaic solar	Operational
Kamuthi Solar Plant	India	648	Photovoltaic solar	Operational
Solar Star	USA	579	Photovoltaic solar	Operational
Longyangxia Dam	China	1280	Hydroelectric	Operational
Belo Monte Dam	Brazil	11,233	Hydroelectric	Operational
Alta Wind Energy	USA	1548	Wind	Operational
Gansu Wind Farm	China	7965	Wind	Operational
London Array	UK	630	Wind	Operational

Sources Data based on national greenhouse gas emissions reports and environmental inventories of power plants

Table 3 Greenhouse gas emissions

Country	GHG emissions (Mt CO_2e)
Morocco	71.8
India	2654.4
USA	6511.3
China	11,858.1
Brazil	2049.6
UK	435.2

Sources Data based on national greenhouse gas emissions reports and environmental inventories of power plants

India's progress in solar energy is evident through the Kamuthi Solar Plant. However, India's heavy reliance on coal poses a significant obstacle to a smooth transition to renewable energy sources.

Both the USA and China have made significant investments in renewable energy. However, their high emissions levels indicate a continued dependence on fossil fuels. Scaling up renewable energy sources to meet their growing energy demands remains a considerable challenge for these nations.

In Brazil, the Belo Monte Dam represents a significant source of renewable energy. However, it has faced criticism due to its environmental impact. Sustainable development of hydropower is vital to mitigate such concerns.

The UK has achieved success in offshore wind energy with the London Array. Nevertheless, challenges like intermittency and high costs hinder the wider adoption of this renewable energy source in the country.

- Future Trends and Approach

To reduce the impact of climate change, future trends in energy transition focus on several key areas. Firstly, advancements in energy storage are crucial for managing the intermittency of renewable energy sources El Amri et al. [30]. Developing efficient and cost-effective storage solutions will enable a more reliable and sustainable energy supply.

Secondly, grid modernization plays a vital role in the integration of renewable energy [31]. Upgrading grids to accommodate the variable nature of renewable sources and implementing smart grid functionalities will enhance the flexibility and efficiency of energy distribution.

Thirdly, policy support is essential for driving the transition to renewable energy. This includes implementing carbon pricing mechanisms, setting ambitious renewable energy targets, and providing incentives for the adoption of clean energy technologies [32]. These policies can stimulate investment in renewable energy and accelerate the transition away from fossil fuels.

Lastly, international cooperation is key to addressing climate change. Collaborating on a global scale allows for the sharing of best practices, technologies, and resources, facilitating a more coordinated and effective approach to achieving a sustainable energy future.

7 Conclusion

This chapter has explored various aspects of the energy transition, highlighting its critical importance for the sustainable future of our planet. We have discussed technological innovations, economic and social challenges, and global strategies that shape this transition. As highlighted by Aghion et al. [33], the transition to renewable energies is not only imperative for reducing greenhouse gas emissions but also offers substantial economic opportunities in terms of job creation and technological development. However, the challenges remain considerable, particularly in terms of financing, political resistance, and the integration of intermittent energy sources.

To overcome these challenges and accelerate the energy transition, coordinated action between governments, the private sector, and communities is essential. Policymakers must strengthen regulatory and policy frameworks that encourage investments in renewable energies and the development of sustainable energy infrastructures. As indicated by IRENA [34], the implementation of incentive policies, such as subsidies and tax credits for renewable technologies, can stimulate innovation and large-scale adoption. Researchers must continue to explore new energy technologies and the socio-economic implications of their deployment. Finally, it is crucial that citizens are better informed and involved in decision-making processes related to energy to ensure an inclusive and equitable transition [35].

Future research should focus on improving energy storage technologies, a key area for managing the intermittency of renewable energies. According to Acker et al. (2021), exploring advanced solutions such as flow batteries or green hydrogen may provide promising avenues for large-scale energy storage. Additionally, studies on the social impact of the energy transition, particularly in terms of energy justice and the distribution of costs and benefits, remain crucial for ensuring an equitable transition (Jenkins et al. 2016). Finally, increased attention should be given to policies that can effectively integrate new entrants into the energy market, such as microgrids and distributed technologies, into the global energy system.

References

1. IPCC (2021) Climate change 2021: the physical science basis. In: Contribution of working group I to the sixth assessment report of the intergovernmental panel on climate change. Cambridge University Press. https://doi.org/10.1017/9781009157896
2. Masson-Delmotte V et al (eds) (2021) IPCC 2021: climate change 2021: the physical science basis. In: Contribution of working group I to the sixth assessment report of the intergovernmental panel on climate change. Cambridge University Press. https://doi.org/10.1017/9781009157896
3. United Nations (2015) Paris agreement. Retrieved from UNFCCC website
4. Creutzig F et al (2017) The underestimated potential of solar energy to mitigate climate change. Env Res Lett 12(034022). https://doi.org/10.1088/1748-9326/aa7e6b
5. Geels FW (2002) Technological transitions as evolutionary reconfiguration processes: a multi-level perspective and a case-study. Res Policy 31(8–9):1257–1274. https://doi.org/10.1016/S0048-7333(02)00062-8
6. Geels FW, Schot J (2007) Typology of sociotechnical transition pathways. Res Policy 36(3):399–417. https://doi.org/10.1016/j.respol.2007.01.003
7. Carlsson B, Stankiewicz R (1991) On the nature, function and composition of technological systems. J Evol Econ 1(2):93–118. https://doi.org/10.1007/BF01224915
8. Markard J, Truffer B (2008) Technological innovation systems and the multi-level perspective: towards an integrated framework. Res Policy 37(4):596–615. https://doi.org/10.1016/j.respol.2008.01.004
9. Rogers EM (2003) Diffusion of innovations, 5th edn. Free Press, New York, NY
10. Stavins RN (1995) Transaction costs and tradeable permits. J Environ Econ Manag 29(2):133–148. https://doi.org/10.1006/jeem.1995.1036
11. Williamson OE (1981) The economics of organization: the transaction cost approach. Am J Soc 87(3):548–577
12. Smil V (2010) Energy transitions: history, requirements, prospects. Praeger
13. REN21. (2021). Renewables 2021 global status report. REN21 Secretariat
14. World Commission on Environment and Development (WCED) (1987) Our common future. Oxford University Press
15. International Energy Agency (IEA) (2021) World energy outlook 2021. https://www.iea.org/reports/world-energy-outlook-2021
16. International Renewable Energy Agency (IRENA) (2020) Global renewables outlook: energy transformation 2050. https://www.irena.org/publications/2020/Apr/Global-Renewables-Outlook-2020
17. World Economic Forum (2021) Fostering effective energy transition 2021 edition. https://www.weforum.org/reports/fostering-effective-energy-transition-2021
18. Intergovernmental Panel on Climate Change (IPCC) (2018) Global warming of 1.5 °C. https://www.ipcc.ch/sr15/
19. Zhao Y, Liu B, Wu W, Wang L (2021) Recent advances in floating solar photovoltaic systems: a review. Sol Energy 219:389–411
20. Sørensen JD (2018) Wind energy systems: optimising design and construction for safe and reliable operation. Woodhead Publishing

21. Tan X, Xu Z, Zhang W, Cui G (2020) A review of solar paint: progress and prospects. Sol Energy 197:661–676
22. Zeng K, Zhang D, Zhang Y, Zhang Y (2021) Recent advances in hydrogen storage materials for fuel cell applications. Int J Hydrogen Energy 46(7):4990–5016
23. Yao L, Yu W, Sheng W, Wang B (2019) A review of smart grid development in major economies: policies, pilot projects and technologies. Renew Sustain Energy Rev 108:234–245
24. Ketterer J, Eichhammer W, Orasch W (2020) Renewable energy auctions: a review. Renew Sustain Energy Rev 134:110358
25. Razykov TM, Ferekides CS, Morel D, Stefanakos E, Ullal HS, Upadhyaya HM (2011) Solar photovoltaic electricity: current status and future prospects. Sol Energy 85(8):1580–1608. https://doi.org/10.1016/j.solener.2010.12.002
26. Burton T, Sharpe D, Jenkins N, Bossanyi E (2011) Wind energy handbook, 2nd edn. Wiley
27. Paish O (2002) Small hydro power: technology and current status. Renew Sustain Energy Rev 6(6):537–556. https://doi.org/10.1016/S1364-0321(02)00006-0
28. Park N-G (2015) Perovskite solar cells: an emerging photovoltaic technology. Mater Today 18(2):65–72. https://doi.org/10.1016/j.mattod.2014.07.007
29. Bertuccioli L et al (2019) Study on early business cases for H_2 in energy storage and more broadly power to H_2 applications. In: Fuel cells and hydrogen 2 joint undertaking
30. El Amri A, Oulfarsi S, Eddine AS, El Khamlichi A, Hilmi Y, Ibenrissoul A et al (2022) Carbon financial market: the case of the EU trading scheme. In: Handbook of research on energy and environmental finance 4.0. IGI Global, pp 424–445
31. El Kezazy H, Hilmi Y (2022) Towards more agile management: literature review of information systems as the pillar of management control. Rev Int Chercheur 3:4
32. El Kezazy H, Hilmi Y (2023) Improving good governance through management control in local authorities. Int Rev Manage Comput 7(3). https://doi.org/10.5281/zenodo.10577979. hal-04422229
33. Aghion P, Hepburn C, Teytelboym A, Zenghelis D (2016) Path dependence, innovation and the economics of climate change. Centre for Climate Change Economics and Policy Working Paper No 272
34. IRENA (2021) Renewable energy statistics 2021. International Renewable Energy Agency
35. Abbas S, Yuan Y, Hassan A, Zhou J, Ahmed A, Yang L, Bisengimana E (2023) Effect of the concentration ratio on the thermal performance of a conical cavity tube receiver for a solar parabolic dish concentrator system. Appl Therm Eng 227:120403. https://doi.org/10.1016/j.applthermaleng.2023.120403
36. Bazilian M et al (2014) Energy access scenarios to 2030 for the power sector in sub-Saharan Africa. Utilities Policy 29:1–16. https://doi.org/10.1016/j.jup.2013.11.002
37. Del Río P, Mir-Artigues P (2012) A cautionary tale: Spain's solar PV investment bubble. Glob Environ Chang 22(1):237–248. https://doi.org/10.1016/j.gloenvcha.2011.10.002
38. Diesendorf M (2016) Subject to change without notice: a review of Australia's renewable electricity policies. BioPhys Econ Resour Qual 1(2):11. https://doi.org/10.1007/s41247-016-0008-6
39. Fouquet R (2016) Historical energy transitions: speed, prices and system transformation. Energy Res Soc Sci 22:7–12. https://doi.org/10.1016/j.erss.2016.08.014
40. Harris A, Martinez L, Brown G (2019) Environmental monitoring using IoT and drone technologies. J Environ Manage 236:22–30. https://doi.org/10.1016/j.jenvman.2019.01.045
41. Jacobsson S, Lauber V (2006) The politics and policy of energy system transformation—explaining the German diffusion of renewable energy technology. Energy Policy 34(3):256–276. https://doi.org/10.1016/j.enpol.2004.08.029
42. Kammen DM, Sunter DA (2016) City-integrated renewable energy for urban sustainability. Science 352(6288):922–928. https://doi.org/10.1126/science.aad9302
43. Lund H, Mathiesen BV (2015) Smart energy systems for coherent 100% renewable energy and transport solutions. Appl Energy 145:139–154. https://doi.org/10.1016/j.apenergy.2015.01.075

44. Nykvist B, Nilsson M (2015) Rapidly falling costs of battery packs for electric vehicles. Nat Clim Chang 5:329–332. https://doi.org/10.1038/nclimate2564
45. Rogelj J, Den Elzen M, Höhne N, Fransen T, Fekete H, Winkler H, Schaeffer R, Sha F, Riahi K, Meinshausen M (2016) Paris agreement climate proposals need a boost to keep warming well below 2 °C. Nature 534(7609):631–639. https://doi.org/10.1038/nature18307
46. Sovacool BK, Dworkin MH (2015) Global energy justice: principles, problems, and practices. Cambridge University Press
47. Zhang S, Andrews-Speed P, Zhao X (2019) Political and institutional analysis of the successes and failures of China's wind power policy. Energy Policy 130:312–319. https://doi.org/10.1016/j.enpol.2019.03.066

Energy, Economic, and Thermal Assessment of Insulating a Moroccan Building Envelope with a Biobased Material for Promoting Sustainability

Hicham Kaddouri, Abderrahim Abidouche, Mohamed Saidi Hassani Alaoui, Ismael Driouch, Abderrahim Ed-Daoui, Firyal Latrache, and Abdelhafid Elouassidi

Abstract Biobased materials and thermal insulation are essential levers for sustainable, energy-efficient construction. A material like hemp offers an innovative alternative to traditional materials, thanks to its renewability, low embodied energy, and potential to reduce carbon emissions. Its thermal performance translates into effective insulation, essential for stabilizing conditions inside buildings and reducing their energy consumption. This efficiency is particularly relevant in contexts such as that of a residential building in Al-Hoceima, Morocco, where the use of hemp concrete for roof insulation has led to a 30% reduction in energy demand while improving thermal comfort by 8%. The study of this type of material, integrated with an energy and economic approach, highlights their potential to satisfy the challenges of the construction sector, which accounts for a significant proportion of energy consumption and greenhouse gas emissions. Although promising, biobased materials such as hemp concrete still face challenges in terms of standardization and widespread adoption. However, the numerical simulation performed with TRNSYS software clearly highlights the benefits of an integrated approach combining sustainability and thermal performance.

H. Kaddouri (✉) · M. S. H. Alaoui · I. Driouch · A. Elouassidi
Experimentation and Modelling Team in Mechanics and Energy Systems, Abdelmalek Essaadi University, National School of Applied Sciences, Al-Hoceima, Morocco
e-mail: hiichamkaddouri@gmail.com

A. Abidouche
Innovative Technologies Laboratory, Sidi Mohammed Ben Abdellah University, High school of technology, Fez, Morocco

A. Ed-Daoui
Laboratory of Physics of Matter and Radiations, Mohamed First University, Faculty of Sciences, Oujda, Morocco

F. Latrache
Laboratory of Modelling and Science Calculations, Mechanics Energetics Systems and Signals Team, Mohamed First University, National School of Applied Sciences, Oujda, Morocco

F. Latrache et al. (eds.), *Digital Technology for an Innovative Energy Transition*, Green Energy and Technology, https://doi.org/10.1007/978-3-031-89657-6_5

Keywords Biobased materials · Thermal insulation · Hemp concrete · Energy efficiency · Thermal comfort · Numerical simulation

1 Introduction

Morocco relies heavily on imported fossil fuels to satisfy its growing energy needs. More than 90% of the energy consumed in the country comes from imported nonrenewable sources, such as oil, natural gas, and coal [1]. This energy dependence poses several major problems, and geopolitical risks, due to energy supply, are one of them. Furthermore, fluctuations in the price of fossil fuels on the international market can lead to high and unpredictable costs for the national economy. In fact, the combustion of these fossil fuels is a major source of greenhouse gas (GHG) emissions, contributing then to climate change and air pollution. Faced with these challenges, it is becoming imperative for Morocco to diversify its energy mix by promoting renewable energies and energy efficiency solutions [2].

The building sector in Morocco accounts for around a third of the country's total energy consumption [3]. This consumption is largely due to the high demand for heating and air conditioning, exacerbated by regional climatic variations. In winter, heating is crucial, especially in mountainous regions, while in summer, demand for air conditioning explodes, mainly in arid and semi-arid areas. This energy demand for thermal comfort places a significant burden on Morocco's final energy consumption, underlining the urgent need to improve energy efficiency in the building sector to reduce dependence on imported fossil fuels.

The crucial issue of energy efficiency in the building sector in Morocco has aroused growing interest among researchers, who have been looking into the impact of various thermal insulation measures on the energy performance of buildings. Bouchark et al. [4] perform a comparative study to assess the effect of thermal insulation on the energy requirements of a tertiary building in Meknes. Their research highlights the crucial importance of choosing the insulation materials to reduce the energy consumption in building sector in Morocco, which is one of the largest consumers of energy. Using energy simulations with TRNSYS software, they demonstrated that integrating efficient insulation materials into the structure of a building lead to substantial reductions in energy requirements. Their results indicate an overall decrease in energy requirements, with a reduction of 40.1% through using polystyrene, followed by hemp concrete with 27.4% and phase change materials PCM with 18%.

Kaddouri et al. [5] conducted a study on the use of biobased insulating materials to enhance the thermal and energy performance of a standard residential building located in Al-Hoceima. This research focused on the integration of three types of natural insulators: hemp wool, wood fiber, and expanded cork, into the building's walls. Using numerical simulations with the TRNSYS software, the researchers evaluated various insulation scenarios to optimize thermal comfort while reducing energy consumption. The findings revealed that insulating the roof with 8 cm of hemp wool achieved energy savings of up to 36.7% for cooling needs and 35.2% for heating, while significantly improving indoor thermal comfort. Moreover, in another study [6], the authors attempted to assess the energy and thermal performance of two biocomposites, hemp concrete and hemp plaster, in four Moroccan locations: Al-Hoceima, Oujda, Marrakech, and Ouarzazate, thereby representing a variety of the country's climates. The results of this research revealed that the most significant energy savings were observed in the cold semi-arid climate of Oujda, with a gain of approximately 24%.

A complementary study by Kaddouri et al. [7] examined how the incorporation of hemp concrete combined with passive design strategies influences the energy consumption of a residential building in Al-Hoceima. The findings showed that using hemp concrete with an optimal thickness of around 10 cm allowed compliance with the performance thresholds required by the Moroccan Thermal Building Regulations (RTCM). Simulations also revealed that this thickness could be reduced to 4 cm with double-glazing, or even to 1 cm by implementing controlled natural ventilation during both the summer and winter seasons.

Overall, the literature highlights the effectiveness of different insulation solutions in reducing the energy consumption in buildings in Morocco, thereby contributing to the overall effort to improve energy efficiency and reduce the carbon footprint of the building sector.

The building envelope plays a central role in initiatives to enhance energy efficiency. It is recognized as a highly effective passive solution for optimizing thermal properties and reducing the energy requirements of structures, whether residential or commercial [8, 9]. Numerous studies have focused on improving thermal insulation [10–13]. Furthermore, thermal modeling techniques, particularly in the context of heat exchangers and solar concentrators [14, 15], can be leveraged to refine the design of insulating materials.

Previous studies have not discussed the economic aspects of insulation approaches. For this purpose, the present chapter aims to identify the most suitable insulation scenario for the Mediterranean climate in Morocco through energetic and economic visions. To this end, this study assesses the impact of insulating with a bio-sourced material for five different insulation scenarios on the energy and thermal performance of a building using thermodynamic simulations. It also examines the cost of insulation and the payback time for each case. The overall objective is to determine the ideal energy and thermal insulation scenario with the lowest cost.

2 Materials and Methods

2.1 Description of the Building Studied

A residential building in the Moroccan city of Al-Hoceima was chosen, with a surface area of 108 m^2. The building has two façades exposed to the outside, with the main façade facing north. Figure 1 shows the 3D plan of this building.

Table 1 shows the materials used to construct the building envelope and their thermophysical properties.

Thermal insulation represents a physical barrier against heat flow through the building envelope via these elements: external walls, roofs, and floors on the ground.

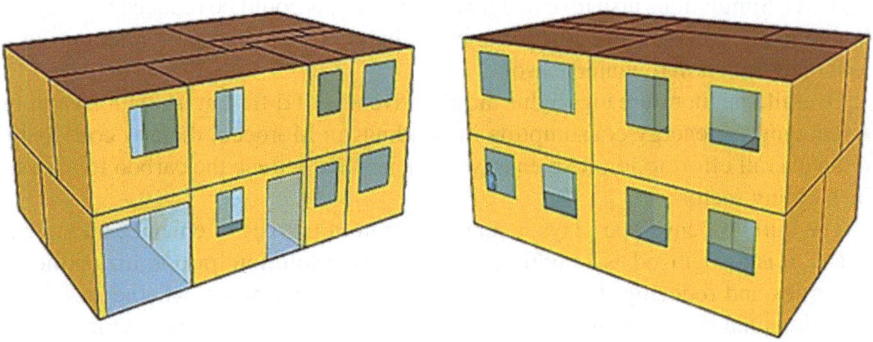

Fig. 1 3D plan of the building

Table 1 Composition of the building envelope

Envelope elements	Materials	Thickness (cm)	Density (Kg/m^3)	Thermal conductivity (W/m K)	Heat capacity (kJ/kg K)
Exterior wall	Mortar	2	1900	1.30	1.00
	Brick	20	664	0.22	0.74
	Mortar	2	1900	1.30	1.00
Interior wall	Mortar	2	1900	1.30	1.00
	Brick	7	918	0.19	0.74
	Mortar	2	1900	1.30	1.00
Floors	Floor tiles	2	2300	1.30	0.84
	Concrete	7	2450	2.00	1.00
	Slab	16	1513	1.04	1.00
	Plaster	1	1350	0.56	1.00

- Single glazing ($U = 5.72 \text{W/m}^2$ K—g = 0.84)
- The window/wall ratio (WWR): 12%

Table 2 Different insulation scenarios

Scenarios	Description
0	Without insulation
1	Roof insulation with 5 cm of hemp concrete
2	Internal insulation of external walls with 5 cm of hemp concrete
3	External insulation of external walls with 5 cm of hemp concrete
4	Roof insulation and internal insulation of external walls with 5 cm of hemp concrete
5	Roof insulation and external insulation of external walls with 5 cm of hemp concrete

Heat loss through these elements is estimated at the following respective percentages: roof 30%, external walls 25%, glazed surfaces 13%, thermal bridges 5%, and floor 3% [16]. The incorporation of an insulating material in the wall layer of a building serves to attenuate heat exchange between the outside and the inside of the building, in other words, to ensure a state of liveable thermal comfort for the occupants and to reduce heating and cooling energy loads.

We propose to study insulation scenarios that will enable us to make a relevant choice to improve comfort in our building. The insulation chosen is hemp concrete with a thermal conductivity of 0.082 W m^{-1} K^{-1} [17] Table 2.

2.2 Meteorological Data

According to the Moroccan Thermal Building Regulations RTCM [18], Morocco is divided into six climate zones (Fig. 2). Our approach studies the city of Al-Hoceima in the second climatic zone of the RTCM. Al-Hoceima city has a Mediterranean climate, with hot, dry summers and wet, mild winters.

The meteorological data used in the simulation comes from the typical meteorological year TMY file generated using Meteonorm software for the city of Al-Hoceima. Table 3 shows the geographical coordinates of the weather station used and Fig. 3 illustrates the ambient temperature distribution in the chosen location.

2.3 Physical and Mathematical Modeling

The shell of a building is the site of three heat transfer phenomena, i.e., in the vicinity of a building wall, heat is propagated in three ways: by conduction, by convection, and by radiation (Eqs. 1, 2, and 3):

$$\dot{Q}_{\text{conduction}} = UA\left(T_{\text{surf,in}} - T_{\text{Surf,out}}\right) \tag{1}$$

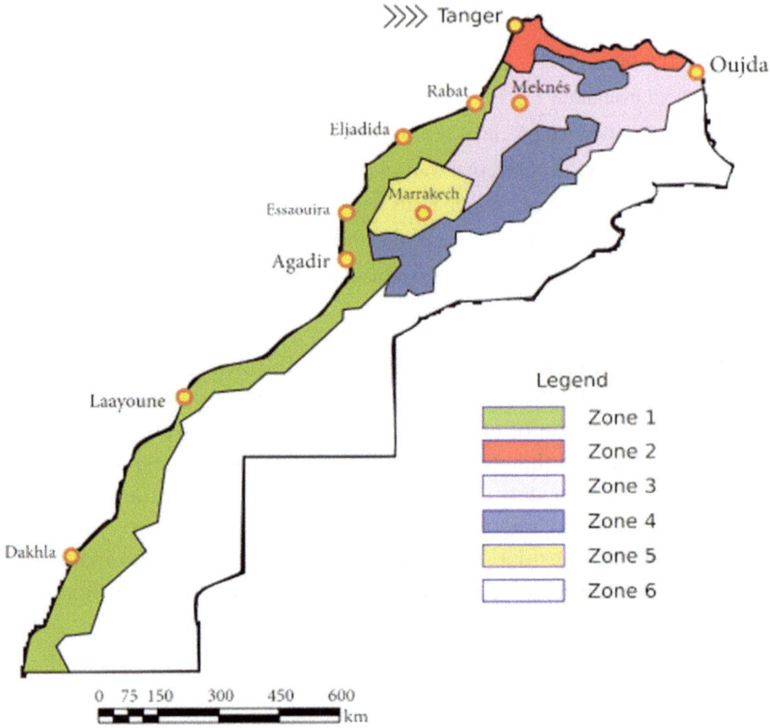

Fig. 2 RTCM-compliant climate map [19]

Table 3 Location of the meteorological station

Location	Latitude	Longitude	Altitude
Al-Hoceima	35.18°N	– 3.85°E	14 m

$$\dot{Q}_{\text{convection}} = h\left(T_{\text{air}} - T_{\text{surf,out}}\right) \tag{2}$$

$$\dot{Q}_{\text{radiation}} = \varepsilon\sigma\left(T^4_{\text{surf,out}} - T^4_{\text{sky}}\right) \tag{3}$$

Heat exchange by convection and radiation occurs on both sides of a wall, while heat exchange by conduction occurs through the layers of the wall envelope. The building modeling of 16 thermal zones describes the heat balance as follows (Eq. 4):

$$\begin{aligned}\dot{Q}_{\text{total}} &= \dot{Q}_{\text{inf,air}} + \dot{Q}_{\text{vent,air}} + \dot{Q}_{\text{cplg}} + \dot{Q}_{\text{surf}} \\ &\quad + \dot{Q}_{\text{int,gain}} + \dot{Q}_{\text{solar}} + \dot{Q}_{\text{solar,int,sha}}\end{aligned} \tag{4}$$

where:

$\dot{Q}_{\text{inf,air}}$ Heat flow due to air infiltration in the thermal zone

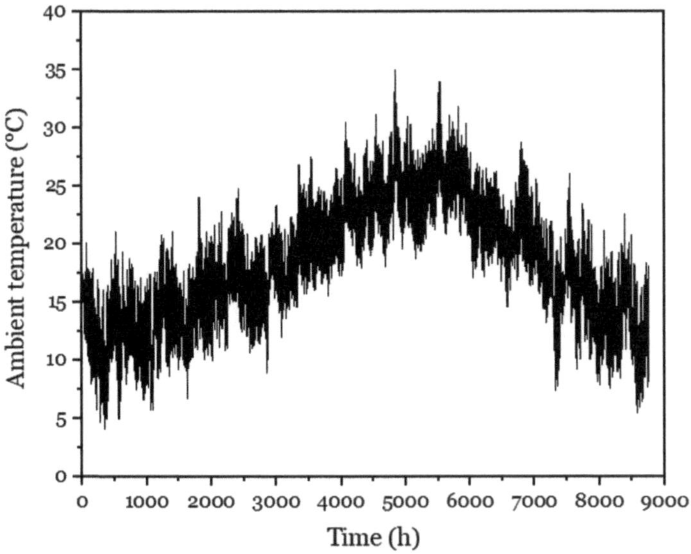

Fig. 3 Ambient temperature distribution of Al-Hoceima city

$\dot{Q}_{vent,air}$	Heat flow due to air ventilation in the thermal zone under study
$\dot{Q}_{cplg}$	Heat flow due to the phenomenon of air coupling between neighboring thermal zones
$\dot{Q}_{surf}$	Heat flow d to convection near the walls of the thermal zone
$\dot{Q}_{int,gain}$	Heat flow due to internal gains from equipment and occupants in the thermal zone
$\dot{Q}_{solar}$	Heat flow due to solar gain entering the thermal zone via the glazing
$\dot{Q}_{solar,int,sha}$	Heat flow due to solar gain captured by shading devices in the thermal zone.

2.4 Numerical Simulation

The thermal study is based on transient dynamic energy simulations using TRNSYS software. The principle of the simulation is to subdivide the building into thermal zones, each of which is considered an air node.

The software calculates the building's energy demand to compare it with the building performance requirements set out in Morocco's Thermal Building Regulations. The calculation is based on determining winter heating loads and summer cooling loads for summer and winter comfort temperatures of 26 °C and 20 °C, respectively, under Moroccan standard NM ISO 7730.

The following assumptions were taken into consideration during the simulation:

- The transient energy simulation was performed with a time step of 1 h.

- The rate of air infiltration through the envelope is estimated by 0.6 vol h^{-1}.
- The solar absorption coefficient of the building's walls is estimated at 0.5.
- Thermal bridges are not taken into account.
- Type 77 based on the Kusuda correlation is used to predict ground temperature [20].

$$T = T_{\text{mean}} - T_{\text{amp}} \times \exp\left[-\text{depth} \times \left(\frac{\pi \alpha_{\text{Soil}}}{365}\right)^{0.5}\right] \times \cos\left[\frac{2\pi}{365} \times \left(t_{\text{now}} - t_{\text{shift}} - \frac{\text{depth}}{2} \times \left(\frac{365\alpha}{\pi}\right)^{0.5}\right)\right] \quad (5)$$

2.5 The Discomfort Percentage Model

To quantify the effect of insulation on thermal comfort, the annual percentage of hours of discomfort experienced by occupants is calculated by:

$$\%\text{discomfort} = \frac{\text{Number of hours of discomfort}}{8760} \times 100 \quad (6)$$

3 Results and Discussion

3.1 Comparison with the Moroccan Thermal Regulations

Thermal regulations set minimum technical requirements for limiting the energy consumption of buildings. They encourage construction industrialists to benefit from architectural features and improve the quality of the building envelope to maximize thermal comfort and reduce heat loss.

Figure 4 shows a comparison, which has been realized between the heating and cooling energy requirements of the reference building and the technical requirements of the Moroccan Thermal Building Regulations RTCM.

A comparison of the annual energy requirements obtained by our numerical calculation code (TRNSYS software) with those required by the RTCM indicates that the results are almost equal. For heating, the difference is no more than 1.1 kWh m^{-2}/year, while for cooling the difference is around 0.8 kWh m^{-2}/year. We therefore conclude that our numerical results are reliable.

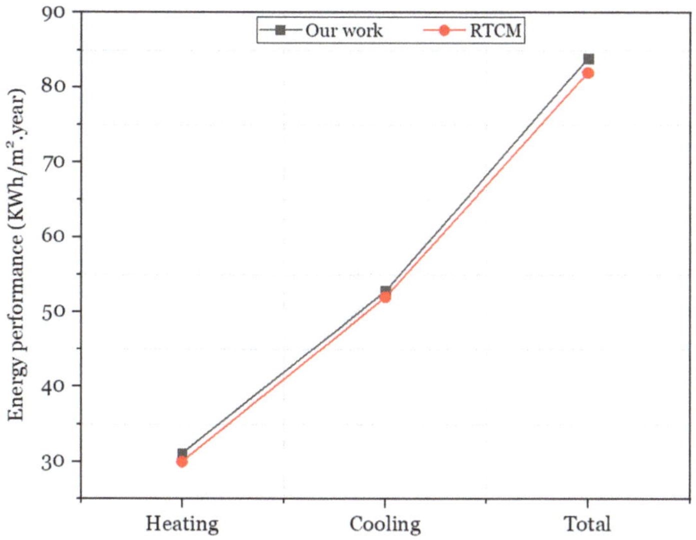

Fig. 4 Comparison of the building's energy performance with RTCM requirements

3.2 *Energy and Economic Study*

In this section, a series of simulations are done to examine different insulation scenarios to determine the most appropriate solution in terms of energy requirements and thermal comfort. Figure 5 compares the annual heating, cooling, and total energy consumption for the building and for the different insulation scenarios.

As a result, insulating with hemp concrete significantly reduces the building's energy requirements compared with the reference scenario. Then:

- Insulating the roof with 5 cm of hemp concrete (scenario 1) reduces heating and cooling energy requirements by 24% and 26%, respectively.
- Insulating external walls with 5 cm of hemp concrete (scenarios 2 and 3) reduces heating and cooling energy requirements by around 7% and 3.5%, respectively.
- Scenarios 4 and 5 reduce the heating energy by around 31%, respectively. At the same time, cooling requirements are reduced by 29% for the 2 scenarios (4 and 5), respectively.

Through these results, it is clear that scenarios 1, 4, and 5 deliver reductions in energy consumption of 25%, 29.7%, and 29.9%, respectively. A combination of roof and external wall insulation is the best solution from an energetic aspect. Furthermore, external wall insulation is preferable to internal insulation.

Choosing the most optimal insulation scenario takes into consideration not only its energy efficiency but also the cost and return on investment of the insulation chosen. To this end, it is necessary to estimate the additional cost of insulation. Table 4 summarizes the cost of insulation for the five scenarios studied (Fig. 6).

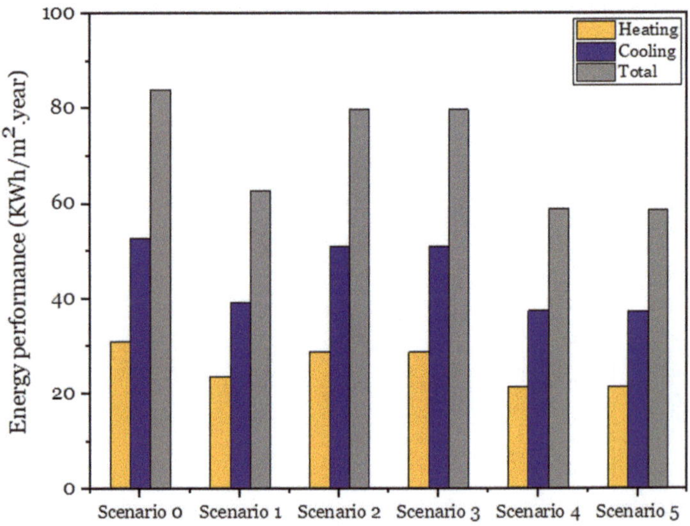

Fig. 5 Comparison of energy consumption for the different insulation scenarios proposed

Table 4 Cost of insulation for the five scenarios

Scenarios	Insulated surface (m^2)	Cost of insulation (MAD/m^2) [21]	Cost of insulation scenario (MAD)
1	108.4	1950.5	211,434.2
2	194		378,397
3	194		378,397
4	302.4		589,831.2
5	302.4		589,831.2

Where: 1MAD = 0.1USD

The results in Fig. 5 show that insulation scenario 5 represents the most significant reduction compared to the reference scenario (scenario 0), with an energy saving of around 30%. Scenario 1, on the other hand, represents the scenario with the lowest insulation cost, estimated at 211,434.2MAD. However, scenario 5 represents the best solution from an energetic aspect, but this solution is expensive due to the cost of insulation, which is why it is necessary to choose the least expensive scenario with a reasonable payback period and energy savings. Hence, the most appropriate scenario is scenario 1.

The comparison of the annual energy consumption for the different areas of the building for the reference case (scenario 0) and for scenario 1 is illustrated in Fig. 7.

Figure 7 highlights that the optimal energy savings are achieved in the rooms on the first floor because their roofs are exposed to solar radiation. Therefore, the roof insulation reduces cooling requirements. On the ground floor, bedroom 2 and the dining room are the areas where energy savings are optimum. This is explained

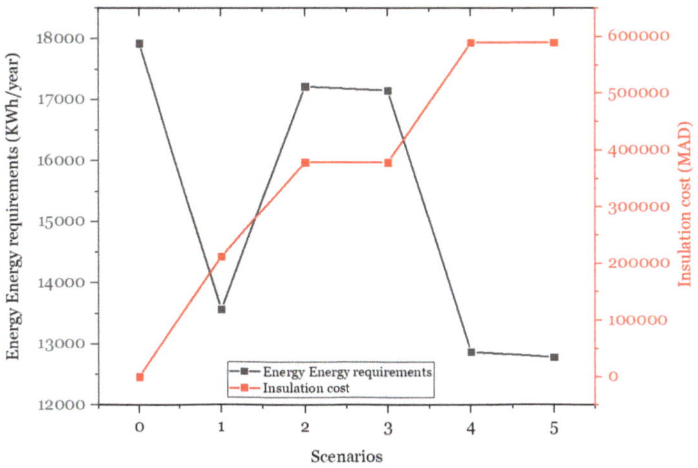

Fig. 6 Comparison of energy demand and insulation costs for the five scenarios

that these rooms are directed to south, they have high solar gains compared with north-facing rooms, and with the application of insulation, and thermal loads are reduced.

3.3 Thermal Study

Through the thermal aspect, we evaluate the influence of insulation on indoor thermal comfort. For reasons of visibility, Fig. 8 shows the variation in indoor temperature in room 4 for the reference scenario (scenario 0) and scenario 1 and for two typical days in summer and winter.

According to Fig. 8a, the indoor temperature varies significantly in summer, especially for the reference building scenario (which does not include any energy efficiency measures). This result indicates that external heat gains have an impact in this case. In addition, the indoor temperature fluctuation for scenario 1 oscillates around the summer comfort range of 26 °C. As a result, scenario 1 shows considerable damping of the temperature profile.

The amplitudes of the indoor temperature in winter are minimal, as shown in Fig. 8b for both days. A reduction of 1 °C is also visible in this figure between the indoor temperature for scenario 1 and the temperature curve for the reference scenario (scenario 0).

Figure 9 compares the percentage of annual discomfort (Eq. 6) in room 4 for scenarios 0 and 1 in order to quantify the effect of insulation on thermal comfort.

The application of insulation according to scenario 1 can lead to an improvement in the thermal comfort of the building's interior environment of around 8%.

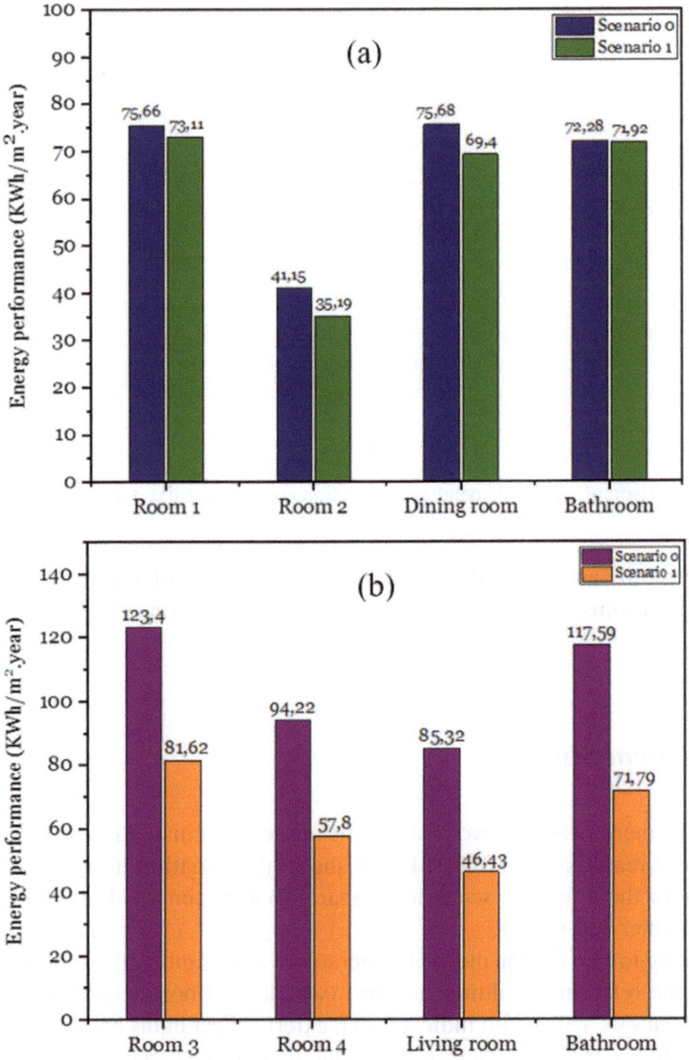

Fig. 7 Comparison of energy consumption in certain areas of the building for the reference scenario (scenario 0) and the scenario with insulation (scenario 1): **a** ground floor—**b** first floor

4 Conclusion and Future Research Work

This study presents a solution for minimizing energy demand and improving the level of thermal comfort inside a residential building in Al-Hoceima by using hemp concrete on the envelope as an insulation support. The study was performed in transient energy simulation of TRNSYS numerical software, using the type 56 multizone model. This study leads to the following conclusions:

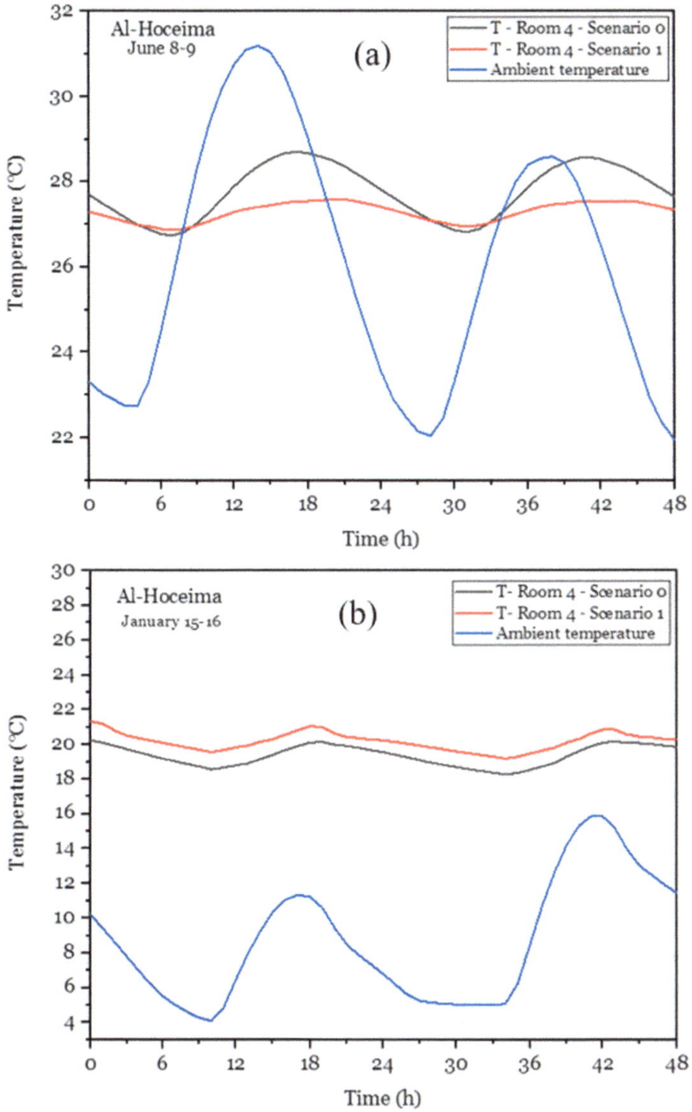

Fig. 8 Variation in outdoor and indoor temperature for the proposed insulation scenarios over two typical days of each season: **a** summer (June 8 and 9) and **b** winter (January 15 and 16)

- All the insulation solutions offer energy consumption reduction.
- Roof insulation has a considerable impact on energy demand than external wall insulation.
- Combined roof and external wall insulation has resulted in maximum energy savings.
- External wall insulation is preferable to internal insulation

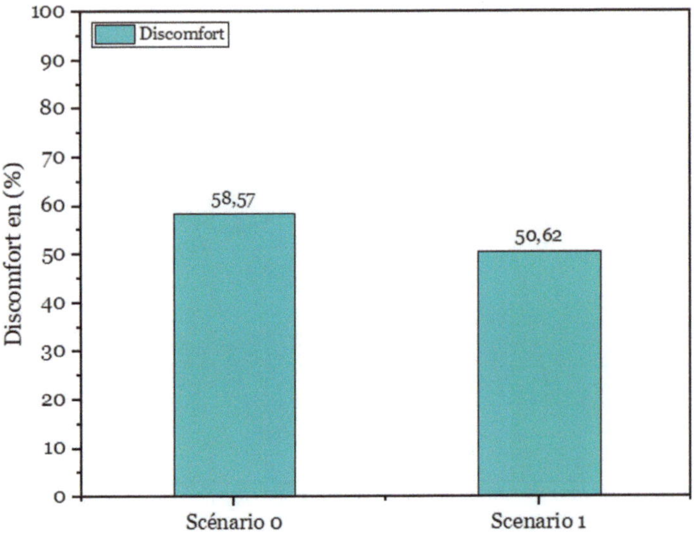

Fig. 9 Comparison of the percentage of annual discomfort in room 4 for scenarios 0 and 1

- Scenario 1 represents the optimum solution in terms of reducing energy demand by 25% at a lower cost than scenario 5.
- Scenario 1 improves the level of thermal comfort inside the building.

In conclusion, the thermal insulation with 5 cm of hemp concrete on the roof of a residential building located in the city of Al-Hoceima in Morocco, leads to an energy reduction of approximately 30%. This thermal insulation improves the thermal comfort of approximately 8% with the most reduced cost compared to the insulation of the external walls, which combines the roof and the external walls.

As an objective to promote sustainability, the insulation is considered as an optimal alternative solution instead of using cooling and heating equipment. The insulation, then, contribute to the reduction in the carbon footprint and therefore decarbonization by 2050. As a perspective and in order to improve the use of digital technologies in the energy sector, researchers and experts are invited to apply sensors for controlling and monitoring the thermal comfort in buildings to guarantee a sustainable insulators. They must investigate validation approaches through AI algorithms to better track this digitalization wave.

References

1. Kousksou T, Allouhi A, Belattar M, Jamil A, El Rhafiki T, Arid A, Zeraouli Y (2015) Renewable energy potential and national policy directions for sustainable development in Morocco. Renew Sustain Energy Rev 47:46–57. https://doi.org/10.1016/j.rser.2015.02.056

2. Élaboration du projet de la stratégie nationale de l'efficacité énergétique à l'horizon 2030 (s. d.). Consulté 27 février 2024, à l'adresse https://www.mem.gov.ma/Pages/secteur.aspx?e=3&prj=38
3. AMEE|Energy efficiency in buildings (s. d.) Consulté 15 octobre 2024, à l'adresse https://www.amee.ma/en/node/118
4. Bouchark et al (s. d.) Comparative analysis of the effect of thermal insulation on the energy requirements of a tertiary building in meknes. Consulté 15 octobre 2024, à l'adresse https://mrforum.com/product/9781644903117-16/
5. Kaddouri H, Abidouche A, Alaoui MSH, Driouch I, Hamdaoui S (2024) Impact of insulation using bio-sourced materials on the thermal and energy performance of a typical residential building in Morocco. J Adv Res Fluid Mech Therm Sci 117(1):43–59. https://doi.org/10.37934/arfmts.117.1.4359
6. Kaddouri H, Abidouche A, Alaoui MSH, Driouch I, Hamdaoui S, Msaad AA (2024) Study of the energy, economic, environmental, and thermal comfort impact of the integration of hemp concrete and hemp plaster in a residential building envelope in Morocco. J Adv Res Numer Heat Trans 23(1):1–27. https://doi.org/10.37934/arnht.23.1.127
7. Kaddouri H, Abidouche A, Saidi Hassani Alaoui M, Ismael D, Msaad A, Hamdaoui S (2024) Numerical simulation study of the effect of integrating hemp concrete and passive strategies on the energy consumption of a residential building in Al-Hoceima. Stat Optim Inf Comput 12(3):727–736. https://doi.org/10.19139/soic-2310-5070-1933
8. Abidouche A, Kaddouri H, Hamdaoui S, Msaad AA, Driouch I, Hassani AMS (2024) The analysis of occupants' thermal comfort in a residential building in Tangier, Morocco. Math Model Comput 752. https://doi.org/10.23939/mmc2024.03.752
9. Meng X, Huang Y, Cao Y, Gao Y, Hou C, Zhang L, Shen Q (2018) Optimization of the wall thermal insulation characteristics based on the intermittent heating operation. Case Stud Constr Mater 9:e00188. https://doi.org/10.1016/j.cscm.2018.e00188
10. Abu-Jdayil B, Mourad A-H, Hittini W, Hassan M, Hameedi S (2019) Traditional, state-of-the-art and renewable thermal building insulation materials: an overview. Constr Build Mater 214:709–735. https://doi.org/10.1016/j.conbuildmat.2019.04.102
11. Aditya L, Mahlia TMI, Rismanchi B, Ng HM, Hasan MH, Metselaar HSC, Muraza O, Aditiya HB (2017) A review on insulation materials for energy conservation in buildings. Renew Sustain Energy Rev 73:1352–1365. https://doi.org/10.1016/j.rser.2017.02.034
12. Asdrubali F, D'Alessandro F, Schiavoni S (2015) A review of unconventional sustainable building insulation materials. Sustain Mater Technol 4:1–17. https://doi.org/10.1016/j.susmat.2015.05.002
13. Kumar D, Alam M, Zou PXW, Sanjayan JG, Memon RA (2020) Comparative analysis of building insulation material properties and performance. Renew Sustain Energy Rev 131:110038. https://doi.org/10.1016/j.rser.2020.110038
14. Latrache F, Hammouch Z, Lamnaouar K, Bellach B, Ghammouri M (2023a) Development of geometrical parameters for a conical solar concentrator—application for vapor generation.https://doi.org/10.1007/978-3-031-29857-8_68
15. Latrache F, Hammouch Z, Lamnaouar K, Bellach B, Ghammouri M (2023b) Mathematical modeling of the heat generated through an evaporator—absorber accompanied by thermal storage for the solar energy applications. Int J Eng Res Afr 64.https://doi.org/10.4028/p-3m6c59
16. Inard C, Depecker P, Roux J-J (1997) Un modèle simplifié pour la prédiction du champ de température dans les bâtiments. Rev Gén Therm 36(2):113–123. https://doi.org/10.1016/S0035-3159(99)80056-7
17. Samri D (2008) Analyse physique et caractérisation hygrothermique des matériaux de construction. These de doctorat, Lyon, INSA. https://theses.fr/2008ISAL0067
18. AMEE|la Règlementation Thermique de Construction au Maroc (RTCM) (s. d.) Consulté 15 octobre 2024, à l'adresse https://www.amee.ma/reglementation-thermique
19. Analysis and comparison of energy efficiency code requirements for buildings: A Morocco–Spain case study (s. d.). Consulté 15 octobre 2024, à l'adresse https://www.mdpi.com/1996-1073/13/22/5979

20. Earth temperature and thermal diffusivity at selected stations in the United States (s. d.) Consulté 15 octobre 2024, à l'adresse https://apps.dtic.mil/sti/citations/AD0472916
21. Béton de chanvre en brique multichanvre|Construire|Alsabrico (s. d.) Consulté 15 octobre 2024, à l'adresse https://www.alsabrico.fr/construire/gros-oeuvre/briques-ecologiques/beton-de-chanvre-en-brique-multichanvre-de-vieille-materiaux.html?srsltid=AfmBOopaP586YVT10ZqLUnIHMLL9GcDkbsworwq8GH_VUYqtDIvxOQ7t

Deep Learning Method for Evaluating Morocco's Urban Photovoltaic Potential

Chouaib Benqlilou, Sarah Kahlaoui, Nabila Zrira, Youssouf Maazou Manomi, and Abdesadeq Elbayad

Abstract The goal of this project is to develop an energy management method for cities, specifically for managing electrical energy in buildings. The proposed method involves detecting roofs using deep learning techniques and determining their photovoltaic solar potential. The estimated theoretical solar potential is 363.33 GWh/year, with a technical potential of 291 GWh/year that can be exploited by installing solar panels on the roofs. A sample of 100 roofs provides a technical photovoltaic power of 3 MW equivalent to 24 MWh over eight hours. The study includes an optimization of energy consumption for the sample, comparing two scenarios and resulting in energy savings of 117,720 DH/day (11,735.62 USD/day) in electricity purchases from the National Office of Electricity and Water (ONEE Electricity Branch). The optimization produces an electrical energy management program, which can be uploaded onto an embedded system such as an Arduino card. As a proof of concept, the operation of the Arduino board was simulated on two buildings. To further improve the project, an economic analysis is necessary to assess its profitability. The analysis shows an initial investment of 24.7 million DH (2.40 million USD) and a payback time of 7 months for the scenario with intelligent energy management. Taking into account the Carbon tax avoided, the net present value over a 20-year project duration is estimated to be 682 million DH (68.15 million USD).

Keywords Deep learning · Mask RCNN · ArcGIS · Potential · PV · Arduino · Energy · Neural networks · Simulation · Model · Colab · Annotation · Rooftop · Detection · Images · Solar panels

---

C. Benqlilou (✉) · S. Kahlaoui · N. Zrira · Y. M. Manomi · A. Elbayad
Ecole Nationale Supérieure des Mines de Rabat, Rabat, Morocco
e-mail: benqlilou@enim.ac.ma

F. Latrache et al. (eds.), *Digital Technology for an Innovative Energy Transition*, Green Energy and Technology, https://doi.org/10.1007/978-3-031-89657-6_6

1 Introduction and Literature Review

1.1 Introduction

Buildings in Morocco consume 33% of the nation's electrical energy, with 25.4% in the residential sector and 8.2% in the tertiary sector. This makes the building sector the second most energy-consuming sector in Morocco, according to the CTEFMAROC (Company of Electricity and Cold Works in Morocco). It is important to note that 65% of the Moroccan population resides in urban areas, and this figure is expected to reach 75% by 2050, resulting in a significant increase in the number of buildings. To manage this rising energy consumption, new energy management strategies must be implemented. These strategies may include the installation of suitable solar panels on rooftop structures to maximize the sun's energy and optimize photovoltaic solar potential [1–3].

Evaluating this potential is a crucial step toward harnessing its benefits. Several techniques have been developed for this purpose, including the use of Deep Learning and Geographic Information Systems (GIS) to detect rooftops in satellite imagery and assess their solar photovoltaic (PV) potential [4–11]. Tapping into this potential can inform the development of a management program and reduce the need for purchasing electrical energy. For improved management, the electrical energy management program can be integrated into an onboard system. This approach can optimize energy consumption and promote sustainability in the ever-growing building sector in Morocco.

1.1.1 Context

Despite a centralized energy system and the high cost of existing solar infrastructure, there are limitations in large available surfaces for solar installations. Additionally, drawbacks associated with wind turbines and the underutilized solar potential underscore the need for more effective energy strategies.

1.1.2 Objectives

In this context, our study aims to estimate the photovoltaic solar potential of roofs in the city of Rabat. Our objectives include implementing a detection system based on deep learning, initially tested in Rabat and subsequently extended to other Moroccan areas such as Tétouan, Agadir, and Ifrane. We plan to evaluate the detected surfaces in square meters using ArcGIS, and quantify the theoretically and actually captured solar energy on the roofs. Furthermore, we aim to assess the photovoltaic energy generated by the installation of PV modules on the roofs, optimize building consumption using a sample of 100 roofs, and implement a smart electrical consumption management system with Arduino. Finally, a comprehensive economic evaluation of the project will be conducted, including a net present value (NPV) analysis.

1.1.3 Hypothesis

For our calculations, we assume a daily sunlight duration of 8 h, randomly generated building consumptions, and an electricity cost of 0.9 DH/kWh (0.090 USD/kWh). We use solar panels with a power of 400 Wc, with a correction factor to determine the actual potential of the roofs. Additionally, a carbon tax of 0.5 DH/kWh (0.050 USD/kWh) is considered, and the project duration is set at 20 years.

This study is organized into six sections. The first section provides an overview of the project and highlights the importance of installing photovoltaic systems on rooftops in Morocco. The second section introduces the deep learning model employed for rooftop detection. The third section outlines the steps involved in developing the rooftop detection model and calculating photovoltaic solar potential. In the fourth section, we examine the optimization problem concerning electrical energy purchases. The fifth section discusses the implementation of an intelligent management system using Arduino. Finally, the sixth section presents an economic analysis of the project. We will conclude our chapter with a summary of our findings, followed by recommendations and an examination of future prospects.

1.2 Literature Review

Various approaches and methodologies have been developed by researchers to assess the photovoltaic potential in different regions. These techniques encompass a wide range of methods, from simple satellite image analysis to more complex geospatial models. In this section, we will explore several of these techniques and discuss their effectiveness and applicability in detecting the solar potential of photovoltaic systems.

To estimate the solar potential of PV systems, Ninsawat and Hossain [2] employed a methodology that assumes rooftops are the only locations for solar panel installation in urban areas. Consequently, they considered the impact of shading on solar PV electricity production. Their goal was to evaluate the extraction of roof types from high-resolution satellite images using object-based image analysis (OBIA). By generating a 3D structure from height and roof type data, they simulated shadows throughout the year and identified potential and financial prospects.

The researchers applied this methodology to the city of Bangkok, utilizing WorldView2 satellite images from the Geo-Informatics and Space Technology Development Agency in Thailand. The system proved to be effective in calculating the financial benefits of solar photovoltaics, concluding that the payback period ranged from 7 to 13 years depending on the roof type, direction, and impact of shading. Moreover, the potentiality analysis revealed that homeowners could see a profit of up to 200%.

Similarly, Luqman et al. [8] addressed the issue of electricity load shedding in Pakistan and discussed the government's policies promoting alternative energy sources. Given this context, they sought to determine the potential of solar energy as an alternative energy source using rooftops of residential areas in Lahore district. They studied and described the types of roofs in the district before scanning portions of them. The uniform structure of the houses and the district's well-planned urban layout simplified their study. They employed the ArcGIS Solar Analyst tool with a georeferenced Google Earth image to create the GIS model for roof detection.

As a result, the estimated solar energy potential was 0.96×10^7 kWh/year, while the total monthly energy consumption of a nearby company was 3.47×10^5 kWh. This means that the energy production from solar PV panels could cover only 11% of the company's energy consumption.

However, the authors, led by Wang et al. [12] explore the crucial role of rooftop photovoltaic systems in solar energy generation, especially in urban environments. Their approach involves an evaluation method for the photovoltaic power generation potential from rooftops in China. They adopted a methodology that integrates a machine learning model to analyze a substantial amount of data, including gross domestic product, building footprint, road length, and population, at a high geographic resolution of 10 km by 10 km. The results indicate that the rooftop generation potential in China amounts to 3.27×10^9 megawatt-hours annually, contributing to an annual reduction of 2.41×109 tons of CO_2. Significant monthly variability is observed in Shandong, ranging from 18.89 TWh in November to 27.41 TWh in May. The eastern region exhibits a high accumulated potential for photovoltaic electricity due to the abundance of rooftop resources, exceeding that of the western region by 3.21 times. At the municipal level, Beijing has a potential of 45.11 TWh/year, and Tianjin has 34.62 TWh/year, making them suitable for rooftop solar photovoltaic development.

Furthermore, in this seminal work, Cui et al. [13] have devised an efficacious system to estimate the potential of Chinese rural photovoltaic rooftops on a large scale. The assessment is grounded in three facets: geographical potential, physical potential, and technical potential. An enhanced SegNeXt deep learning network was proposed to extract rooftop images, achieving an average overlap rate of approximately 87% and an average pixel accuracy of about 96%.

Leizhou City exhibits a geographical potential of 1500 kWh/m^2, a physical potential encompassing 25 186 181.7 m^2, and a technological potential of 442.4 MW. Innovatively, the authors employed the Bass Demand Diffusion Model to project the installed capacity over the next 35 years. The integration of the Commodity Diffusion Model with the installed capacity yielded commendable results, aligning with the dual-carbon "3060" plan for China's future.

Nevertheless, Kozlovas et al. [14] emphasize the significance of rooftop photovoltaic (PV) systems in Lithuanian urban areas concerning the achievement of energy and climate targets. They conducted a four-stage analysis of residential (AP), commercial (COM), and public (PUB) building characteristics regarding their suitability for the installation of solar PV systems, demonstrating that the identified building categories have varying impacts on the expansion of solar PV systems on their roofs. The authors utilized LiDAR and photogrammetry data on building topologies, geographic information system (GIS), digital maps for roof measurements, factual modeling of PV performance, average PV system profiles, energy consumption statistics in buildings, levelized cost of energy (LCOE) models, and a literature review.

The authors' estimated result indicates that the roof area per installed PV unit ranges from 14.1 to 18.3 m^2/kW in AP buildings, followed by 18.0–18.6 m^2/kW in COM buildings and 6.1–18.3 m^2/kW in PUB buildings. The roof suitability factor for solar PV system installation varied in ranges of 31.5–85.4%, 48.9–99.4%, and 50.1–88.2%, respectively. Rooftop solar energy is consumed locally, with only small amounts of energy injected into the grid.

The technical potential analysis results show that on AP building roofs, 272 MW could be installed, followed by 78 MW on PUB building roofs and 67 MW on COM building roofs. If the technical potential is realized, an investment of 332.7 million euros is possible. The researchers estimated the cost of energy produced by solar PV systems at 2.34–5.25 EURct/kWh, significantly lower than electricity prices on the market and retail, which are 23 EURct/kWh and 24 EURct/kWh, respectively (with government support) in 2022.

On the other hand, Fakhraian et al. [1] identified urban environments as having high potential for electricity production using rooftop-mounted photovoltaic systems. They conducted a semantic review of various methodologies developed and published in the current state of the art. Their aim was to pinpoint the critical factors for solar PV potential in urban rooftops and to determine the most effective methods, thus providing a comprehensive foundation for future studies. Their review encompassed a wide range of techniques for estimating rooftop solar potential, from GIS methods utilizing LIDAR data to 3D modeling approaches. They offered a visual comparison of different methodologies and recommended the creation of an accurate, consistent multifactor method that relies on high-resolution open data sources. This would enable the determination of photovoltaic potential maps for urban rooftops on a global scale.

Therefore, in a significant review conducted by Zhang et al. [10], the authors explored the relationship between urban rooftop photovoltaic potential and urban land use. They utilized a U-NET model to detect rooftops and two types of datasets to train the model. The first dataset was used to pre-train the model with limited data, while the second, larger dataset was employed for the main training process. Additionally, they used the VGG16 model to classify land use types in Wuhan.

With a total solar irradiation of 1241 kWh/m^2/year and considering only three land use types, the annual photovoltaic potential reached approximately 5800 GWh/year in total.

As a result; in his study, Afia [4] focused on comparing the estimated technical solar energy potential with the photovoltaic roof production based on the area of installed PV panels in Jagstfeld—Bad Friedrichshall, Germany. He used ArcGIS and OpenStreetMap to carry out this comparison. The comparison aimed to fill the information gap on how much of the estimated solar potential was actually utilized for electricity production through installed PV panels and assess the energy data of the estimated solar potential.

The method he employed to determine the appropriate roof area was based on solar panel areas, taking into account the actual situation to calculate the PV production. He used solar potential data from LUBW and OSM/Bing Basemap data to extract the study area. Concurrently, he digitized the solar panels on the roofs to determine the panel area and exported all this data, along with the solar potential of the study area and the OSM building footprint, into geoprocessing tools. The first tool provided the photovoltaic area of the roofs, while the last two types of data gave the solar potential of the roofs. He then reintegrated the data into the geoprocessing tools to calculate the electricity production and made a comparison between the solar potential area and the electricity production area on one hand, and the electricity production of the entire solar potential on the other hand.

The results showed 43 buildings with photovoltaic systems on their roofs, generating a total annual electricity production of 0.42 GWh. In the city, 198 systems were installed, producing 2.15 GWh/year.

Another study by Lan et al. [15] uses the developed mixed-use city of Adelaide as a case study to illustrate a comprehensive workflow employing machine learning algorithms. The process includes data collection, data pairing, model training, model interpretation, and model application to better understand the relationships between urban morphology and solar potential at the urban neighborhood scale. Among the machine learning models, Lan et al. [15] choose the artificial neural network (ANN), which exhibits the highest accuracy at around 86.1%. The results of the artificial neural network model reveal six key urban morphology indicators that negatively impact the solar potential of urban blocks. Among these indicators, building density is the most influential in the model, followed by plot ratio, building height, building floors, and variance of height and volume. The model also reflects the nonlinear impacts of these urban morphology indicators on urban solar potential, as well as the complex interactions between these morphology indicators and urban solar potential. Additionally, five typical morphological prototypes are identified. The average solar potential of Cluster 1 is 1402.53 kWh/m^2/year, 4.3% higher than Cluster 3, 11.2% higher than Cluster 4, 21.2% higher than Cluster 2, and 36.8% higher than Cluster 5. This study proposes a standardized and simplified workflow in the early stages of planning and design to assess urban form and land use associated with harvesting solar energy.

However, Li et al. [16] present a novel methodology and workflow grounded in deep learning for the expeditious assessment of solar energy potential across diverse topographies in rural Wuhan, China. This methodology elucidates the nuanced distribution of solar potential in rural settings and delineates zones conducive to development. The authors conducted a comprehensive survey of all viable surfaces for photovoltaic installations, encompassing building rooftops, water surfaces, and fallow lands. They curated their dataset leveraging publicly available satellite imagery, validating correlations utilizing the classical U-Net convolutional neural network. The findings reveal an average PV potential of 42.25 kWh/year per unit area for rural regions, with the potential to reach 0.09 GWh/km^2/year.

For aquatic areas, the average PV potential per unit area is 0.014 GWh/year, reaching a maximum of 2.48 GWh/km^2/year.

Regarding fallow lands, the average PV potential per unit area is 0.003 GWh/year, with the potential to achieve 56.66 GWh/km^2/year.

These outcomes could significantly contribute to mitigating electrical infrastructure costs and fostering solar energy development in the region.

In addition, Sun et al. [3] conducted an integrated study and feasibility analysis to determine the geographic potential of rooftop PV systems in urban buildings using GIS tools. The proposed approach was validated through a case study in the central area of Wuhan, China, with a spatial resolution of 100 m. The results indicate that the maximum annual PV power generation potential in Wuhan's central area amounts to 10.757 GWh, which could meet 31.83% of local energy consumption. Furthermore, deploying rooftop PV systems at a 100% rate would lead to a significant reduction in carbon emissions.

Moreover, Cadei et al. [6] utilized a supervised method based on U-net convolutional neural networks to outline the available rooftop area for photovoltaic panel installation through per-pixel image segmentation. The model was trained to perform pixel-wise binary segmentation on satellite images of Geneva, Switzerland, with the aim of generalizing the solution to a national scale. As a result, they successfully identified rooftops in residential areas, achieving a test set accuracy of approximately 0.97 and an intersection over union index of 0.77 using only 244 training images.

Additionally, Zhong et al. [11] demonstrated the significant solar rooftop potential in the Nanjing area of China using publicly available high-resolution satellite images. They developed a deep learning method for extracting rooftop areas and estimating solar PV potential. Their primary objective was to create a low-cost, generic framework for city-scale extraction of building rooftops. They developed a spatial optimization sampling strategy to address challenges when training the model, based on prior knowledge of urban and rural spatial layouts and land use. This strategy helped them reduce labor costs associated with preparing the dataset by about 80%.

Utilizing Google Earth Satellite images and the Deeplab v3 semantic segmentation algorithm, their model achieved an accuracy of 92%. The total rooftop surface extracted amounted to 330.46 km^2, and the estimated solar PV potential was approximately 311.853 GWh/year, with a corresponding power generation estimate of 49.897 GWh for 2019.

Thus, Bayrakci Boz et al. [5] utilized LIDAR (light detection and ranging) data and building footprints to develop a rooftop detection model using ArcGIS. Their objective was to identify rooftops suitable for solar energy systems, so they first created rooftop segments using seven slope (tilt), five aspect (azimuth) classes, and six different building types. They took into account the direct beam shading caused by nearby objects and surrounding terrain on a monthly basis for each rooftop. Using these data, they implemented the method as an ArcGIS model in Model Builder and developed a tool. The use of LIDAR data increased the precision of their study. When tested in the city of Philadelphia, PA, USA, the results showed that 33.7%

of the building footprint areas and 48.6% of the rooftop segments identified were suitable for PV systems.

Hence, Muhammed et al. [9] presented a roof extraction model for photovoltaic potential estimation based on two main phases. The first phase involves detecting rooftops from satellite images using preprocessing techniques and a machine learning algorithm. The preprocessing steps include gamma correction, shadowing, vegetation masking, k-means, and connected components, with the support vector machine (SVM) applied to extract rooftops. The second phase employs two GIS-based methods, PVGIS and the Solar Analyst tool in ArcGIS, for PV estimation.

To evaluate their approach, they applied the method to a part of Madinaty city in Egypt and achieved good precision and recall scores, approximately 95.7% and 90%, respectively. They successfully detected 112 roofs in the test image, with a total area of 26.131 m^2. The annual PV potential was estimated to be 9.3 and 8.7 MWh/year using the GIS-based methods. The Solar Analyst tool was less accurate because it relied on a digital elevation model with a 30-m resolution. They were able to estimate that an average of 48% of CO_2 emissions could be offset annually by using solar panels instead of traditional energy sources.

Comforting the subject, Carl [7] aimed to identify the solar photovoltaic potential of residential roofs in the city of Kailua Kona, located on the island of Hawaii. She used Geographic Information Systems (GIS) and Light Detection and Ranging (LIDAR) data, along with a statistical analysis. The study quantified the potential of Solar World's SW260 monocrystalline panels on residential rooftops in the study area, solar radiation modeling, estimation of available rooftop area, and calculation of photovoltaic potential from incoming solar radiation.

To achieve this objective, Carl used high-resolution LIDAR data and Esri's solar modeling tools to calculate incoming solar radiation on a sample of digitized roofs. She then employed equations developed by Šúri et al. [17] to calculate the photovoltaic potential of the sample. Additionally, she used a statistical model to identify the correlation between roof area and lot size. As a result, the least squares multiple linear regression analysis was performed to determine the influence of slope, elevation, roof area, and lot size on the modeled PV potential values.

Carl's study highlighted the use of LIDAR data to estimate the PV potential of roofs. She applied the equations constructed from the statistical analyses of the entire sample to the entire study area to calculate the total roof area and PV potential. Consequently, she estimated a potential of approximately 11.9×10^8 kWh/year, which amounts to about 17% of the total electricity supplied by the utility to the entire island in 2012.

2 Methodology and Research Framework

2.1 Rooftop Detection and PV Potential Estimation

The methodology we adopt in this chapter is divided into three distinct parts:

- First, the preparation of our dataset using ArcGIS.
- The second part involves implementing the deep learning model on Google Colab for training, exporting the model to ArcGIS, and applying it.
- The third part focuses on surface measurements and estimation of the photovoltaic potential (Fig. 1).

Before proceeding with the creation of information, it is important to choose the region on which we will be focusing our efforts (ROI). In this study, we have selected four major regions in Morocco, chosen for their unique characteristics and variety of roof types: Rabat oceanic city in the center of the country roofs that are mainly side-by-side.

- The mountainous snow-covered city of Ifrane, known for its houses with roofs sloping downwards.
- The city of Agadir, located in the south of Morocco.
- The city of Tetouan, located in the north of Morocco.

In all of these cities, the roofs have different shapes, and we apply our deep learning model to test its accuracy. The cities of Agadir and Tetouan have different climates and types of roofs compared to each other. After selecting the region of interest, the next step is to create a dataset. This task can be divided into several subtasks, beginning with downloading the images of the ROI. Google Earth satellite provides high-resolution images that are free to download and can be used to train the deep learning model to recognize rooftops accurately. By using Google Earth Pro software, it was simple to download the training images, as illustrated in Fig. 2.

The images downloaded from Google Earth Satellite are 4800 × 2718 pixels, which is the maximum size allowed by GES. Once downloaded, we use ArcGIS Pro to label the rooftops and export the data. To train, test, and validate our model, we downloaded four images from the regions of interest. From this, we conclude that each region requires its own model because they have different roof structures.

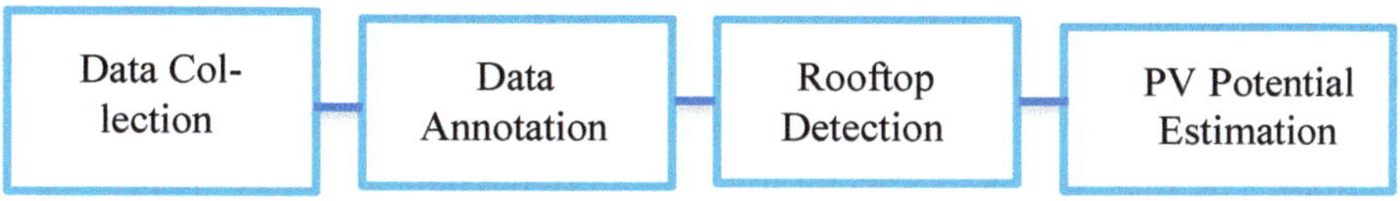

Fig. 1 Rooftop detection and photovoltaic potential estimation framework

Fig. 2 Training image downloaded from Google Earth satellite

2.2 Importing, Labeling, and Exporting Training Images Using ArcGIS

ArcGIS Pro is a software that provides a plugin module for testing and experimenting with the concepts of deep learning. Additionally, ArcGIS has a specific module in Python that facilitates the coding of our detection model. In any deep learning project, the first and most crucial thing is the database. In our case, the data will consist of roof images taken from GES for the four ROIs. ArcGIS provides a tool dedicated to creating data for deep learning, which is accessible through the Imaging section. This section also has a classification tools section that we will use to create our data.

After accessing the Label Objects for Deep Learning section, we can begin annotating our images and creating the data in the form of objects that the model requires. Since the roofs in the city have different shapes, it is recommended to use the polygonal shape to annotate them.

After annotating all the roofs on the training image, we export the dataset. ArcGIS provides several options for exporting training data, such as the chip size (default is 256) or the gap between two chips (default is half the chip size, which is 128). We can also choose the dataset format, such as Pascal Visual, RCNN Masks, or KITTI Labels. In our project, we studied chip sizes of 512 and 256 to compare and determine which size provides the best accuracy (see the Experiments and Results section). The final step is to select a folder to store the data and start the process in ArcGIS.

2.3 Detection

Due to hardware limitations, we used Google Colab to train the model instead of the Jupyter notebook provided by ArcGIS. Google Colab is a Google Research product that allows users to write and execute Python code of their choice via their browser. It provides free access to powerful GPUs (Graphics Processing Units), which is crucial for deep learning tasks.

To import data into Google Colab, we can connect to the drive associated with our Google account. Therefore, we first need to import the data into the hard drive to access it on Colab. To read the data, we need to use the ArcGIS library.

2.4 Model Training and Its Use in ArcGIS

The ArcGIS library provides different object detection models such as Mask RCNN, YOLO, and Single Shot Detector. In our project, we used Mask RCNN as it suits our needs to calculate the surface area of roofs. To train the model, we determine the learning rate using a predefined function in the ArcGIS library called ".lr_find()". Then, we use this function in the fit function to train the model for a specified number of epochs, which is 100 in our case. Finally, we save the current model to the hard drive and download it to export it to ArcGIS. Once the model is exported, we access the "Detect Objects Using Deep Learning" tool to import the model.

Next, we select the image on which we want to apply the model and the import path of the previously downloaded template after training. The model will be imported with all the parameters that we defined during the training, and it is possible to modify some of them, such as the batch size or the threshold. The tool consists of two parts. The first part is to view the model inputs and select the image in which we want to detect the roofs. The second part is the environment, where we can select the coordinate system of the image, the extent of our model's application, and the type of processor we should use in the simulation (CPU or GPU).

GPU is faster than CPU. To apply the model and calculate the surface of the roof, we need georeferenced images. The surface calculated by ArcGIS is based on the pixels present in the image. However, to measure the actual surface, we need a geographic coordinate system. In Morocco, the "North Morocco Degree" coordinate system is used. There are many websites or software available to download georeferenced images, and we have downloaded them using SAS Planet software. Initially, the images were provided in the Mercator geographic coordinate system, but we changed it to North Morocco Degrees in ArcGIS to ensure we obtained the correct area.

2.5 Model Training and Its Use in ArcGIS

After detecting the roofs, we need to calculate their surface and photovoltaic potential. To calculate the total area, we can access the attribute table of the detection frame. This table displays all the polygons (roofs) detected in the image, and each of them has a shape area that models the roof surface. However, the geographic coordinate system on the image is a Mercator system, so we need to convert it to North Morocco Degrees. To do this, we simply add a new field (e.g., real surface) and choose the geometric shape and the coordinate system. We can also choose the unit of the real surface, and ArcGIS will calculate it automatically. The final task in this work is estimating the PV potential, which is straightforward. We need to know the global annual solar irradiation in the city and multiply it by the total area we collect. This data can be obtained from PVGIS or other meteorological websites, using Eq. (1).

$$P = A \times R \tag{1}$$

A represents the total rooftop area (m^2)
R represents the global irradiation ($kWh/m^2/year$)
P represents the total solar PV potential (kWh/an)

In reality, the total area of the roofs cannot be fully exploited due to obstacles that may be present on them. In this study, we assume that only 80% of the total surface area of each roof is exploitable. Therefore, the real potential will be expressed as follows Eq. (2).

$$P_0 = P \times 0.8 \tag{2}$$

P_0 represents the new potential that takes into account the shadowing from rooftops.

We can use ArcGIS conversion tools to export the attribute table as an Excel table. Then, after determining the exploitable potential, all rooftops without significant potential can be deleted. In this study, we set the exploitable potential limit to 200 kWh. After performing this task, we obtain the technical PV potential: Eq. (3).

$$\begin{aligned} P_t &= P_0 - \sum P \\ P &\le 200 \end{aligned} \tag{3}$$

P_t is the technical potential.
P_0 is the potential that takes in account the obstacles.
P represents all the potential lesser than 200 kWh

3 Experiments and Results

To validate the effectiveness of our methodology, we selected three cities with different types of roofs, layouts, and climates: Tetouan, Agadir, and Ifrane. Before testing the model on these cities, we need to define some data parameters to optimize the detection. The primary parameter that affects the detection is the size of the images in the dataset. In this case, we created 512-pixel images and trained the model. These images were exported in different formats to determine which format works best for detecting surfaces such as roofs.

After fixing the size of the dataset images on the Agdal samples in Rabat, we applied the optimal size to the other regions. The "average_precision_score" is the parameter that indicates the accuracy of the model, so we selected the best value. The results are listed in the Table 1 below:

4 Study Results and Discussion

4.1 Discussion

Both the YOLO and SSD models use the same data types. However, we did not use the SSD model in this study due to its low "average_precision_score." We found that the best image size for the dataset is 256 pixels with a stride of 128, which provides the best accuracy. When we applied these values to the datasets of the other regions, we obtained an 81% accuracy score for Ifrane, 67% for Tetouan, and 86% for Agadir. In conclusion, we can say that the number of training data is crucial for properly training the deep learning model, and the size of the data chips can impact the detection accuracy. Additionally, the shape and configuration of the roofs also affect the detection of the roofs. Therefore, Agadir, which has the best roof configuration, obtained the best accuracy score. However, Rabat and Tetouan, whose roofs are almost identical to each other, obtained low accuracy scores.

The detection system does not perform well with images of closely spaced rooftops, as observed in the Ifrane and Agadir areas where the detection is more accurate due to the clear and distinguishable roof structures. In contrast, in the Agdal region, the precision is lower because the roofs are close together with indistinct boundaries. This issue stems from the quality of the input images, which should be higher, as with LIDAR systems used in some of the studies mentioned above. Therefore, detection is more effective for sharp images with well-spaced roofs. Another limitation is that the system requires further training to refine its performance and achieve better detection results.

Table 1 The evaluation of detection models using different parameters

Modèles	Images size	Stride	Learning rate (LR)	Batch size	Average precision score	Threshold	Early stopping	Epoch
RABAT								
SSD	256	128	0.001202264	64	0.096318043	0.9	6	100
	512	256	0.0003113	64	0.097956167	0.9	6	100
MaskRCNN	256	128	3.01995E−05	16	0.794091713	0.9	10	100
	512	256	6.30957E−05	4	0.69177778	0.9	22	100
YOLO	256	128	–	–	–	–	–	–
	512	256	–	–	–	–	–	–
IFRANE								
SSD	Not defined	Not defined	Not defined	Not defined	Not defined	Not defined	Not defined	Not defined
MaskRCNN	256	128	2.0893E−05	16	0.812993158	0.9	44	100
YOLO	Not defined	Not defined	Not defined	Not defined	Not defined	Not defined	Not defined	Not defined
TÉTOUAN								
SSD	Not defined	Not defined	Not defined	Not defined	Not defined	Not defined	Not defined	Not defined
MaskRCNN	256	128	0.000025119	4	0.670192217	0.9	27	100
YOLO	Not defined	Not defined	Not defined	Not defined	Not defined	Not defined	Not defined	Not defined
AGADIR								
SSD	Not defined	Not defined	Not defined	Not defined	Not defined	Not defined	Not defined	Not defined
MaskRCNN	256	128	0.000036308	4	0.86027299	0.9	31	100
YOLO	Not defined	Not defined	Not defined	Not defined	Not defined	Not defined	Not defined	Not defined

4.2 *Results After Applying the PV Potential Estimation*

In this section, we estimated the photovoltaic potential of the roofs in the Agdal region. After detecting the roofs, we used ArcGIS to estimate the total area detected by the model using the North Morocco Degrees geographic coordinate system. The model detected about 722 roofs in the ROI, corresponding to an area of 185×10^3 m^2. Using the previous formulas, we calculated a total theoretical potential of 364 GWh/year and a technical potential of 291 GWh/year. The solar radiation factor of the city is 1760 kWh/m^2/year.

4.2.1 Estimation of PV Power Generation

For the proof of concept, SUNPOWER 400 Wc panels were selected as examples. They were chosen based on their ranking as one of the best solar panels in Morocco in terms of performance. We obtained their characteristics from PVSYST, which contains a large database on photovoltaic panels. To evaluate the photovoltaic capacity installed on the roofs, we needed to calculate the photovoltaic area. The photovoltaic surface represents the actual surface occupied by the solar panels, taking into account their inclination. For our study, the tilt angle of a solar panel is 34° (according to PVSYST). To avoid the effects of shadows, signs must be placed at a certain distance to ensure that the panels receive the maximum amount of solar irradiation. The distances and angles required to achieve this configuration are shown in Fig. 3:

- α corresponds to the angle of inclination of the panels, it's equal to 34° in Rabat;
- l represents the width of the panel;
- L represents the length of the panel
- x represents the distance between two parallel modules

Fig. 3 Rooftop panel configuration

The first parameter to be determined is the distance x between two PV modules. This distance allows us to calculate the corrective factor for the actual available PV surface, which is needed to determine the PV power and the energy delivered by each installation. We calculate the energy delivered based on the average insolation time in the city of Rabat, which is 8 h. The following Eqs. (4, 5, 6, 7, 8, 9, 10) represent the expressions for the three parameters required for the calculations.

$$x = 2 \times L \times \cos(\alpha) \tag{4}$$

$$s = 2x \times L = x \times l \tag{5}$$

$$N = \frac{\text{Roof area}}{s} \tag{6}$$

$$\text{PV area} = N \times L \times l \tag{7}$$

$$\text{FC} = \frac{\text{Roof area}}{2x \times L} = \frac{l}{2x} \tag{8}$$

$$P = \text{PV area} \times l \tag{9}$$

$$E = P \times h \tag{10}$$

With:

S	existing surface between two strings (parallel modules)
N	number of panels on the roof
Area PV	area occupied by panels FC: the corrective factor
FC	the corrective factor
P	the power of the installation (kW)
I	solar irradiation on Rabat (in kWh/m^2 h, is equal to 0.22 Wh/m^2 h)
h	number of hours of use per day
Roof area	the roof area (determined with ArcGIS).

The equations provided above give a corrective factor of 0.302326, which will be applied to the different surfaces in our sample of 100 roofs. This results in a PV power of 3 MW.

4.2.2 Energy Purchases Optimization by Studying Two Consumption Cases

This study is not limited to estimating the solar photovoltaic potential; it also looks at minimizing energy procurement from external suppliers. Optimizing external energy procurement is beneficial in terms of energy efficiency. To achieve this, we need to solve an optimization problem with the objective of minimizing procurement while ensuring that each building meets its energy requirements.

We conducted a comparative study between two scenarios for building energy consumption. In the first scenario, each building will consume its own production, and we consider two cases.

- If a building satisfies its energy needs and there is surplus energy, it will go unused and therefore wasted. This is not energy-efficient as the surplus energy could have been used elsewhere.
- In the second case, if a building is unable to meet its energy requirements, it will be required to buy energy from the distributor.

The second scenario aims to satisfy the energy needs of all the buildings collectively. This means that we can distribute the excess energy of a building that meets its needs to another building that has an energy deficit. The problem defined previously consists of two main variables: the energy procured from the distributor and the energy surplus for each building. We used a sample of 100 buildings with a production duration of 8 h per day, resulting in matrices of size 100×8 representing variables that we denote as A for the purchased energy and E for the surplus. The data used includes the energy production of the panels and the energy demand of the buildings, represented by the matrices P and D, respectively. In the equations, the indices b and h represent the buildings and the hours, respectively. $A(b, h)$ represents the energy purchase of building b at hour h. From the matrices P and D, we can determine the initial matrix of A noted A_0 and the surplus matrix noted E_0 with the following Eqs. (11, 12).

$$E_0 = P - D \tag{11}$$

$$A_0 = D - P = -E_0 \tag{12}$$

The negative values of the matrices are updated to zero. Indeed, when the surplus is negative it would mean that there is a procurement of energy and when the procurement is negative it would mean that there is a surplus of energy. The goal of this optimization is to minimize the cost of energy that is purchased, a cost factor of energy sold in is defined equal to 0.9 DH/kWh- (0.090 USD/kWh). This optimization problem is defined by Eq. (13).

$$min\left(z = \sum b, hA(b, h) \times 0.9\right) \tag{13}$$

This objective function is subject to constraints; in this case, there is only one: for each building the demand must be satisfied every hour. This constraint can be defined by: Eq. (14).

$$(\forall h, \sum b, E(b, h) + A(b, h) - D(b, h) \geq 0 \quad (14)$$

Our aim is finally defined as a production and surplus affectation problem.

4.2.3 Optimization Results

In order to solve the optimization problem, we used MATLAB's optimization tool. After 70 iterations we obtained the results of the optimization. For the first scenario, the results show a total purchase of 117,720 DH/day (11,735.62 USD/day) for the 100 selected buildings making an average of 1177.2 DH/day (117.36 USD/day) per building. The second scenario on the other hand gives a procurement of 136 DH/day (13.56 USD/day) for the 100 buildings.

From these results, we can see that the implementation of the intelligent management system will have a significant impact on the energy that will be saved. This management system will make it possible to make considerable gains in energy procurement. The final impact of this system will be to reduce the production of energy by the thermal power plants which work with fossil energies and thus the emission of CO_2 will be strongly reduced (34,370 tons annually).

Indeed, the results obtained from the optimization problem show the potential of an intelligent management system in reducing energy procurement costs for buildings. By distributing excess energy from one building to another, the system can achieve a more efficient use of the photovoltaic potential and minimize the need for external energy procurement. This not only leads to cost savings for the building owners but also reduces the reliance on fossil fuels and contributes to a reduction in greenhouse gas emissions. The implementation of such a system can have a significant impact on energy sustainability and environmental preservation.

5 Conclusion and Perspectives

Resource management in urban areas is crucial for the sustainable future of cities. This project addresses energy management in buildings by harnessing rooftop spaces for decentralized electric power generation through solar panels. The strategy involved assessing the solar PV potential of rooftops using satellite imagery and a model trained on neural networks, revealing a technical solar PV potential estimated at 291 GWh per year.

Furthermore, we demonstrated how rooftop photovoltaic systems could be implemented to optimize energy consumption and production, using Arduino boards to manage the distribution of surplus energy. This management system aligns perfectly with the Smart Cities concept and is projected to save 42 million Moroccan dirhams annually on energy purchased from the National Office of Electricity and Water

(ONEE) and reduce CO_2 emissions by 34,370 tons per year, with an initial investment of 24.7 million Moroccan dirhams (2.40 million USD). The project's internal rate of return is 7 months, making it highly profitable.

Moving forward, the project could be expanded to other Moroccan cities, thereby increasing energy savings and contributing more substantially to national CO_2 emission reductions. The integration of more advanced artificial intelligence techniques could also refine roof detection and enhance the efficiency of energy management systems.

Considering partnerships with other sectors, such as transportation and public infrastructure, could lead to a more holistic approach to urban energy management. Furthermore, the ongoing adoption of advanced technologies to improve the efficiency of solar panels and energy storage solutions could further reduce the payback time and increase the sustainability of the project.

These initiatives will extend the relevance of the project and allow it to adapt to emerging challenges and opportunities in the field of urban energy management.

6 Competing Interests

The authors declare that they have no known competing financial interests or personal relationships that could have appeared to influence the work reported in this paper.

References

1. Fakhraian E, Forment MA, Dalmau FV, Nameni A, Guerrero MJC (2021) Determination of the urban rooftop photovoltaic potential: a state of the art. Energy Rep 7:176–185. https://doi.org/10.1016/j.egyr.2021.06.031
2. Ninsawat S, Hossain M (2016) Identifying potential area and financial prospects of rooftop solar photovoltaics (PV). Sustainability 8(10):1068. https://doi.org/10.3390/su8101068
3. Sun L, Chang Y, Wu Y, Sun Y, Su D (2022) Potential estimation of rooftop photovoltaic with the spatialization of energy self-sufficiency in urban areas. Energy Rep 8:3982–3994. https://doi.org/10.1016/j.egyr.2022.03.035
4. Afia A (2017) Evaluation of solar energy potential and electricity output using OpenStreetMap and ArcGIS. https://doi.org/10.13140/RG.2.2.11968.97285
5. Bayrakci Boz M, Calvert K, Brownson JRS 2015 An automated model for rooftop PV systems assessment in ArcGIS using LIDAR. AIMS Energy 3(3):401–20.https://doi.org/10.3934/energy.2015.3.401
6. Cadei R, Attias R, Jiang S (2020) Detecting rooftop available surface for installing PV modules in aerial images using deep learning
7. Carl C (2014) Calculating solar photovoltaic potential on residential rooftops in Kailua Kona, Hawaii. PhD thesis. https://ui.adsabs.harvard.edu/abs/2014PhDT........95C. 16 July 2024
8. Luqman M, Ahmad SR, Khan S, Ahmad U, Raza A, Akmal F (2015) Estimation of solar energy potential from rooftop of Punjab government servants cooperative housing society lahore using GIS. Smart Grid Renewable Energy 6(5):128–139. https://doi.org/10.4236/sgre.2015.65012

9. Muhammed E, Morsy S, El-Shazly A (2021) Building rooftops extraction for solar PV potential estimation using gis-based methods. Int Arch Photogramm Remote Sens Spatial Inf Sci XLIV-M-3–2021:119–25. https://doi.org/10.5194/isprs-archives-XLIV-M-3-2021-119-2021
10. Zhang C, Li Z, Jiang H, Luo Y, Xu S (2021) Deep learning method for evaluating photovoltaic potential of urban land-use: a case study of Wuhan, China. Appl Energy 283:116329. https://doi.org/10.1016/j.apenergy.2020.116329
11. Zhong T, Zhang Z, Chen M, Zhang K, Zhou Z, Zhu R, Wang Y, Lü G, Yan J (2021) A city-scale estimation of rooftop solar photovoltaic potential based on deep learning. Appl Energy 298:117132. https://doi.org/10.1016/j.apenergy.2021.117132
12. Wang L, Xu S, Gong Y, Ning J, Zhang X, Zhao Y (2022) High resolution photovoltaic power generation potential assessments of rooftop in China. Energy Rep 8:14545–14553. https://doi.org/10.1016/j.egyr.2022.10.396
13. Cui W, Peng X, Yang J, Yuan H, Lai LL (2023) Evaluation of rooftop photovoltaic power generation potential based on deep learning and high-definition map image. Energies 16(18):6563. https://doi.org/10.3390/en16186563
14. Kozlovas P, Gudzius S, Ciurlionis J, Jonaitis A, Konstantinaviciute I, Bobinaite V (2023) Assessment of technical and economic potential of urban rooftop solar photovoltaic systems in Lithuania. Energies 16(14):5410. https://doi.org/10.3390/en16145410
15. Lan H, Gou Z, Hou C (2022) Understanding the relationship between urban morphology and solar potential in mixed-use neighborhoods using machine learning algorithms. Sustain Cities Soc 87:104225. https://doi.org/10.1016/j.scs.2022.104225
16. Li Z, Zhang C, Yu Z, Zhang H, Jiang H (2023) Deep learning method for evaluating photovoltaic potential of rural land use types. Sustainability 15(14):10798. https://doi.org/10.3390/su151410798
17. Šúri M, Huld TA, Dunlop ED (2005) PV-GIS: a web-based solar radiation database for the calculation of PV potential in Europe. Int J Sustain Energ 24(2):55–67. https://doi.org/10.1080/14786450512331329556

Implementation of AI Techniques on Efficiency Enhancement of Solar PV Relating to Performance Parameter

Syed Saad Farooq, Amnah Kamran, Mumtaz A. Qaisrani, Naveed Ahmed, Muhammad Umer Farooq, and Faheem Abbas

Abstract AI-driven technologies are using embedded evolutionary algorithms, fuzzy and neural logics to extract maximum solar energy. It is the need of an hour to optimize the Solar PV performance variables like aligning with the sun's trajectory in intelligent tracking mechanisms with an addition of regular cleaning and substantial cooling of panels simultaneously. The chapter headings have aimed to give insight of the inter dependency amongst these Solar PV performance variables in the context of multi-objective optimization. By implementation of these intelligent algorithms, it is concluded later in the chapter that it has significantly reduced the levelized cost of energy (LCOE) by minimizing the need for unnecessary panel replacements and by controlling variables like temperature and contamination ratios. Furthermore, it is also shown that by optimization, solar PV electrical and thermal efficiencies have increased and hence solar energy is being declared as a sustainable alternative.

Keywords Solar tracking · Soiling ratio · PVT cooling · Degradation rate · LCOE · Multi-objective optimization

1 Introduction

1.1 Generalities of Solar PV Optimization

Nowadays energy produced by fossil fuels has a risk of shortage and the world is now seeking renewable energy as a vital energy generation and utilization source, solar photovoltaic (PV) is one sustainable solution to the cause. This energy is now

S. S. Farooq · A. Kamran · M. A. Qaisrani (✉) · M. U. Farooq · F. Abbas
Institute of Mechanical and Manufacturing Engineering, Khwaja Fareed University of Engineering and Information Technology, Rahim Yar Khan, Pakistan
e-mail: mumtazqaisrani@yahoo.com

N. Ahmed
U.S.-Pakistan Center for Advanced Studies in Energy National Univsersity of Sciences and Technology, Islamabad, Pakistan

F. Latrache et al. (eds.), *Digital Technology for an Innovative Energy Transition*, Green Energy and Technology, https://doi.org/10.1007/978-3-031-89657-6_7

anticipated by automation. To enhance the performance of solar power, optimality in solar tracking, PVT systems, and cost functions need to be addressed. The optimization algorithm can be implemented through MATLAB, LabVIEW, C++, python or Arduino IDE and with Sensory integration to solar panels. Besides this, an intelligent scheduling of soiling can be done; degradation rate can also be altered as corresponding objective variables. In this way, multi-objective optimization has placed special importance on developing automation strategies and process planning.

Optimization techniques are categorized as deterministic and stochastic approaches [1]. Linear analysis can be done using traditional methods like newton Raphson, secant, and bisection methods while evolutionary algorithms and AI methods are under the umbrella of non-linear analysis as shown in Fig. 1. Data mining and data learning are now are being used extensively today for solar photovoltaic tracking. People are interested now in using one of the strategies either machine learning (ML) or deep learning (DL.) DL methods work on non-linear data and multi-layered evaluation [2] and [3]. It has attractive features in image processing. Deep learning can be either supervised semi or unsupervised versions.

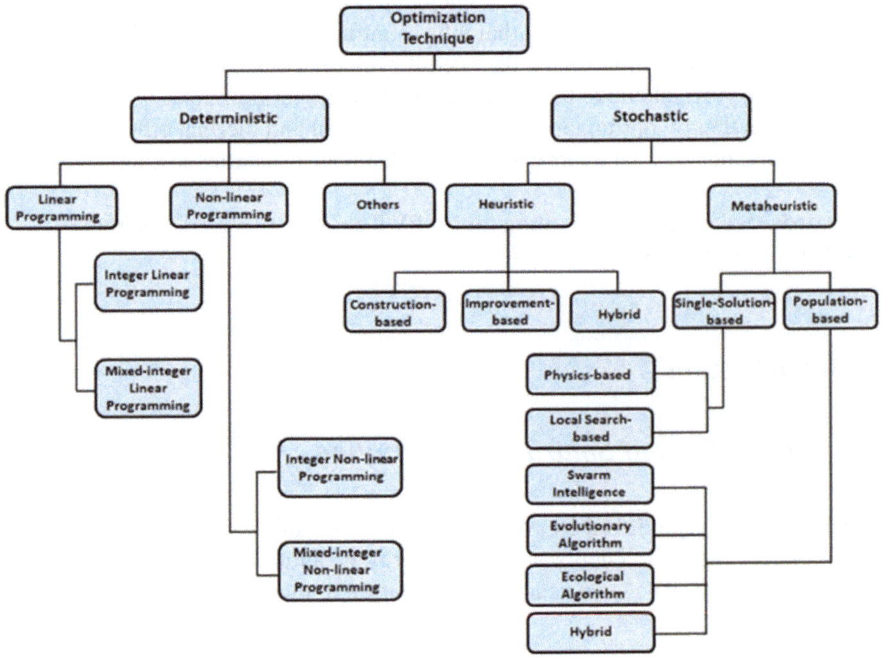

Fig. 1 Schematic diagram of optimization procedures [1]

1.2 Literature Survey

Usually, solar photovoltaic with less efficiency has traditional stand-alone arrangements. With the advent of dual-axis tracking as shown in Fig. 2, the efficiency of solar photovoltaic was however increased but this arrangement is still costly for a consumer to afford [4] but its design can be optimized to get maximum power point tracking with less LDRs [5, 6] and [7]. LDRs develop the logic for the motors as shown in Fig. 3. This LDR arrangement is being replaced now with a novel UV sensor for precise tracking even in overcast conditions [8]. A comparison of Meta-heuristic and Pareto front algorithms like particle swarm optimization, and genetic and other predictive models like artificial neural networks or fuzzy logic is being used for data mining and power enhancement in partially shaded conditions [9] and [10]. However, these traditional optimization procedures have a common issue of prolonged settling times and bad transient responses for the global optimum values. To serve this purpose, hybrid meta-heuristic techniques were adopted [11]. There was still a need to address the transient response in plausibly finding extreme global values. This work closely related the effects of hybrid algorithms for better transient responses numerically [12]. The dual axis trackers are being controlled through PID controllers for saving of electrical energy which previously was being utilized by motors in inaccurate positioning relative to sun [13]. These traditional optimization positions the serial framework is now being advanced with parallel actuators operated through servo motors [14]. Servomotors are specialized for continuous feedback as it has a gear reduction unit with an optical encoder. The actuation with parallel actuators is complex, architecture is bulky but is being used for precise movements with better stiffness [15]. Its use is recommended for long array of solar panels in solar power plants. The levelized cost of energy in this arrangement can only be justified if operated in solar power plants.

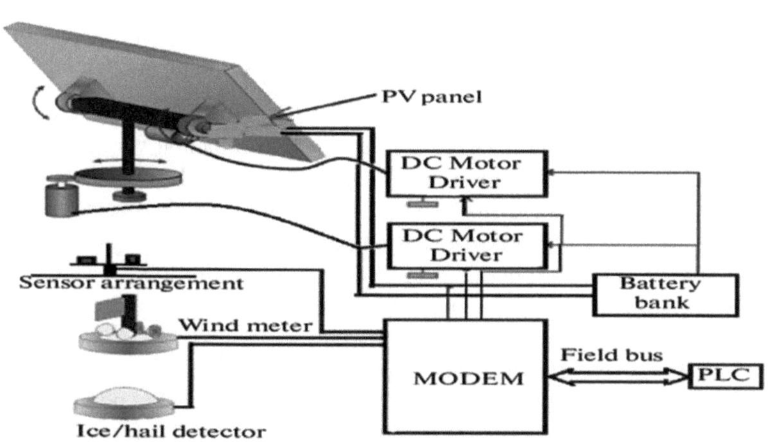

Fig. 2 Sensor based dual axis solar tracker [16]

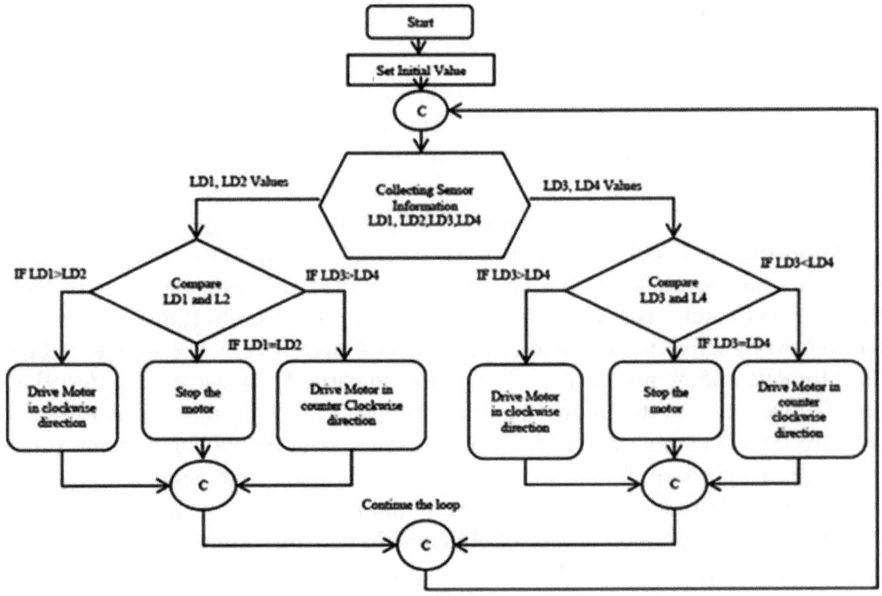

Fig. 3 Motor Logic through LDRs [17]

Another major issue is the rise in temperature of solar panels which significantly lowers the solar PV efficiency. PVT systems have different arrangements of copper piping beneath the surface of solar panels that has contributed for efficiency enhancement. A combination of water and air circulation is also an option. It is recommended that this cooling mechanism must be operated itself whenever it faces rise in temperature [18]. An optimization evolutionary algorithm can be implemented where objective variables like irradiation models are to be set according to the cell efficiency objective function [19]. This is further explained by obtaining optimized tilt angles to get maximum solar energy [20]. Points below are some of the extended reasons behind power fluctuations.

1. Primary changes in power factor are due to environmental variations. Environmental changes include sand storms, extensive rain fall, over cast conditions and snowy weather. Sand storms can cause soiling which is a major source of reducing the panel efficiency. It is quite difficult to examine the contaminants being attached on panel in the form of dust and bird droppings. There is a need for an intelligent system for panel to address the soiling and an autonomous system for its regular cleaning.
2. Secondary changes are due to an inappropriate solar panel placement has shading effects in afternoon which is also a vital cause of decrement in power. Fault diagnosis is sometimes too difficult because there is no feedback response systems which can help engineers find faulty panels or wiring around the solar park [21].

In the context of the above discussion and problem identification, solar power plants needs an effective intelligence model which can detect these power changes and extract optimal values of tracking, cleaning, PVT cooling systems to attain maximum efficiency yet lowering lCOE and degradation rate at the same time. Later section of this chapter will have emphasis on the optimization and variations of these important solar PV performance parameters.

2 Optimization Parameters of Solar PV Technology

2.1 Optimization with Solar Tracking

In solar tracking, it is required that the positions of the solar panel must be adjusted in a way that maximum irradiations can strike the panel orthogonal to the panel plane to get maximum efficiency. Normal beam irradiation (BI) is the one which is already incident 90 degrees to the panel. Reflected Irradiation (RI) comes to the panel after striking multiple things. Diffused irradiation is the part of irradiance (DI) that is absorbed or scattered in the atmosphere while reaching the panel. Adjusting to get the maximum irradiance is achievable if the system has a feedback mechanism i.e. panel must track the sun automatically with the movement of the sun and the motors have to change their orientation accordingly. Besides sensors, the system can have an algorithm based on mathematical calculations that can record the Greenwich time and season-wise sun terrestrial movement. The algorithm will also tend to observe the placement of the solar panel (latitude or longitude). The objective variables will be the tilt and azimuthal angles. The algorithm can use multiple iterations in seconds to find the optimal global tilt and azimuthal angles. This numerical strategy can be achieved using evolutionary optimization. This system can reduce the settling time of the motors as they will face less feedback from the sensors. This can save energy and improve efficiency. The system has to act according to the variance in seasons and optimal tilt azimuthal angles are the only solution to cover this uncertainty. The modular BN method was adopted to see the probabilistic casual interference by seasonal and abrupt weather changes [22]. One of the works has used online methods instead of mathematical embedded data to detect the possible movement of the sun. Besides the evolutionary algorithm, it has used neural networks with an ingredient of regression plots [7]. An Arduino or PLC based hardware control unit is used to compute the objective function dependent on the voltage changes in LDRs as shown in Fig. 2.

Tracking in heliostats and concentrated solar power plants is important and it must be dealt with care. A recent development is made in which feedback computer vision is introduced. It is done by observing the presence of light from the light sensor and after capturing an image of the stick shadow from the webcam. In this procedure, solar angles are approximated, and this data is sent to the microcontroller that gives signal to the motors [23].

Tracking also facilitates controlling soiling ratio which is an aside source for dust cleaning. This autonomous tracking algorithm is following the sun rotation. In the morning, the solar panel is aligned at 90 degrees of tilt angle; the dust may fall due to gravity. It should be noted that dust should not mix with humidity all night. Otherwise, this strategy may fail, and the solar panel will now only be cleaned with wet cleaning. This multi-tasking can reduce the effort to clean panels manually and integration of robotic mechanism.

2.2 *Optimization with PVT Cooling Systems*

PVT systems hold significance in providing sufficient cooling to solar panels. Continuous incident rays of light increase the temperature of the panel when exposed to sunlight during the day. The cells reduce their efficiencies and power production decreases. Several approaches have been adopted for years to enhance its efficiency by solar panel interaction with water, air, phase change materials, or nanoparticles. It is still a vacant portion of research to get optimal solutions for a sufficient amount of cooling. Being thermal efficiency objective function, it is necessary to get the optimal flow rates of water or air [24]. Secondly, there is a need to observe the best possible combination of water flow through the different pipe arrangements beneath the solar panel surface as shown in Fig. 4. There are commonly oscillatory, parallel, and serpentine arrangements and each one of the arrangements provides various conduction rates to the solar panel [25]. This later be analyzed on CFD to check the temperature distribution around its contact surface [26].

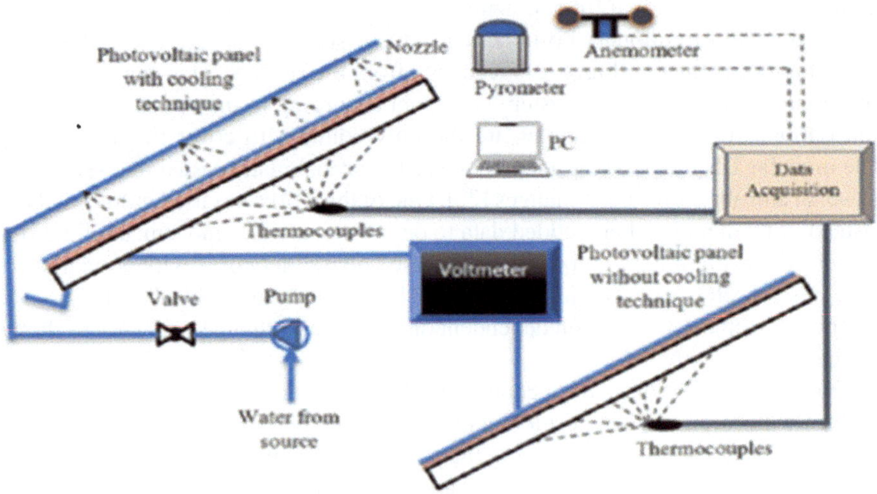

Fig. 4 PVT controlled cooling system [27]

PV cooling can also be implemented in another way. The placement of water nozzles above and beneath the surface of solar panels can also serve a cooling purpose. However, the exact location of water nozzles to get the maximum power needs optimal solutions for its design. Flow rate, droplet size, and flow distribution will be a major factor. It must be set in a way to get water on the maximum panel surface. Water pumps must also operate with low power. If it cannot be done, at least pumps must activate only when there is a rise in temperature. All must be done if there is some multi-objective algorithm being implemented for an autonomous cooling system. The main disadvantage of this type of cooling is the wastage of water. Sprinklers are also a possible solution to this problem. This type of cooling has the advantage of simultaneous wet cleaning of solar panels.

2.3 *Optimization with Cleaning*

One of the corresponding optimal parameters is a soiling ratio, it is necessary to remove the dust particles or contaminants from the solar panel. Solar panels need continuous monitoring and regular scheduling of cleaning to always get maximum power. It is usually a tiresome job if this task is being done by labor if there are many solar modules. There must be a mechanism to detect the amount of dust and then clean it. Two methods for solar panel cleaning are dry cleaning and wet cleaning. Many researchers have tried multiple strategies to rectify the problem. Robotic strategy is being used for cleaning tasks in terms of intelligent automation [28]. This robotic mechanism cannot move all day for cleaning purposes. It therefore requires the automatic dust detection and scheduling algorithm. That algorithm will get feedback from the pyrometers and the signal will trigger the motor if it crosses some dust level. The dust will reduce the irradiations to panel. A variation of voltages with cleaning is best described in Fig. 5. The pyranometer is an indirect method for the calculation of the soiling ratio. The soiling ratio depends on dust accumulations rates which will be evaluated through tilt and azimuthal configurations of angles.

2.4 *Optimization with Cost*

Cost is the most important aspect for the implementation of solar photovoltaic technology. Levelized cost of energy (LCOE) is one parameter which is being used extensively. It is required that the solar photovoltaic system must have optimum LCOE with all the features of tracking, cleaning and cooling systems. LCOE is dependent on different components like installation cost operational maintenance, system lifespan, discount rate, payback period and revenue generation. The LCOE can be obtained by PVsyst i.e. software to calculate the feasibility by calculating LCOE and all its dependencies. These corresponding components need to be optimized to get an optimum LCOE with respect to the features of tracking, cooling and

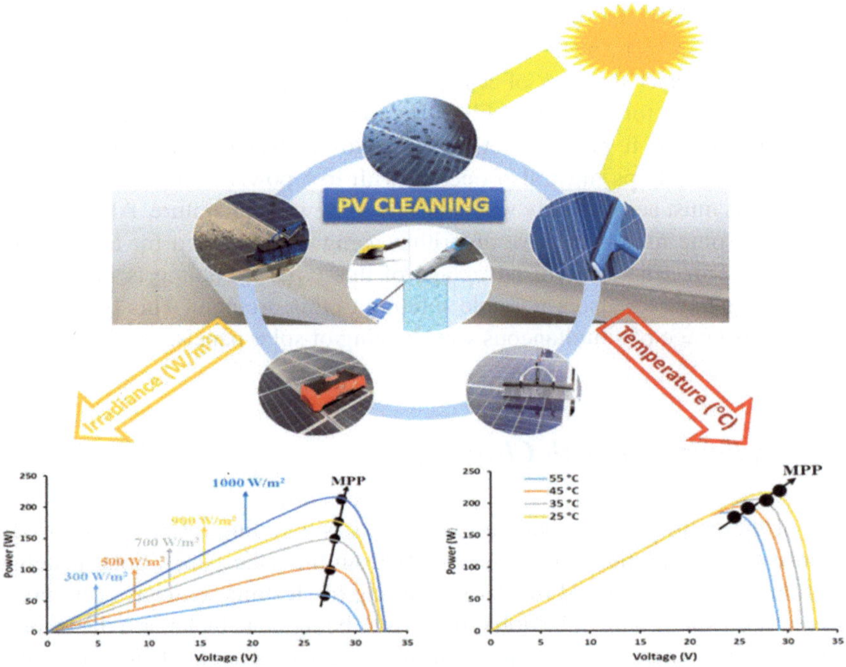

Fig. 5 Voltage variations with respect to cleaning [29]

cleaning so that an investor or shareholders can declare it as viable solution. With an addition of autonomous cleaning, cooling and tracking, it can save a lot of money with an enhanced solar PV performance. Further other parametric optimization like yield rating, capacity factor and reference yield can also be a source of efficiency enhancement. The study [30] has detailed back reflector optimization of irradiations with LCOE.

2.5 *Optimization with Degradation Rate*

Degradation rate (DR) is a suitable illustration to describe the operating life of solar panels. Solar panels always expose to environments changes like high temperatures, dust storms, hailstorms. This reduces the life of the panel. Corrosion, hot spots, discoloration, deterioration of frame and short circuiting between the diodes etc. are the effects. Degradation rate may increase in non-linear environment trends. This can be reduced if a proper mechanism of cooking and cleaning be adopted. The weightages for temperatures, humidity and incident solar irradiation and wind must be adjusted in a way if a controlled architecture of cleaning and cooling be set. The LCOE is heavily dependent on degradation rate of the solar panel. If DR increases, there often needs to be replaced by solar panels. Usually, the life span of the panel is 25 years, if proper manual maintenance is not done regularly, its span may reduce to 22 years or so. Automatic systems of tracking, cooling and cleaning will save the solar panel from deteriorations and hence improved degradation rate. Although tracking alone will increase the temperature, it is negative for degradation rate. Solar panels must be coupled with cooling systems by keeping in view of LCOE. Degradation rate is well coped in the recent research [31] with solar tracking by fuzzy logic algorithms.

3 Multi Objective Optimization of Tracking, Cooling and Cleaning

It is evident from the above literature that automation in solar photovoltaic needs some algorithms or systematic feedback. Majorly tracking, cooling and cleaning will be the major concern for high power whole day. There must be multi-objective function which will cater to all these phenomena simultaneously. Function will be comprised of cell efficiency for tracking, thermal efficiency for cooling, LCOE for cost and Degradation rate (DR). In tracking, corresponding tilt and azimuthal values need to be optimized in order to maximize the cell efficiency. In cooling, maximized objective function being thermal efficiency, flow rates must be optimized together with the best possible arrangement of the flow lines. In cleaning, dust accumulation rates (DAR) will be checked for the optimal values of tilt and azimuthal angles. Cost function depends on all the above parameters separately. With an aim to reduce the cost and degradation rate, these tracking, cooling and cleaning are set. It is important to note here that all parameters will be in some constraint and pareto optimal solutions will be obtained. Here lies a question that how the above tracking, cooling, cost and degradation rate will be evaluated as these are distinct quantities with different nature of constraints. Procedure illustrates that these all will be normalized. Secondly each one of the parameters must set priority levels to reach some decision to finalize optimal values. TOPSIS (technique for order preference by similarity to ideal solution) may be adopted to prefer of one parameter onto other [32]. Epsilon constraint strategy may also be adopted for this cause in which one parameter in constraint with respect to other [15]. Figure 6 is the complete inter dependency between the parameters for the optimization process.

Optimization as being an adventurous procedure for finding the desired values yet has some limitations of heavy computations and complex numerical procedures. This work has exhibited four main performance parameters to be treated as objective functions i.e. cell efficiency, thermal efficiency, degradation rate and Levelised cost of energy which further have many underlying interdependent variables like tilt angle, mass flow rate, soiling ratio etc. as shown in Fig. 6. Optimization algorithms which run with this scenario has time consuming iterative process in order to reach for global optimum solutions [33]. Some optimization techniques usually stuck in local optimum solutions and declare them as final solutions. This must be dealt

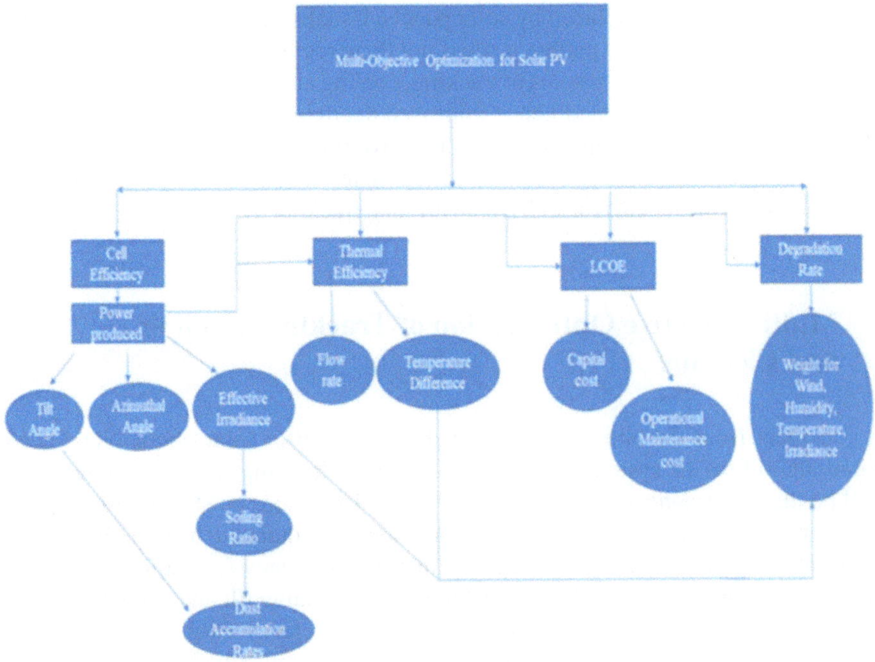

Fig. 6 Multi objective functions and corresponding variables dependency layout

carefully and much iteration should be performed to reach a steady state result. It is important to select a reliable set amongst the optimum set and decision technique like "TOPSIS" is used here which normalizes yet assign ranking to optimum sets to get the most fit solutions.

4 Conclusion and Perspectives

This work has highlighted the key aspects for the performance enhancement of solar PV. Before intelligent automation, it seems nearly impossible to manually monitor and act accordingly. Tackling solar panel efficiency is a difficult task as it is dependent on many internal design parameters as explained in figure '1'. Intelligent working demands useful optimization to perform the tasks simultaneously.

Firstly, this work has overcome the problems of un-effective irradiance, excessive heat generation and dust deposition. Flow rate is being set to a value which will have better conduction rates for lowering of panel temperature. Soiling ratio is being approached to one which means the panel is clean enough. Consequently, cell efficiency and thermal efficiencies are being increased.

With automation, there must always be cost saving ideas. Secondly this study has introduced a cost function as a part of multi-objective optimization procedure; it is now easier to get the economic feasibility with an insight of increase in lifespan of solar panels. This multi-objective function is also ensuring that degradation rate must not be high by controlling the air humidity, wind speed, temperature of panel and effective irradiations.

Hence these optimal values are giving us the tradeoff between performance parameters. Its HMI (human machine interface) can also be developed so that any individual can get the optimized data in a short time. This study can play a crucial role towards the sustainability of solar PV and encourage common people, governments and investors to adopt and promote as useful alternative forms of energy.

Moreover this algorithm can add the results by the variation of solar panel types like monocrystalline, polycrystalline and thin. Also the variations in efficiency by wind effects for the solar PVT systems can also be added. Parametric optimization can also help for the design of solar panel tracking, cooling and cleaning systems. Hybrid water and air cooling system can be installed by the different pipe or slot configurations. Further advanced AI and vision based robotic cleaning methods can also be introduced. This algorithm can serve this purpose too and can help to select the best possible combinations of design parametric values but the main objective like LCOE should be kept in mind for all necessary alterations in solar PV systems.

References

1. Ogundoyin SO, Kamil IA (2021) Optimization techniques and applications in fog computing: an exhaustive survey. Swarm Evol Comput 66:100937. https://doi.org/10.1016/J.SWEVO.2021.100937
2. Ahmad T, Chen H (2019) Deep learning for multi-scale smart energy forecasting. Energy 175:98–112. https://doi.org/10.1016/J.ENERGY.2019.03.080
3. Deng L, Yu D (2013) Deep learning: methods and applications. Foundations and Trends in Signal Processing 7(3–4):197–387. https://doi.org/10.1561/2000000039
4. Farooq SS, Attique J, Ahmad SA, Farooq MU, Qaisrani MA (2022) Performance analysis of dual axis solar tracking system actuated through serial manipulators. In: 2022 Proceedings of international conference on emerging trends in electrical, control, and telecommunication engineering, ETECTE 2022. https://doi.org/10.1109/ETECTE55893.2022.10007364
5. Zheng S, Shahzad M, Asif HM, Gao J, Muqeet HA (2023) Advanced optimizer for maximum power point tracking of photovoltaic systems in smart grid: a roadmap towards clean energy technologies. Renew Energy 206:1326–1335. https://doi.org/10.1016/J.RENENE.2023.01.023
6. Aguila-Leon J, Vargas-Salgado C, Chiñas-Palacios C, Díaz-Bello D (2023) Solar photovoltaic maximum power point tracking controller optimization using grey wolf optimizer: a performance comparison between bio-inspired and traditional algorithms. Expert Syst Appl 211:118700. https://doi.org/10.1016/J.ESWA.2022.118700
7. Masoumi AP, Bagherian V, Tavakolpour-Saleh AR, Masoomi E (2023) A new two-axis solar tracker based on the online optimization method: experimental investigation and neural network modeling. Energy and AI 14:100284. https://doi.org/10.1016/J.EGYAI.2023.100284
8. Jamroen C, Fongkerd C, Krongpha W, Komkum P, Pirayawaraporn A, Chindakham N (2021) A novel UV sensor-based dual-axis solar tracking system: Implementation and performance analysis. Appl Energy 299:117295. https://doi.org/10.1016/J.APENERGY.2021.117295

9. Wasim MS, Amjad M, Habib S, Abbasi MA, Bhatti AR, Muyeen SM (2022) A critical review and performance comparisons of swarm-based optimization algorithms in maximum power point tracking of photovoltaic systems under partial shading conditions. Energy Rep 8:4871–4898. https://doi.org/10.1016/J.EGYR.2022.03.175
10. Subiyanto S, Mohamed A, Hannan MA (2012) Intelligent maximum power point tracking for PV system using Hopfield neural network optimized fuzzy logic controller. Energy Build 51:29–38. https://doi.org/10.1016/J.ENBUILD.2012.04.012
11. Li L, Zhang Y, Li H, Liu R, Guo P (2024) An optimized approach for solar concentrating parabolic dish based on particle swarm optimization-genetic algorithm. Heliyon 10(4):e26165. https://doi.org/10.1016/J.HELIYON.2024.E26165
12. Liu G, Bai Y, Zhu L, Wang Q, Zhang W (2024) A sequential excitation and simplified ant colony optimization based global extreme seeking control method for performance improvement. Swarm Evol Comput 86:101522. https://doi.org/10.1016/J.SWEVO.2024.101522
13. Lu W, Ajay P (2024) Solar PV tracking system using arithmetic optimization with dual axis and sensor. Measurement: Sensors 33:101089. https://doi.org/10.1016/J.MEASEN.2024.101089
14. Jeng S-L, Chieng W-H, Lue B-H (2014) Design and analysis of spatial parallel manipulator for dual axis solar tracking. J Chinese Soc Mech Eng 35(3):221. Accessed: 23 May 2024. [Online]. Available: https://www.researchgate.net/publication/298423753
15. Farooq SS, Baqai AA, Shah MF (2021) The multi objective optimization of a tricept parallel manipulator using evolutionary algorithm. J Eng Res 9(2):278–295. https://doi.org/10.36909/JER.V9I2.9073
16. Samanta A, Dutta A, Neogi S (2012) A simple and efficient sun tracking mechanism using programmable logic controller 1. Appl Solar Energy (English translation of Geliotekhnika) 48(3):218–227. https://doi.org/10.3103/S0003701X12030140/METRICS
17. Rajesh T, Tamilselvan KS, Vijayalakshmi A, Kumar NC, Reddy KA (2021) Design and implementation of an automatic solar tracking system for a monocrystalline silicon material panel using MPPT algorithm. Mater Today Proc 45:1783–1789. https://doi.org/10.1016/J.MATPR.2020.08.635
18. Shaban WM, Kabeel AE, El Hadi Attia M, Talaat FM (2024) Optimizing photovoltaic thermal solar systems efficiency through advanced artificial intelligence driven thermal management techniques. Appl Therm Eng 247:123029. https://doi.org/10.1016/J.APPLTHERMALENG.2024.123029
19. Al-Ghussain L, Taylan O, Abujubbeh M, Hassan MA (2023) Optimizing the orientation of solar photovoltaic systems considering the effects of irradiation and cell temperature models with dust accumulation. Sol Energy 249:67–80. https://doi.org/10.1016/J.SOLENER.2022.11.029
20. Gharakhani Siraki A, Pillay P (2012) Study of optimum tilt angles for solar panels in different latitudes for urban applications. Sol Energy 86(6):1920–1928. https://doi.org/10.1016/J.SOLENER.2012.02.030
21. Balachandran GB, Devisridhivyadharshini M, Ramachandran ME, Santhiya R (2024) Comparative investigation of imaging techniques, pre-processing and visual fault diagnosis using artificial intelligence models for solar photovoltaic system—a comprehensive review. Measurement 232:114683. https://doi.org/10.1016/J.MEASUREMENT.2024.114683
22. Kim KH, Cho SB (2019) An efficient concentrative photovoltaic solar system with Bayesian selection of optimal solar tracking algorithms. Appl Soft Comput 83:105618. https://doi.org/10.1016/J.ASOC.2019.105618
23. Tchao ET, Wu G, Xu J (2022) Solar photovoltaic optimization using a deep reinforcement learning approach. Energy 238:121654. https://doi.org/10.1016/J.ENERGY.2021.121654
24. Podder B, Rahman A, Islam SM (2021) Performance enhancement of solar photovoltaic systems through optimization algorithms: a comprehensive review. Renew Sustain Energy Rev 144:110916. https://doi.org/10.1016/J.RSER.2021.110916
25. Abdin ZU, Rachid A (2021) A survey on applications of hybrid PV/T panels. Energies (Basel) 14(4):1205. https://doi.org/10.3390/EN14041205
26. Aste N, Del Pero C, Leonforte F (2012) Thermal-electrical optimization of the configuration of a liquid PVT collector. Energy Procedia 30:1–7. https://doi.org/10.1016/J.EGYPRO.2012.11.002

27. Al-Doori GF, Mahmood RA, Al-Janabi A, Hassan AM, Chala GT (2023) Impact of surface temperature of a photovoltaic solar panel on voltage production. Lecture Notes in Energy 92:81–93. https://doi.org/10.1007/978-981-19-6688-0_6
28. Antonelli MG, Beomonte Zobel P, De Marcellis A, Palange E (2020) Autonomous robot for cleaning photovoltaic panels in desert zones. Mechatronics 68:102372. https://doi.org/10.1016/J.MECHATRONICS.2020.102372
29. Salamah T et al (2022) Effect of dust and methods of cleaning on the performance of solar PV module for different climate regions: comprehensive review. Sci Total Environ 827:154050. https://doi.org/10.1016/J.SCITOTENV.2022.154050
30. Hamed A, Al-Ghussain L, Hassan MA, Annuk A (2022) Techno-economic analysis for optimal configurations of PV systems with back reflectors. Energy Rep 8:14979–14996
31. Piotrowski LJ, Farret FA (2022) Feasibility of solar tracking and fixed topologies considering the estimated degradation and performance of photovoltaic panels. Sol Energy Mater Sol Cells 244:111834. https://doi.org/10.1016/J.SOLMAT.2022.111834
32. Podder B, Biswas A, Saha S (2021) Multi-objective optimization of a small sized solar PV-T water collector using controlled elitist NSGA-II coupled with TOPSIS. Sol Energy 230:688–702. https://doi.org/10.1016/J.SOLENER.2021.10.078
33. Carissimo C, Korecki M (2024) Limits of Optimization. Mind Mach 34(Suppl 1):117–137. https://doi.org/10.1007/s11023-024-09764-4

The Study of Direct Torque Controller (DTC) with the Proportional Integral (PI) Controller to Promote the Energy Transition

Ranjit Bindal and Akhil Nigam

Abstract There are various types of electrical machines which are currently utilizing with some emerging techniques for achieving reliable operation. So this book chapter delves in the study of AC machine like induction motor which is integrated with an intelligent controller to enhance energy stability and its reliability. Within the realm of proposed controller system, this manuscript explores about proportional integral (PI) controller for monitoring and regulating speed and its torque. By considering many advanced techniques in enhancing with optimization of energy efficiency across various AC and DC electrical machines, the utilization of PI controller enables a comparative study of some key aspects which are key essential for an intelligent controller-based induction motor functioning. This manuscript scrutinizes productiveness and efficiency of numerous parameters which are relevant to induction motor operation which includes like settling time, rotor speed, stator current and electromagnetic torque etc. By exploiting an intelligent controller, the objective is to attain advanced control over induction motor characteristics. Further, it includes discussion of existing controllers using emerging techniques to ensure reliable electricity production. Then this chapter concludes with intricacies in regulating speed and torque of induction motor, graveling the way for achieving advancement in efficiency of reliable and stable electricity. This chapter approaches with using proposed intelligent controller for optimizing speed and torque in addition to comparison with existing type of DTC controller.

Keywords Induction motor · PI controller · DTC technique · Electromagnetic torque · Settling time · Rise time · Transient time

R. Bindal · A. Nigam (✉)
EED, Chandigarh University, Mohali, Punjab, India
e-mail: akhilnigam03@gmail.com

F. Latrache et al. (eds.), *Digital Technology for an Innovative Energy Transition*,
Green Energy and Technology, https://doi.org/10.1007/978-3-031-89657-6_8

1 Introduction

In 1834, electrical machines have been utilized and they have expanded their operation day by day. The electrical machines have been categorized into static and dynamic machines which may further explained in terms of transformers and motors. In the role of electric motor, it has reached its milestone through using emerging techniques since year 1834. They have tried to replace diesel and gas-based engines while converting into an optimized version of electrical machines. Electric motors have been widely used in aeroplane system, trains, cars, ships etc. From the point of achieving clean energy, an electric motor can provide reliable operation. Multiple aspects have been overcome through the use of electric motor. At each different power applications, electric motor has been differentiated. From the point of low power applications, induction motor can be utilized. Similarly for high-power applications, synchronous motor can be utilized. From the point of power supply, further it has been classified into next section.

In conventional times, there was huge problem of gas emission and high maintenance cost by using diesel-based energy source. After the introduction of induction motor, it has reduced these existing problems especially in terms of fuel economy. Similarly, from the point of speed operation of motor, it depends upon parameters defined for any proposed electric system. Most of the times for high-speed operation, induction motor, reluctance motor and permanent magnet motors may be preferred. The construction and their working principle also depend upon the material used in electric motors. Now these days in electric vehicles, electric motors play vital role due to its high efficiency and high-power density. In industrial purpose, around 65% electricity is consumed rest energy will be lost in terms of heat or loss which affect on the performance of electric system. The rotational speed/torque is controlled or affected by change in temperature and some internal heat losses like stator and rotor core losses, winding losses, eddy current loss, etc.

The electrical motors play an essential role in factories, agricultural, traction, commercial and domestic applications. The applications of electrical motors are increasing continuously in today's technology era, for example, fans (ceiling/exhaust/wall), mixer juicer, cooler, lath machine, grinders, cutters, mills, lifts, elevators, ships, aeroplanes and many more. Electrical motors are the backbone of each and every industrial application. The invention of electric motors changed the wheels of development and advancement which formally commenced the second industrial revolution [1–3].

1.1 Classification of Electric Machines

The electric motors are mainly classified as DC motors and AC motors:

DC Motors: DC motors are referred to as those that run on a DC source. A DC motor is a machine that changes electrical energy into mechanical energy. The

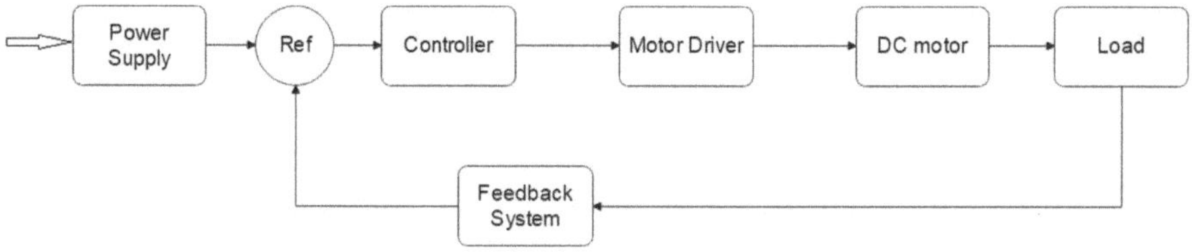

Fig. 1 Block diagram of DC motor

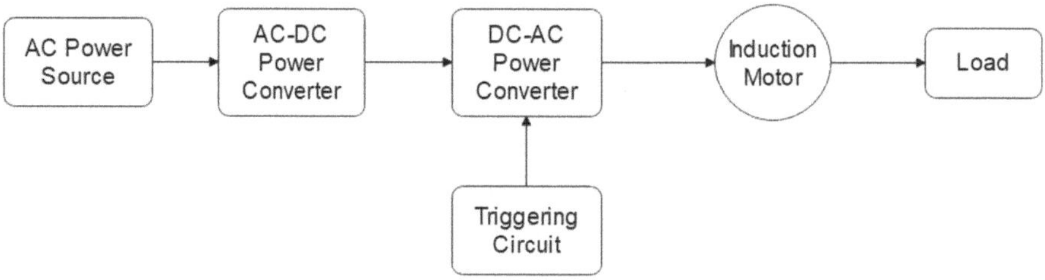

Fig. 2 Block diagram of AC motor

Lorentz law, which indicates that 'the current carrying conductor put in a magnetic and electric field experiences a force', is the fundamental tenet of a DC motor [4] (Fig. 1).

AC Motors: The electric motor that runs on AC supply is known as AC motor. It is divided into two main categories such as induction motor and synchronous motor (Fig. 2).

1.2 DTC with PI Controller

PI controller is an arrangement of the proportional controller and integral controller. It is a feedback controller used to control the feedback and calculate the error signal which is the variation between output and required output. Due to its simple design and improved performance under demanding working circumstances, the PI-type controller is the most used controller in industrial applications. The fundamental flaw of this controller is choosing the appropriate PI gain values by using constant gains. With changes in values and power plant operating circumstances, the PI controller does not give the required performance. The diagram of the proportional integral controller is displayed in Fig. 3 [5].

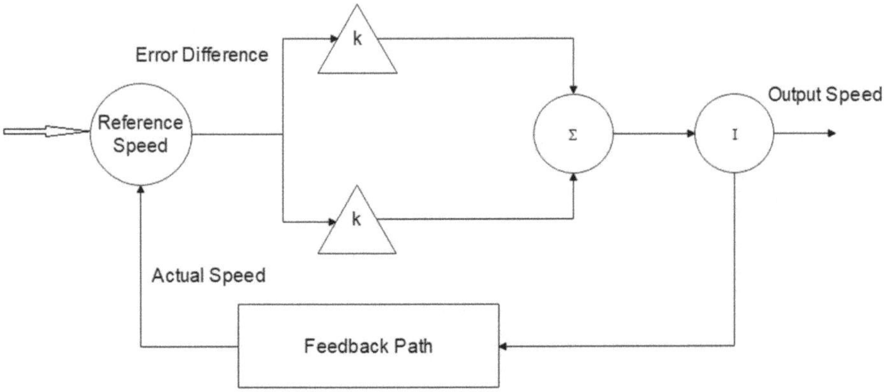

Fig. 3 Block diagram of PI controller [5]

1.3 Literature Review

Koo and Ki [3] presented a new DTC method for AC motor to minimize the torque ripples at fixed switching frequency. In the experiments, the control time interval is 20 s and the hysteresis bands of both torque and flux controllers are set at 0.6% and 0.5% of the rated value. Casadei et al. [1] differentiated and discussed the benefits and drawbacks of two control methods in this paper conventional DTC with PI. The concert of the two control techniques was discussed on the bases of current ripples, torque and transient retort to step change of the torque command. From the result response, it is found that this method needs only a small improvement (25–30%) of the computational period needs by basic DTC technique. The settling time of 3Ø IM is reduced from 3.8 ms to 1.8 ms at a speed of 1200 and 1.7 ms to 0.5 ms at a speed of 100 rpm. Jalluri and Ram [2] proposed the combination DTC using with of space vector modulation technique to limit the speed of AC motor and differentiated with predictable proportional integral (PI) controller. Based on comparative analysis, it is observed that at a torque of 2 Nm, the torque ripples are 50% by conventional method, 6% by PI controller and DTC-SVM it is reduced to 5%. Similarly at a torque of 10 Nm it is 10% for conventional, 1.2% for PI and 1% for DTC-SVM. Salem and Derbel [4] compared the DTC-SVM using PI controller and DTC-SVM using sliding mode controller techniques for induction motor on the basis of constant switching frequency and performance of AC motor. From the comparison, it is analysed that the behaviour at low speeds and robustness to parameters is improved in DTC-SVM using sliding mode controller compared to DTC-SVM using PI controller. The 10 kW induction motor is used for simulation.

Alagappan and Dhanasekaran [5] proposed the comparison between direct torque control with propositional integral controller and direct torque control with anti-windup PI technique for a 3Ø IM for speed and torque control. Anti-windup closed-loop saturation compared the input and output of speed controller. From the comparison, it is found that anti-windup technique is better than PI controller based on the parameters, i.e. minimum overshoot rate, minimum settling time, speed decrease because of variation in load and setting time after load varies. From the result examination, it is found that the overshoot time is minimized from 0.33% to 0.27%, settling time is reduced from 2.15 s to 1.8 s, steady-state error is reduced from 0.033% to 0.02%, speed drop is reduced from 0.43% to 0.02% and settling time after changing the load varies from 0.16 s to 0.08 s. Mehra et al. [6] proposed the comparison between DTC with proportional controller and DTC with PI controller on 3Ø IM on the basis speed regulation and torque ripples. In this paper, DTC with variable frequency drive is used to control the torque of 3Ø IM. From the comparative analysis on the parameters of torque, speed and stator flux, it is experimental that the stator flux and torque ripples are minimized in case of PI controller as compared to proportional controller. Lokriti et al. [7] proposed to improve the direct torque control characteristics by controlling stator flux components and by using hysteresis controllers. A novel switching table is introduced that is independent of sector change and reduced the size of the table as compared to Takahash's one. PI controller is implemented to reduce the present static error and minimize the flux and current harmonics. From the result examination, it is seen that the total harmonic distortion (THD) values is reduced from 11.72% to 5.02% for flux and 84.32% to 38.92% for the current. Venkadesan and Sedhuraman [8] proposed the real and reactive power to minimize error and settling time of 3Ø IM. From the result analysis, it is examined that by increasing the stator resistance from 10 to 50%, the settling time is reducing from 0.5 s to 0.37 s, At the same time, percentage error is improving from 0.33% to 0.36%. Pimkumwong and Wang [9] proposed a simple DTC controller and compared with PI controller on the parameters of speed, torque and stator flux peak value of 3Ø AC motor. In this research, the concept of projected controller is adopted as the idea of changeable step size most power point tracking for photovoltaic system. From the virtual result analysis, it is found that there is no change in previous and proposed PI controller but there is the steady-state error in speed response under loading period (Table 1).

1.4 Problem Identification

The past decades have witnessed significant development in the performance of Induction motors. According to the literature survey, it has been observed that the DTC is used with different controllers, i.e. DTC with Pl, DTC with neural network, DTC with fuzzy logic and DTC with space vector modulation.

The following points have been observed from the literature survey:

1. The previous researcher used motors of different specifications for different torque control techniques.
2. No data is available for different DTC techniques on a single motor for analysis.
3. Further, DTC system still requires smooth torque control and improved system parameters (like rise time, settling time and transient time).

Table 1 Comparison among different controllers in optimizing energy efficiency of induction motors

Reference and year	Technique used	Power	No. of poles	Supply voltage (V)	Torque (Nm)	Phases	N (rpm)	R_s (ohm)	R_r (ohm)	L_s (H)	L_r (H)	L_m (H)	J (Kgm^2)	Remarks
Koo et al. (1999)	DTC	10HP	4	220	–	NM	1750	0.15	0.17	0.035	0.035	0.0338	0.14	Torque ripples are reduced from 15 to 3%
[1]	DTC	4 kW	4	220	26.5	NM	1440	NM	NM	NM	NM	NM	NM	Settling time is reduced from 3.8 ms to 1.8 ms at a speed of 1200 and 1.7 to 0.5 ms at a speed of 100
Jalluri et al. (2012)	DTC-SVM using with flux observer	3 kW	4	380	6.8	3	1420	1.635	1.9	0.086	0.086	0.243	0.8	Torque ripples are reduced from 50 to 5% at a torque of 2 Nm
Salem and Derbel [4]	DTC-SVM PI	10 kW	4	220	NM	3	1470	0.29	0.38	50 mh	50 mh	47.3mh	0.5	Behaviour at low speeds. And robustness to parameters, variance is good as compared to SM and PI

(continued)

Table 1 (continued)

Reference and year	Technique used	Power	No. of poles	Supply voltage (V)	Torque (Nm)	Phases	N (rpm)	R_s (ohm)	R_r (ohm)	L_s (H)	L_r (H)	L_m (H)	J (Kgm^2)	Remarks
Pandian et al. (2014)	DTC with PI Vs anti-wind up PI	4 kW	4	400	NM	3	1430	1.405	1.395	0.005839	0.005839	0.001722	0.0131	Overshoot time is reduced from 0.33% to 0.27%, settling time is reduced from 2.15 s to 1.8 s, steady-state error is reduced from 0.033% to 0.02%, speed drop is reduced from 0.43% to 0.02%
Mehra et al. (2014)	DTC and flux control using P and PI	5 kW	4	440	NM	3	1440	1.77	1.34	0.00139	0.00121	0.00369	0.025	Stator flux and torque ripples are reduced by PI
Lokriti et al. (2015)	DTC	1.5 kW	4	380	NM	3	175 wm	4.75	1.2	0.4	0.072	0.163	0.025	THD values is reduced from 11.72% to 5.02% for flux and 84.32% to 38.92% for the current

(continued)

Table 1 (continued)

Reference and year	Technique used	Power	No. of poles	Supply voltage (V)	Torque (Nm)	Phases	N (rpm)	R_s (ohm)	R_r (ohm)	L_s (H)	L_r (H)	L_m (H)	J (Kgm^2)	Remarks
Venkadesan and Sedhuraman [8]	Real and reactive power for DTC	1.1 kW	4	415	NM	3	1415	6.03	6.085	0.5192	0.5192	0.4893	0.01178	Settling time is reducing from 0.5 to 0.37 and percentage error is improving from 0.33% to 0.36%
Pimkumwong and Wang [9]	DTC and DTC with PI	1 kW	4	220	6.8	3	1400	8.96	8.22	569.32	569.32	535.52	0.035	The steady-state error in speed response under loading period as compared to PI controller

2 Methodology

In space vector position via transients is not affected by the scalar torque control. In contrast, although the immediate location of current, voltage and space vectors flux are controlled in vector torque control, neither the amplitude nor the frequency (angular speed). In order to precise control the directions of the space vectors in a steady state and as well as in transient state, the vector control operates on the position of the space vectors. There is a col-linear relationship between the control parameters and torque. The proposed flowchart of DTC with PI controller has been shown below (Fig. 4).

2.1 *Practical Implications on PI Controller for Improving Efficiency in Electrical Machine*

There are numerous implications for using PI controller for improvement of electric machines as discussed below:

- An adaptive PI controller is able to control its tuning parameters in real time-based system. Such adaptive tuning process enhances robustness and performance either working in time varying or non-linear system.
- By adaptive such intelligent controller, it is able to incorporate with predictive models and anticipate with future prediction if there is any change in system behaviour.
- From the point of safety purpose, intelligent controller can be equipped with fault detection system from protecting any abnormal conditions. These sensors are able to sense any fault and able to take prevent actions.
- An intelligent controller is able to coordinate with control actions for achieving optimal performance. This can be performed by analysing feedback loop system or interdependencies.

Voltage source inverter (VSI): The electromagnetic torque (T_e) of a 3Ø IM can be written as:

$$T_e = \frac{3}{2}\frac{P}{2}\frac{L_m}{\sigma L_r L_s}\Psi_r \Psi_s \sin\eta, \tag{1}$$

whereas η is the angle lie betwixt the stator flux linkage space vector (Ψ_s) and rotor flux linkage space vector (Ψ_r) is displayed in Eq. 1, L_m is a mutual inductance, L_r is inductance of rotor, L_s is a stator inductance and σ is a leakage coefficient given by Eq. 2 [10].

$$\sigma = 1 - \frac{L_m^2}{L_s L_r}. \tag{2}$$

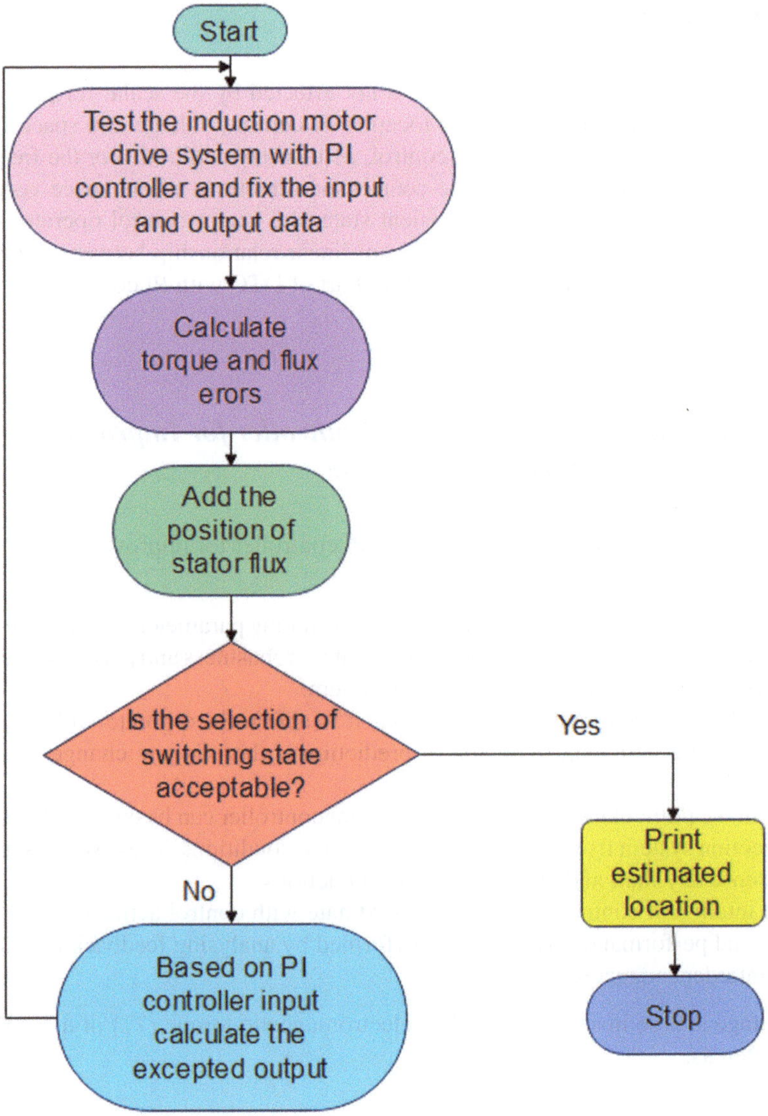

Fig. 4 Flowchart of DTC with PI controller

In both steady state and transient state, situations may be described by Eq. 1. Although these double vectors do not revolve at the same speed during transients, the rotor flux lags behind the stator flux by a torque position. Since the amount of the stator flux is typically constant, it is demonstrated by Eq. 1 that the induction motor torque rotated by rotating the rotor or stator flux linkage vectors. By adjusting the stator supply voltages, the stator flux linkage space vector may be altered. The stator resistance decrease, which is determined by Eq. 3, is assumed to be neglected for simplicity [11].

$$\overline{V} = \frac{\mathrm{d}\overline{\Psi}_s}{\mathrm{d}t} \tag{3}$$

When a frontward dynamic voltage vector is used formerly, it origins fast rotation of $\overline{\Psi}_s$ and torque improves via 'η', while a zero-voltage vector is utilized. Enabling a 3-phase, 2-state operation with 6 pulse voltage source inverters (VSI), around 6 nonzero lively voltage space vectors and 2 nil voltage space vectors as displayed in Fig. 18. In 6 lively voltages, space vectors can be displayed by Eq. 4.

$$V_k = \frac{2}{3} V_{\mathrm{dc}} e^{j(k-1)\frac{\pi}{3}}, \tag{4}$$

($k = 1, 2, 3 \ldots$).

where V_{dc} is supply DC voltage and K is constant.

To restrict combined stator flux and torque, it is likely to switch the appropriate voltage vectors based arranged as the stator flux linkage space vector is situated. From previous research work, it represents that an inverter voltage space vector consisting of $\vec{V}$d (direct axis voltage) is in space with switching position $\vec{V}$1 (100) and the vector $\vec{V}$q (quadrature axis voltage) which is 90 degrees away from $\vec{V}$d. The $\vec{V}$1 to $\vec{V}$6 are the different vector positions having 60 degree apart from each other. The switching states from S1 to S6 of inverter model are changed according as per the given Table. 2 [12–15].

Table 2 shows the variation of supply voltage vectors varying from $\vec{V}$0 to $\vec{V}$7 and the inverter switching states.

Digital output signals of flux controller can be characterized:

$$V_\Psi = 1 \text{ for } \Psi_{st} < \Psi^* - B_\Psi \tag{5}$$

Table 2 Switching states of the inverter

Supply voltage vectors	Switching state of inverter A	Switching state of inverter B	Switching state of inverter C	Inverter operation
$\vec{V}$0	0	0	0	Freewheeling
$\vec{V}$1	1	0	0	Active
$\vec{V}$2	1	1	0	Active
$\vec{V}$3	0	1	0	Active
$\vec{V}$4	0	1	1	Active
$\vec{V}$5	0	0	1	Active
$\vec{V}$6	1	0	1	Active
$\vec{V}$7	1	1	1	Freewheeling

$$V_{\Psi} = -1 \text{ for } \Psi_{st} < \Psi^* + B_{\Psi}. \tag{6}$$

Therefore, the torque controller output can be characterized as:

$$V_{T_e} = 1 \text{ for } T_e < T_o^* - B_{T_e} \tag{7}$$

$$V_{T_e} = 0 \text{ for } T_e = T_o^* \tag{8}$$

$$V_{T_e} = -1 \text{ for } T_e = T_o^* + B_{T_e} \tag{9}$$

Here, V_{Ψ} is the hysteresis flux controller, Ψ_{st} is the input flux, Ψ^* reference flux, B_{Ψ} is the flux acceptance tool, V_{Te} is the torque controller, $T_0{}^*$ is the reference torque, B_{Te} is the torque acceptance band for the hysteresis flux comparators, A binary numbers (0 or 1) in V_{Ψ}, V_{Te} and the stator flux segment is generated by the angular position, $\delta = \tan^{-1}\frac{\Psi_{\text{stq}}}{\Psi_{\text{std}}}$, whose the value is calculated from voltage vector switching table as shown in Table 3. Here, Ψ_{stq} is the q axis flux controller; Ψ_{std} is the d axis flux controller.

Table 3 represents the voltage vector switching. The flux regulator is a twice stage comparator, whereas the torque regulator is a thrice stage comparator. The projected stator flux and electric torque are then inhibited directly through evaluating them by their individual stipulated parameters of utilizing hysteresis comparators. The outputs of these double comparators are then utilized as input signals for an optimal switching table.

Table 3 Voltage vector switching table

V_{Ψ}	V_{Te}	Sector state 1	Sector state 2	Sector state 3	Sector state 4	Sector state 5	Sector state 6
↑	↑	$\vec{V}2$	$\vec{V}3$	$\vec{V}4$	$\vec{V}5$	$\vec{V}6$	$\vec{V}1$
	Zero	$\vec{V}0$	$\vec{V}0$	$\vec{V}0$	$\vec{V}0$	$\vec{V}0$	$\vec{V}0$
	↓	$\vec{V}6$	$\vec{V}1$	$\vec{V}2$	$\vec{V}3$	$\vec{V}4$	$\vec{V}5$
↓	↑	$\vec{V}3$	$\vec{V}4$	$\vec{V}5$	$\vec{V}6$	$\vec{V}1$	$\vec{V}2$
	Zero	$\vec{V}0$	$\vec{V}0$	$\vec{V}0$	$\vec{V}0$	$\vec{V}0$	$\vec{V}0$
	↓	$\vec{V}5$	$\vec{V}6$	$\vec{V}1$	$\vec{V}2$	$\vec{V}3$	$\vec{V}4$

3 Results and Discussion

3.1 *Design and Analysis of Simulation Model for Conventional DTC Technique*

In this paper, a conventional DTC technique has been demonstrated for speed and torque control of 3Ø IM. The simulation model is displayed in Fig. 5.

Then, Fig. 6 displays the changes of supply voltage with respect to time. This is the output voltage of the three-phase measurement block and connected to 3Ø IM. The voltage measurement block measures the voltage across the IM or known as supply voltage. The voltage measured by voltage measurement block is 400 V. The simulation model is run for 0.5 s for simplicity.

Figure 7 displays the current varies from 38.8 A to 5.5 A. The maximum current taken by 3Ø IM during starting is 38.8 A at the time of 0.152 s which is several times more than the normal current. At time of 0.240 s, the current is 5.5 A, and after that, the current is constant. This constant current indicates that the IM is running in a stable condition.

Concerning the starting current, Fig. 8 displays it the starting, which is nearly several times the rated current. The rated current is nearly 11.76 A at a power factor of 0.85 lagging. The starting current is 36 A during the start of IM. It is decreasing

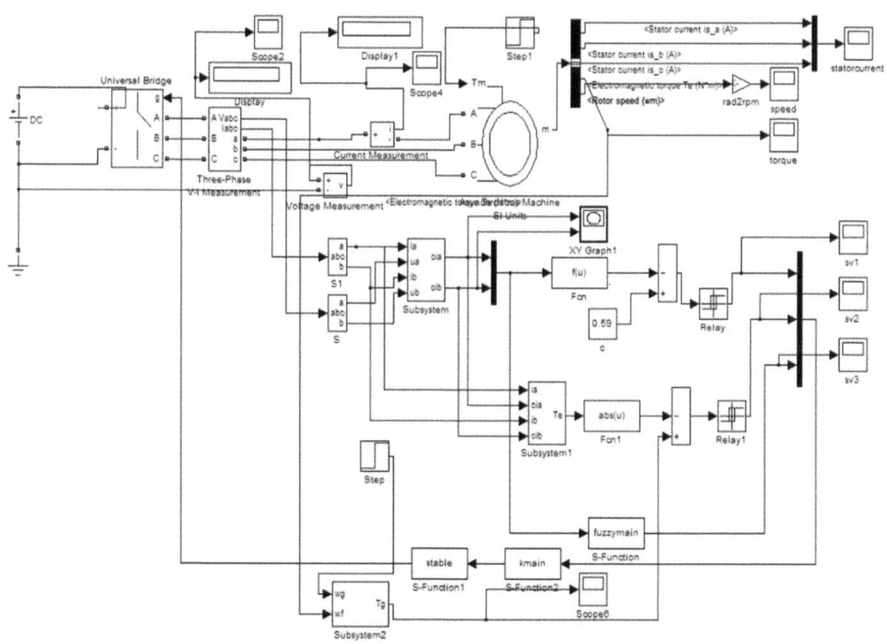

Fig. 5 Simulation model of conventional DTC for 3Ø IM

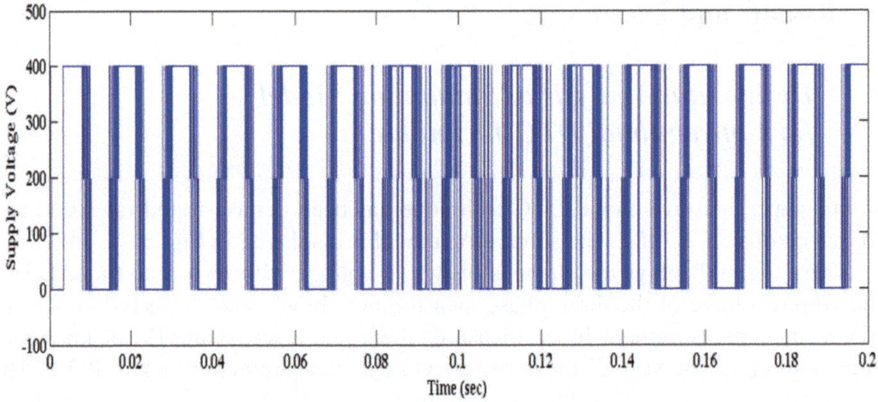

Fig. 6 Supply voltage of conventional DTC for 3Ø IM

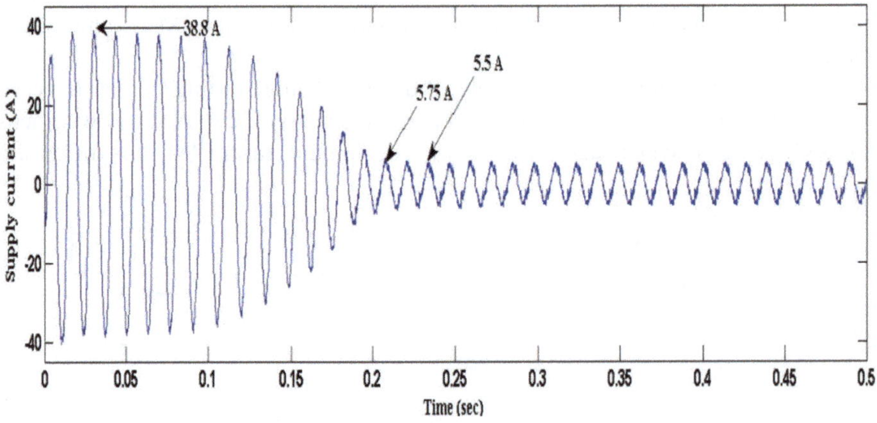

Fig. 7 Supply current of conventional DTC for 3Ø IM

step by step and touches to the value of 4.35 A inside 0.214 s. At the time of 0.23 s, the current touches 4.62 A and remains fixed after that.

Figure 9 demonstrates the simulation model for 0.5 s. The curve is designed by simulation study of conventional DTC technique for 3Ø IM, the curve represents the speed variation of the 3Ø IM is increasing with respect to time and touches a maximum speed of 1178.42 rpm at a time of 0.2923 s. The speed touches 117.842 rpm at a time of 0.279 s; the speed touches 589.21 rpm at a time of 0.1125 s and 1060.578 rpm at a time of 0.1746s. The rise time for 3Ø IM is 0.1467 s, settling time is 0.280 s, transient time is 0.220 s and speed is uniform after a time of 0.2923 s.

In addition, the electromagnetic torque varies from 0 Nm to 24.1 Nm from time 0 s to 0.0824 s (Fig. 10). After 24.1 Nm, the electromagnetic torque starts decreasing,

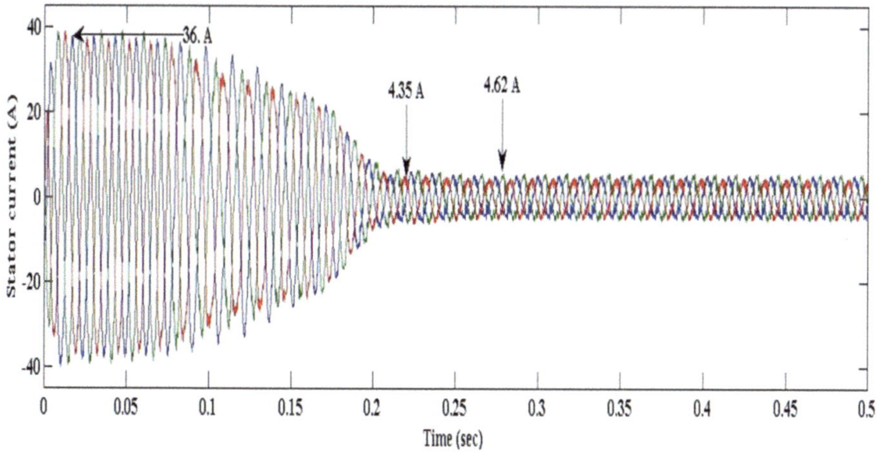

Fig. 8 Stator current of conventional DTC for 3Ø IM

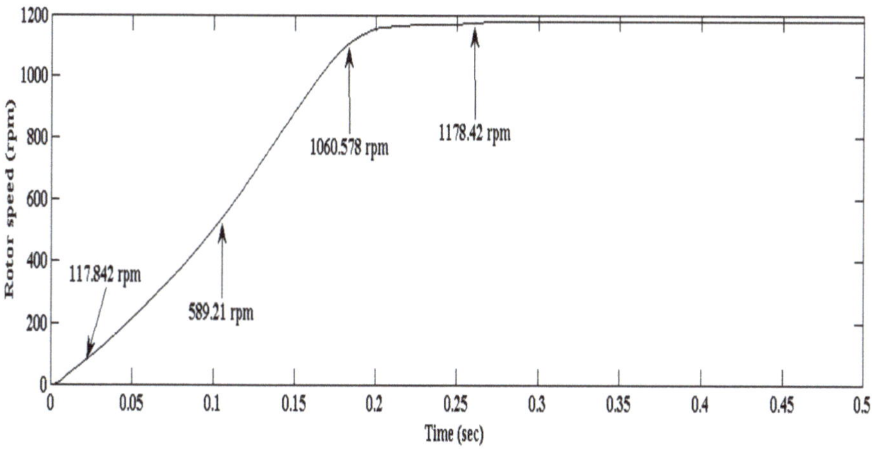

Fig. 9 Rotor speed of conventional DTC for 3Ø IM

at the time of 0.24 s, the electromagnetic torque touches to 5.4 Nm, and after that, it becomes almost uniform, which shows that the IM is running in a stable state.

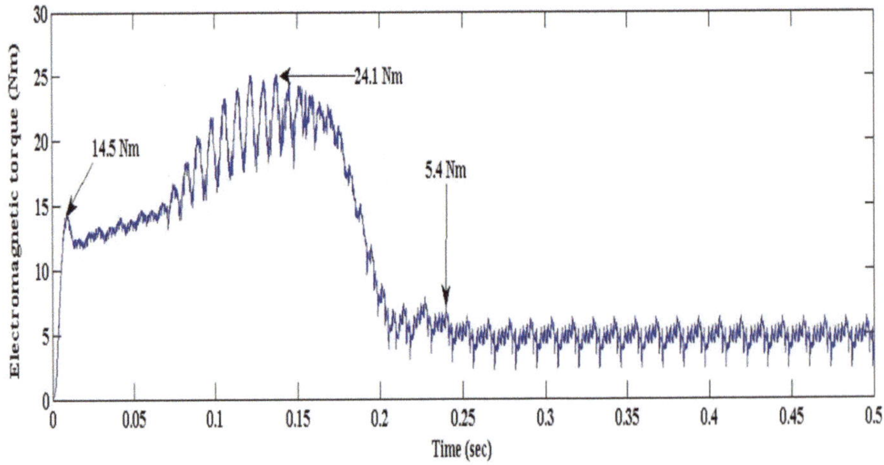

Fig. 10 Electromagnetic torque of conventional DTC for 3Ø IM

3.2 Design and Analysis of Simulation Model for Conventional DTC Technique: DTC with Proportional Integral (PI) Controller

The proportional integral (PI) controller is a modest way of feedback control and is widely utilized in industries. PI controller increases the speed response of the AC motor. It produces very low steady-state error. Different switching positions are selected in the PI controller to limit the rpm and torque of 3Ø IM. Accordingly, the switching position must be rearranged in such a way, the estimations of torque, flux error, error and stator flux angle define a place of stator flux.

3.2.1 System Simulation Model for DTC with PI Controller

The developed simulation model is shown in Fig. 11 having specification as, power rating 5.4 hp(4KW) 3Ø IM, voltage 400 V, frequency 50 Hz, speed 1430 rpm, stator resistance 0.8 Ω, resistance of rotor 1.395 Ω, stator inductance of stator 5.839mH, inductance of rotor 2.6 mH, mutual inductance 172.2 mH, rotor inertia 0.0131 kg/ m^2, pole pair 2.

Figure 12 displays the change in supply voltage in volts with respect to time in sec. The voltage applied is 400 V to 3Ø IM and the simulation model is run for 0.5 s.

Figure 13 displays supply current varies from 38.5 A to 5.27 A. The maximum current taken by 3Ø IM during the start is 38 A at the time of 0.152 s. The 3Ø IM is run at a constant speed of 5.27 A after 0.175 s.

Figure 14 shows the stator current is three times of initial current of 38.5 A. It reduces and touches the point of 4.35 A at a time span 0.142 s. At the time of 0.22 s,

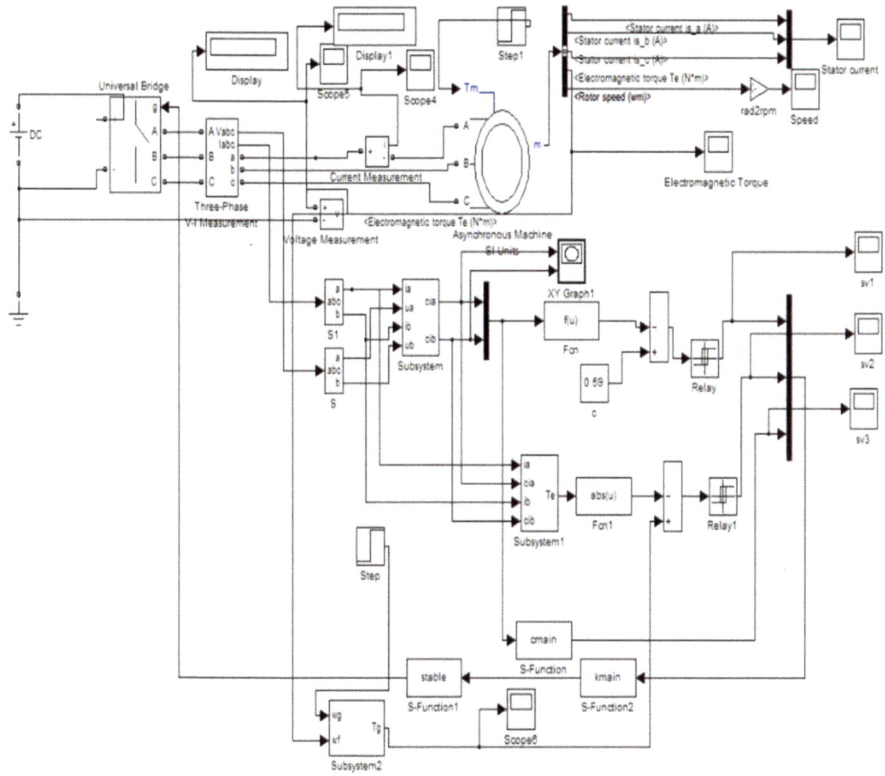

Fig. 11 Simulation model for DTC with PI controller for 3Ø IM

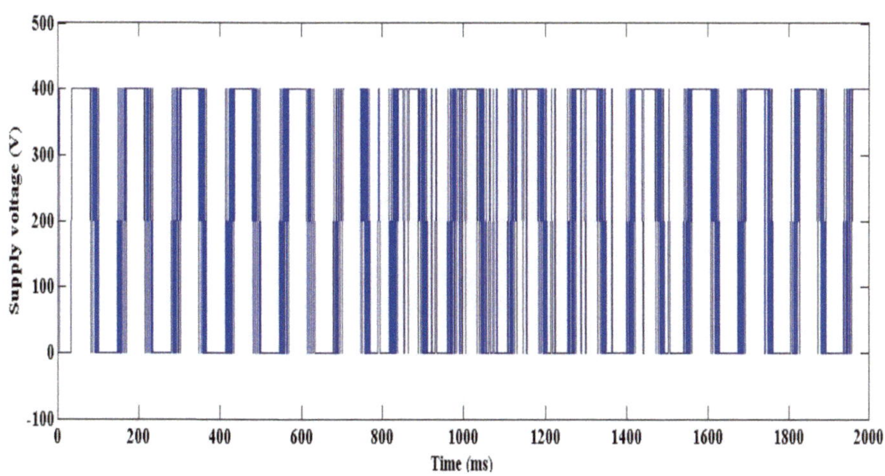

Fig. 12 Supply voltage of DTC with PI controller for 3Ø IM

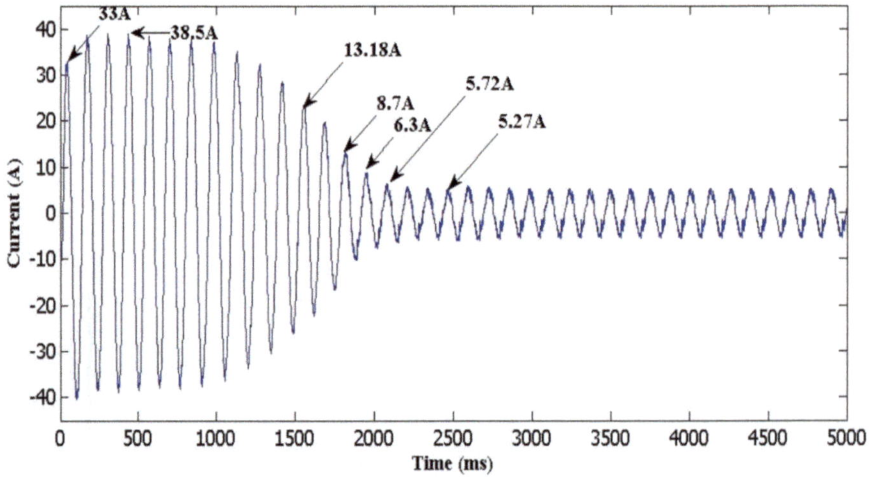

Fig. 13 Supply current of DTC with PI controller for 3Ø IM

the current touches the line of 4.8 A, and after that, it is uniform. The torque ripples are 82.7% generated as per Eq. 1.

Figure 15 displays the rotor speed in rpm with respect to time in sec. The simulation model is run for 0.5 s. The simulation of DTC with PI controller displays the speed of 3Ø IM is increasing and touches the maximum speed of 1129.34 rpm when the time is 0.254 s. The speed is 112.934 rpm at a time of 0.0269 s, the speed is 564.67 rpm at a time of 0.1153 s and the speed is 1016.4 rpm at a time of 0.1713s. The rise time for 3Ø IM is 0.1444 s with a speed of 112.93 rpm, the settling time is 0.240 s and the transient time is 0.201 s.

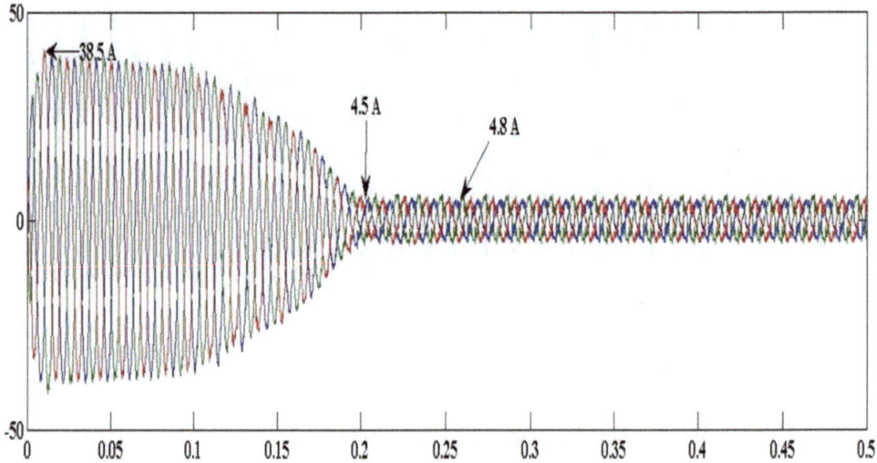

Fig. 14 Stator current of DTC with PI controller for 3Ø IM

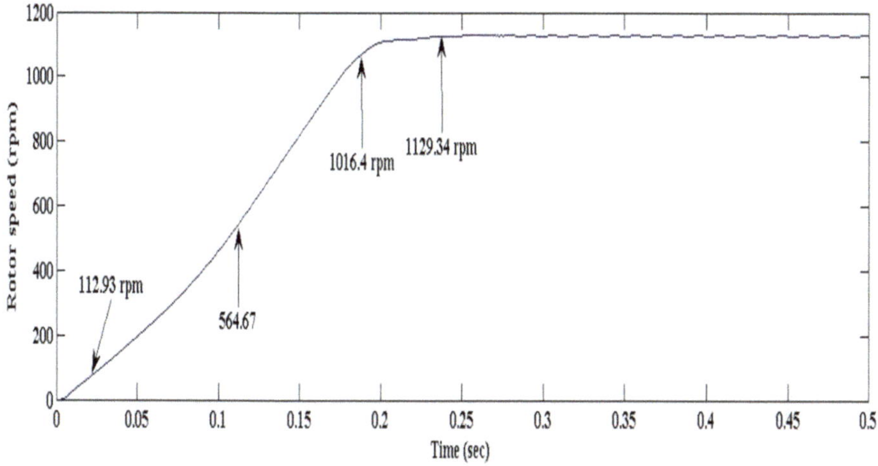

Fig. 15 Rotor speed of DTC with PI controller for 3Ø IM

The electromagnetic torque varies from 0 Nm to 25.2 Nm (Fig. 16) with respect to time from 0 s to 0.0824 s. As the torque reaches maximum position of 25.2 Nm, after that the electromagnetic torque starts decreasing, at the time of 0.240 s, the electromagnetic torque touches the point of 6.15 Nm, and after that, it becomes uniform.

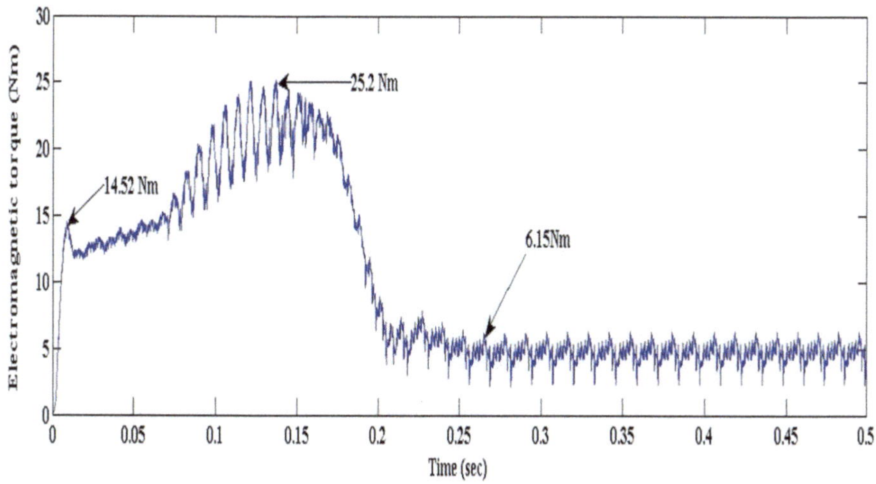

Fig. 16 Electromagnetic torque of DTC with PI controller for 3Ø IM

3.3 Discussion

In this section, comparison between conventional DTC technique with improved DTC technique has been discussed in terms of various parameters.

Table 4 shows the assessment amid the conventional DTC technique and the projected DTC with PI controller method.

Figure 17 shows the graphical comparison of convention DTC with DTC with PI controller on the parameters of rising time, settling time and transient time. As of the bar graph comparative examination, it is observed that rising time is minimized by 0.0023 s or 1.57%, settling time is reduced by 0.040 s or 14.29%, transient time is reduced by 0.019 s or 8.64%.

Figure 18 shows the graphical comparison of convention DTC with DTC with PI controller on the parameters of torque ripples, electromagnetic torque and stator current. As of the bar graph comparative examination, it is observed that torque ripples are reduced by 3.25%, electromagnetic torque is improved by 1.1 Nm or 4.43% and stator current is increased by 2.5 A or 6.94%.

Table 4 Comparison between of conventional DTC technique and DTC with PI controller

Technique	Rise time (sec)	Settling time (sec)	Transient time (sec)	Torque ripples Reduction in (% age)	Electromagnetic torque (Nm)	Stator current (A)	Rotor speed (rpm)
Conventional DTC	0.1467	0.280	0.220	79.45	24.1	36	1178.42
DTC with PI	0.1444	0.240	0.201	82.7	25.2	38.5	1129.42

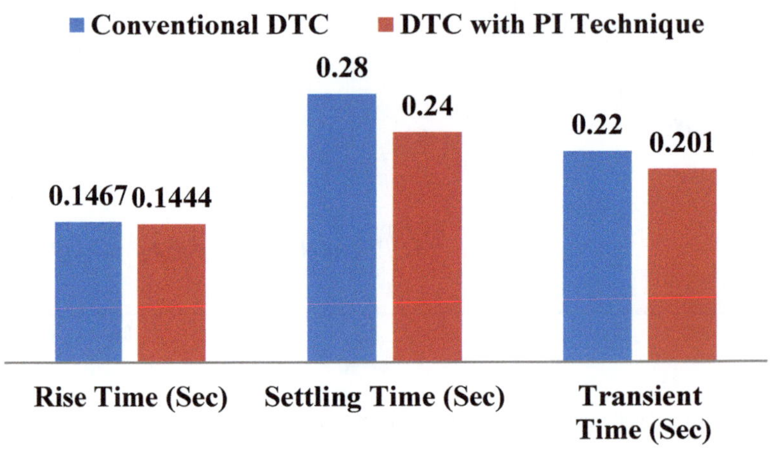

Fig. 17 Comparison of convention DTC with DTC with PI controller on the parameters of rising time, settling time and transient time

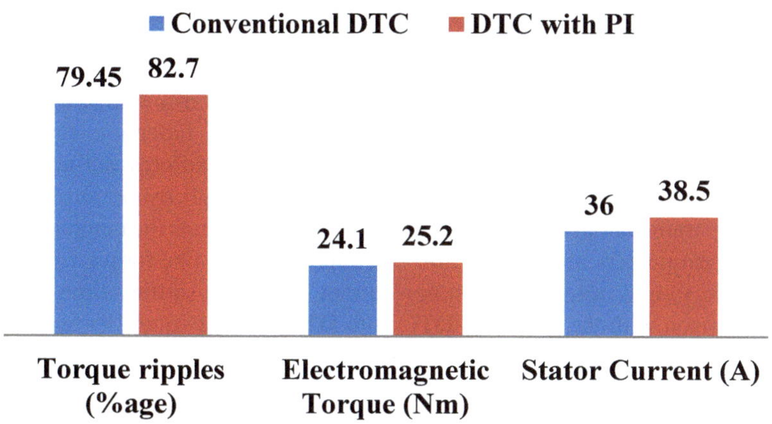

Fig. 18 Comparison of convention DTC with DTC with PI controller on the basis of torque ripples, electromagnetic torque and stator current

Figure 19 shows the graphical comparison of convention DTC with DTC with PI controller on the parameters of rotor speed. As of the bar graph comparative examination, it is observed that the rotor speed is reduced by 48.51 rpm or 4.11%. Hence, the proposed DTC with PI controller gives better results as compared to conventional DTC technique.

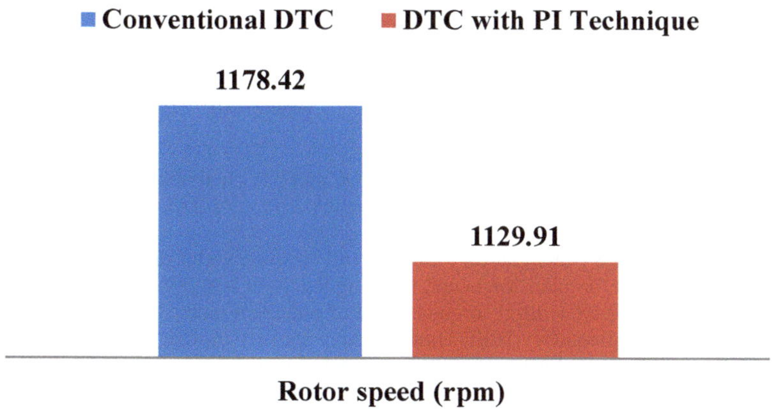

Fig. 19 Comparison of rotor speed with convention DTC with DTC with PI controller

4 Conclusion

This book chapter summarizes basically about integration of intelligent controllers specifically proportional integral type controller with induction machine for enhancing reliable and stable electricity. This chapter explores the ability of PI controller to attain and regulate speed and torque through by optimizing energy efficiency parameters in electrical machines. It also represents the comparison study of various parameters about induction machine operation like torque ripples, electromagnetic torque, rotor speed, stator current, rise time settling time, etc. From the comparison with the proposed DTC with DTC with PI controller technique for three-phase induction motor on different parameters, it is depicted that the rising time is minimized by 0.0023 s or 1.57%, settling time is reduced by 0.040 s or 14.29%, transient time is reduced by 0.019 s or 8.64%, torque ripples are reduced by 3.25%, electromagnetic torque is improved by 1.1 Nm or 4.43%, stator current is increased by 2.5 A or 6.94% and rotor speed is minimized by 48.51 rpm or 4.11%. The key insights highlighted in this book chapter are integration of PI controller with induction machine, scrutinizes various performance parameters such as rise time, settling time, rotor speed, transient time, etc., intricacies of enhancing rotor speed and torque and comparative analysis of conventional DTC and DTC with PI controller which provides insights the better operation of PI controller. The perspective of this research work includes integration of PI controller with emerging techniques for stable operation and emphasizes the importance of induction machine to energy stability, optimizing energy efficiency of electrical machines, advanced monitoring over motor parameters and comparison with advanced controller to obtain optimal solution. Hence, such proposed technique provides a way to pave the path for future emerging trends in applications of induction motors.

References

1. Casadei D, Profumo F, Serra G, Tani A (2002) FOC and DTC: two viable schemes for induction motors torque control. IEEE Trans Power Electron 17(5):779–787. https://doi.org/10.1109/TPEL.2002.802183
2. Jalluri SR, Ram BVS (2012) Direct torque control on space vector modulation with adaptive stator flux observer for induction motor. Int J Eng Res Appl 2(6):297–302
3. Koo J, Ki S (1999) New direct torque control of induction motor for minimum ripple and constant switching frequency. IEEE Trans Ind Appl 35(5):1076–1082
4. Ben Salem F, Derbel N (2014) Direct torque control of induction motors based on discrete space vector modulation using adaptive sliding mode control. Electr Power Compon Syst 42. https://doi.org/10.1080/15325008.2014.927029
5. Pandian A, Dhanasekaran R (2014) Performance analysis of direct torque control of three phase induction motor. J Theor Appl Inf Technol 62(3):819–824
6. Mehra J, Sharma P, Dubey C (2014) Comparative analysis of direct torque control and flux control of induction motor using P and PI controller. Int J Emerg Technol Adv Eng 4(5):887–892
7. Lokriti A, Salhi I, Doubabi S (2015) IM direct torque control with no flux distortion and no static torque error. ISA Trans 59:256–267. https://doi.org/10.1016/j.isatra.2015.08.014

8. Venkadesan A, Sedhuraman K (2015) Stator resistance estimation utilizing real and reactive power for direct torque controlled induction motor drives. Int J Adv Res Electr Electron Instrum Eng 4(6):4920–4926
9. Pimkumwong N, Wang M-S (2018) Direct torque control of three-phase induction motor based on constant voltage per frequency control with simple controller. In: 2018 international ECTI northern section conference on electrical, electronics, computer and telecommunications engineering, pp 15–20. https://doi.org/10.1109/ECTI-NCON.2018.8378273
10. Panda A, Singh SP (2016) A three-level fuzzy-2 DTC of induction motor drive using SVPWM. IEEE Trans Indus Electron63(3):1467–1479. https://doi.org/10.1109/TIE.2015.2504551
11. Fuzzy Predictive DTC of Induction Machines With Reduced Torque Ripple and High-Performance Operation | IEEE Journals & Magazine | IEEE Xplore. (s.d.). Consulté 20 novembre 2024, à l'adresse https://ieeexplore.ieee.org/document/7891647
12. Bellos E (2019) Progress in the design and the applications of linear Fresnel reflectors—a critical review. Therm Sci Eng Prog 10:112–137. https://doi.org/10.1016/j.tsep.2019.01.014
13. Geyer T (2011) Computationally efficient model predictive direct torque control. IEEE Trans Power Electron 26(10):2804-2816. https://doi.org/10.1109/TPEL.2011.2121921
14. Alsofyani IM, Lee KB (2019) Enhanced performance of constant frequency torque controller–based direct torque control of induction machines with increased torque-loop bandwidth. IEEE Trans Indus Electron 67(12). Paper|Power Electronics Lab. (s. d.). Consulté 20 novembre 2024, à l'adresse http://pel.ajou.ac.kr/bbs/board.php?bo_table=paper&wr_id=1343&page=5&device=pc
15. Wu X, Huang W, Lin X, Jiang W, Zhao Y, Zhu S (2021) Direct torque control for induction motors based on minimum voltage vector error. IEEE Trans Industr Electron 68(5):3794–3804. https://doi.org/10.1109/TIE.2020.2987283

The Evolution of the Lithium-Ion Batteries LIBs Technologies and An Overview of Their Contribution in the Global Energy Transition

Mohamed Aklalouch, Asma Ihechmyen, and Firyal Latrache

Abstract Pollution and greenhouse gas emissions from the burning of fossil fuels are causing the planet's average temperature to rise steadily every year. This global warming poses an urgent threat that the United Nations decided, in 2015, to include this issue in its global action plan of the 2030 agenda for sustainable development. Specifically, Goal 13 of this agenda calls for immediate action to combat climate change and mitigate its impacts. In response, countries are transitioning toward sustainable energy by investing in wind and solar power, which can also be used to produce green hydrogen. Then, the use of this energy requires the storage technologies and batteries are the solution for both storing energy and powering modes of transport. An ideal battery is high in energy and power density, environmentally friendly, affordable, and safe. Among current technologies, lithium-ion batteries (LIBs) best meet these criteria. This chapter will focus on the role of LIBs in enabling the energy transition by examining their history, technological improvements, and future prospects. The future of LIBs depends on their inclusion in a circular economy through recycling. This chapter illustrates the major role of the LIBs in promoting energy transition in order to achieve the sustainability and mitigate the climate change. It presents the history of the LIBs and the current progress in cathode and anode and then their applications. In addition, the chapter analyzes the role of recycling the LIBs in order to protect the environment and conserve the biodiversity. At the end, the chapter presents the contributions of scientific research work on the energy storage by LIBs and their applications in electric vehicles and smart grid.

M. Aklalouch (✉) · A. Ihechmyen
Chemistry of Materials & Environment Team, IMED-Laboratory, Department of Chemistry Sciences, Faculty of Sciences and Technologies, University Cadi Ayyad, 40000 Marrakesh, Morocco
e-mail: m.aklalouch@uca.ac.ma

A. Ihechmyen
e-mail: a.ihechmyen.ced@uca.ac.ma

F. Latrache
Mechanics, Energetics, Systems and Signals Team, Laboratory of Modelization and Scientific Calculus, National School of Applied Sciences, University Mohammed I, Oujda, Morocco
e-mail: l.firyal@ump.ac.ma

F. Latrache et al. (eds.), *Digital Technology for an Innovative Energy Transition*, Green Energy and Technology, https://doi.org/10.1007/978-3-031-89657-6_9

Keywords Lithium-ion batteries · Energy transition · Energy storage · Batteries recycling

1 Introduction

Humanity faces three critical environmental challenges: mitigating climate change by reducing CO_2 emissions, improving air quality in major cities, and lessening dependence on oil. Recognizing the urgency of these issues, the United Nations has included them as Goal 13 in its 2030 agenda for sustainable development, calling for immediate action to combat climate change and its impacts. To this end, there is an emergency need to rapidly accelerate the transition to clean energy sources like wind, solar, and hydropower. Since renewable energy sources are intermittent due to weather conditions, they need to be coupled with electrochemical energy storage devices called rechargeable batteries.

A rechargeable battery or accumulator is a device that can store electrical energy through electrochemical oxidation–reduction reactions. These "secondary" electrochemical generators require an initial electricity supply to operate, it is called "charging process". During use, the chemical energy of the active electrode materials is directly transformed into electrical energy through spontaneous reversible redox reactions that release electrons to an external circuit.

The basic unit of a battery is called an electrochemical cell. Depending on the desired voltage and capacity, these cells can be connected in series or parallel. Each cell consists of two electrodes separated by an electrolyte.

The electrodes are typically "composite" materials made up of: (1) an electrochemically active material, (2) an electronic conductor to facilitate electron transfer, and (3) a polymer that enhances mechanical properties. The negative electrode contains the active material with the lowest reduction potential, while the positive electrode contains the material with the highest potential. The negative electrode is the anode and the positive is the cathode.

In most batteries, the electrolyte is a high-ionic-conductivity liquid that wets the separator, this latter prevents direct contact between the two electrodes while allowing ionic transport. In addition to storing energy from clean sources, batteries play a crucial role in powering electric vehicles (EVs) and enabling their mobility. The development of electric vehicles is one of the most promising solutions for reducing air pollution by decreasing our reliance on fossil fuels.

Among the existing technologies, lithium-ion batteries (LIBs) are considered the optimal solution for storing and retrieving energy from renewable sources like solar, wind, and hydropower to power these applications. Since Sony commercialized the first LIB with a layered $LiCoO_2$ structure in 1991, the lithium-ion battery market has experienced dramatic growth. This technology excels due to its high energy and power densities, enabling sufficient autonomy in a lightweight and compact design.

The history of lithium-ion batteries (LIBs) is inextricably linked to the earlier development of nickel–cadmium (Ni–Cd) batteries. The Ni–Cd battery was invented in 1889 by Swedish scientist Waldemar Jungner [1]. It featured a positive electrode made of nickel oxide and a negative electrode of metallic cadmium, using an alkaline electrolyte typically KOH. While French scientist George Neumann succeeded in completely sealing the Ni–Cd accumulator in 1947, leading to the modern sealed nickel/cadmium battery, this technology was eventually phased out due to drawbacks like harmful environmental effects and the "memory effect."

Meanwhile, American chemist Gilbert Newton Lewis was pioneering a potential lithium-based replacement for Ni–Cd batteries. Starting in 1912, Lewis experimented with developing a revolutionary new battery technology using the lightest of all chemical elements—lithium. Lithium offered remarkable properties like extremely high-energy density. However, Lewis faced significant challenges in harnessing the fundamental instability of pure lithium metal. Undeterred, Lewis remained dedicated to unlocking lithium's immense potential as a battery material. His groundbreaking work laid the crucial foundations for the eventual emergence of practical lithium-ion batteries [2].

It was not until the 1970s that the first commercial lithium-ion batteries were finally introduced to the public, marking a major milestone in the evolution of portable power sources. The first commercial lithium-ion batteries were introduced by Exxon. Rechargeable lithium-ion batteries emerged in the 1980s following the discovery by French scientist Michel B. Armand that intercalation materials could be used as electrodes of the battery [3]. A key disadvantage of early lithium-ion batteries was their instability, particularly during charging, which posed a safety risk of explosion due to the use of metallic lithium anodes. To address this issue, Bell Laboratories developed lithium-ion batteries with anodes made of graphite rather than pure lithium metal [4]. Further significant advancements in lithium-ion battery technology came thanks to the groundbreaking work of Professor John B. Goodenough. Goodenough, who would later be awarded the 2019 Nobel Prize in Chemistry alongside Akira Yoshino and Stan Whittingham, made a crucial discovery that paved the way for major improvements in lithium-ion battery. Specifically, he identified and developed the Li_xCoO_2 material as a highly effective cathode for lithium-ion cells. This innovation represented a huge step forward, as the Li_xCoO_2 cathode offered superior energy density, stability, and cyclability compared to earlier lithium battery technologies. The 1990s saw a wave of commercialization and technological advancements that forever transformed the consumer electronics landscape. In 1991, Sony introduced the first commercially available lithium-ion battery, which used a layered $LiCoO_2$ cathode structure. This revolutionary battery technology represented a major step forward from previous rechargeable batteries, offering substantial improvements in energy density, recharge times, and overall performance. Lithium-ion batteries could store far more energy in a much smaller, lighter package compared to lead-acid or Ni–Cd alternatives. This made them ideal for powering the increasingly compact and portable electronic devices rapidly gaining popularity, from mobile phones and laptops to emerging consumer gadgets. Over the past three decades, lithium-ion battery technology has continuously evolved and improved, solidifying its position as the leading power source for a wide range of applications, from consumer electronics to electric vehicles. This battery breakthrough laid the foundation for a new era in portable power, forever changing how we power the devices that shape our digital lives.

This chapter provides a comprehensive overview of lithium-ion batteries (LIBs), tracing their remarkable journey from pioneering research to widespread use in modern energy storage. We'll explore the diverse applications of LIBs, from powering personal electronics to enabling the electrification of transportation and integration of renewable energy. The chapter then delves into the latest progress in cathode and anode materials, examining the innovative approaches researchers are taking to enhance performance, safety, and sustainability. Crucially, we'll highlight the pivotal role LIBs play in the global energy transition, as their efficient energy storage is indispensable for the widespread adoption of clean, renewable power. Additionally, the chapter addresses the growing importance of LIB recycling, outlining strategies and technologies to recover valuable materials and minimize environmental impact. Finally, we will explore how the research community is working to develop truly clean, sustainable LIBs to power the green energy revolution and to successfully implement the energy transition.

2 Current Progress in Cathode and Anode for LIBs

The main components of lithium-ion batteries (LIBs), besides the separator and electrolyte, are the positive and negative electrodes. These electrodes are made of materials that can reversibly insert and remove lithium ions over multiple charge and discharge cycles.

2.1 Positive Electrode Materials for LIBs

2.1.1 Layered Structure Materials

$LiCoO_2$ Layered Material

In 1980, Mizushima et al. conducted pioneering work on the electrochemical cycling of $LiCoO_2$ as a 4 V cathode material [5]. Though this material initially received little attention due to the lack of a stable electrolyte at that potential, it was later commercialized by Sony and became the first widely-used positive electrode for lithium-ion batteries. Lithium cobaltite ($LiCoO_2$) is currently the most commonly used cathode material. Its structure consists of a compact network of oxygen atoms, with lithium and cobalt atoms occupying alternating, ordered layers between the planes (Fig. 1). However, only half of the theoretical energy density of these layered materials can be utilized in practice, as removing more than half the lithium content compromises their structural stability.

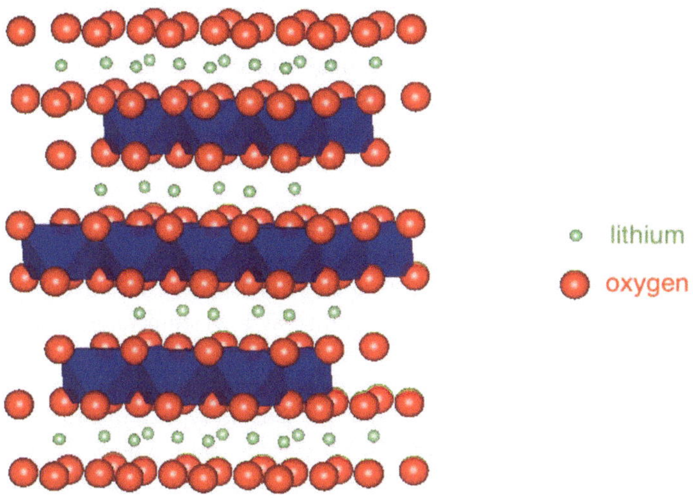

Fig. 1 Crystallographic structure of $LiCoO_2$ lamellar positive electrode where the octahedral positions are occupied by Co (Structure designed in VESTA [6])

The $LiCoO_2$ material has a lamellar structure composed of alternating CoO_2 planes separated by layers of lithium ions in octahedral sites, as shown in Fig. 1. During lithium de-insertion, the material undergoes three distinct phase transitions depending on the Li content (x) in Li_xCoO_2.

The first transition occurs around $x = 0.92$, between two hexagonal structures (hexagonal I and II). This is accompanied by a large increase in the lattice parameter "c" and a slight decrease in "a," attributed to increased repulsions between the CoO_2 sheets.

The second and third transitions occur just before and after $x = 0.5$, where lithium ions are well-ordered with vacancies. These transitions are due to disorder (hexagonal) to order (monoclinic) and order to disorder changes in the Li_xCoO_2 structure, respectively.

In-situ X-ray diffraction has shown that Li_xCoO_2 can be completely and reversibly deintercalated to form CoO_2. An insulator–metal transition also occurs at $x = 0.75$, as $LiCoO_2$ is a semiconductor while Li_xCoO_2 is a conductor for $x < 0.75$.

The $LiCoO_2$ compound has a theoretical capacity of 274 mAh/g as a Li-ion battery cathode. However, practical capacity is limited to 140 mAh/g at $x = 0.5$, corresponding to a 4.2 V cutoff. Charging above 4.2 V can achieve higher capacities, but causes rapid capacity fade due to cobalt dissolution and oxygen loss from the Li_xCoO_2 structure.

To address these issues, coating the $LiCoO_2$ with metal oxides like ZrO_2, Al_2O_3, TiO_2, B_2O_3, and SiO_2 has been implemented. However, this reduces capacity due to the inactive coating mass and can increase internal resistance, causing overvoltages during discharge.

Other Layered Materials

Despite the good electrochemical performance of the $LiCoO_2$ compound, research has continued to explore new materials such as the $LiNiO_2$ compound. The main objectives of this research was to reduce the cost and toxicity. Other work has focused on synthesizing solid solutions with the formula $LiCo_yM_{1-y}O_2$ (where M = Al, Ni, Mn, etc.).

The $LiNiO_2$ material has the same crystalline structure as $LiCoO_2$, but it has the advantage of being less expensive. Additionally, its discharge capacity is greater, around 220 mAh/g, compared to 140 mAh/g for $LiCoO_2$. However, the difficulty in synthesizing $LiNiO_2$ and its low stability at low temperatures have hindered its commercialization. The Li–Ni–O system is characterized by solid solutions of $Li_{1-\alpha}Ni_{1+\alpha}O_2$ (where $0 < \alpha < 1$). At low values of α, the layered structure is maintained but with some Ni^{2+} cations displaced into Li^+ sites (a type of disorder). This cation disorder reduces the capacity at variable flow rates and the reversibility of lithium intercalation. This issue has been addressed by doping with cations such as Co, Al, Mn, which help stabilize the layered structure. For the solid solution $LiNi_{1-x}Co_xO_2$, when $x \geq 0.3$, the interlayer space contains only Li^+ ions, greatly reducing the irreversible capacity compared to $LiNiO_2$, since there is no more oxidizable Ni^{2+} in the Li layers. In addition to cobalt doping, adding excess lithium to $LiNi_{1-x}Co_xO_2$ also helps reduce the cation disorder, as the layers become saturated with Li, which also improves battery life.

Other materials, such as $LiMnO_2$ or $LiFeO_2$, have a cubic structure when prepared at high temperature, rendering them electrochemically inactive. A key scientific and economic challenge in recent years has been to synthesize the isostructural layered $LiMnO_2$ and $LiFeO_2$ phases with the α-$NaFeO_2$ structure, using more moderate synthesis conditions. These materials were prepared by ion exchange from high-temperature sodium precursors ($NaFeO_2$, $NaMnO_2$) that already had the desired two-dimensional layered structure. Since this reaction only causes weak bond breaks, the rigid structure of the layers is maintained. However, the resulting $LiFeO_2$ and $LiMnO_2$ materials are thermodynamically metastable, and their electrochemical properties have so far been unsatisfactory, as it has proven impossible to deintercalate lithium from $LiFeO_2$, and $LiMnO_2$ is unstable during lithium extraction, transforming into a spinel phase during cycling.

2.1.2 Spinel Structures Materials

4 V $LiMn_2O_4$ Spinel

Since the 1980s, spinel-structured $LiMn_2O_4$ materials have garnered significant attention as potential positive electrode materials for lithium-ion batteries (LIBs). These materials are being developed as alternatives to the commonly used layered oxide positive electrodes in LIBs.

The spinel structure material $LiMn_2O_4$ was presented early on as a viable alternative to the layered material $LiCoO_2$. Comparing the two materials (Table 1), their capacity and operating potential values are quite similar. The spinel has a theoretical capacity of 148 mAh/g at an operating potential around 4 V versus Li/Li^+, while the layered oxide has a capacity of 135 mAh/g at a potential of approximately 4.2 V versus Li/Li^+. These comparable energy values make the two materials strong alternatives (Table 1).

The $LiMn_2O_4$ spinel material is an interesting choice for practical applications due to its cost-effectiveness from the natural abundance of manganese and its environmental friendliness compared to other materials. Figure 2 is a schematic representation of the $LiMn_2O_4$ crystallographic structure. This material has a cubic close-packed (ccp) array of oxygen ions, with the manganese ions (Mn^{3+} and Mn^{4+}) occupying the 16d octahedral sites and the lithium ions (Li^+) in the 8a tetrahedral sites. Its space group is Fd $\overline{3}$ m.

The manganese ions have octahedral coordination to the oxygens, and the MnO_6 octahedral share edges in a three-dimensional host for the Li guest ions. The lattice parameter (a_c) generally falls around 8.25 Å, depending on the synthesis method. When lithium ions are inserted into the spinel host structure, they occupy the 16c

Table 1 Comparison of the characteristics of the cathode materials $LiCoO_2$ lamellar and $LiMn_2O_4$ spinel

Cathode material	Specific capacity (mAh/g)	Potential versus Li/Li^+ (V)	Specific energy (Wh/g)
$LiCoO_2$	135	4.2	510
$LiMn_2O_4$	148	4	420

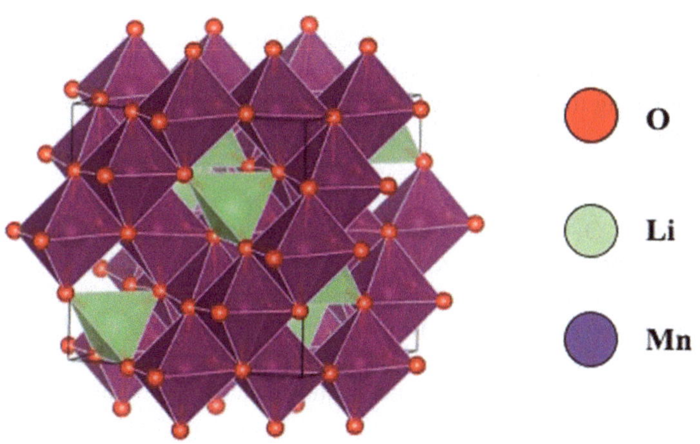

Fig. 2 Crystallographic structure $LiMn_2O_4$ spinel positive electrode where Mn occupies the octahedral positions (Structure designed in VESTA [6])

octahedral sites, causing a first-order phase transition as the lithium ions in the 8a tetrahedral sites are displaced into the vacant 16c sites.

The spinel structure provides 3D pathways for the insertion and de-insertion of lithium ions, enabling outstanding lithium-ion mobility that is beneficial for high-rate applications. The $LiMn_2O_4$ compound is electrochemically active in the 3 V and 4 V regions versus Li/Li^+, with the 3 V region allowing for greater lithium-ion insertion. However, the Jahn–Teller effect due to the d^4 configuration of Mn^{3+} in the 3 V region causes a cubic-to-tetragonal structural distortion, which negatively impacts the electrochemical cycling properties, especially at high discharge rates, resulting in a dramatic decrease in capacity.

In the region of 4 V, the following electrochemical reaction occurs:

$$LiMn_2O_4 \rightleftarrows Li_{1-x} + Mn_2O_4 + x\, Li^+ + x\bar{e} \quad (1)$$

where $0 < x \leq 1$

The extraction of lithium from the spinel structure allows the oxidation of Mn^{3+} to Mn^{4+} at a potential of 4 V versus Li/Li^+. This decrease in Mn^{3+} concentration limits the Jahn–Teller effect. However, the lithium de-intercalation process occurs in multiple stages, leading to phase changes that gradually reduce capacity over repeated cycling. Additionally, the instability of Mn^{3+} ions, which can disproportionate into Mn^{2+} and Mn^{4+}, is another factor that affects capacity depending on the following reaction:

$$2Mn^{3+} \rightleftarrows Mn^{2+} + Mn^{4+} \quad (2)$$

Doping spinel structures with certain atoms can help stabilize them during lithium insertion and extraction, improving electrochemical performance. Doping with non-oxidizable metals like Li^+, Al^{3+}, Mg^{2+}, Zn^{2+}, or Ga^{3+} has demonstrated this ability. However, this approach has the disadvantage of reducing discharge capacities. Alternatively, doping with oxidizable transition metals like Ni^{2+}, Cr^{3+}, Fe^{3+}, Cu^{2+}, or Co^{3+} works by reducing the Mn^{3+} concentration to balance the electronic charges of the doped material.

5 V $LiMn_{1.5}Ni_{0.5}O_4$ Spinel

The $LiNi_{0.5}Mn_{1.5}O_4$ spinel structure material is considered the most promising for lithium-ion batteries compared to other spinel structure $LiM_yMn_{2-y}O_4$ doped compounds. This material exhibits excellent performance in terms of specific discharge capacity (over 130 mAh/g), high-rate capability, and long cycle life. Zhong et al. were the first to use the $LiNi_{0.5}Mn_{1.5}O_4$ compound as a positive electrode material, and it has since become the most extensively studied spinel structure material.

The defining characteristic of this material is its dominant potential plateau around 4.7 V, which results from the electrochemical activity of the Ni^{2+}/Ni^{4+} redox couple. $LiNi_{0.5}Mn_{1.5}O_4$ can crystallize in two different spinel structure space groups: the ordered $P4_332$ structure and the disordered Fd $\overline{3}$ m structure. The disordered Fd $\overline{3}$ m phase is primarily due to a slight formation of oxygen deficiencies, while the ordered structure is stoichiometric. In the Fd $\overline{3}$ m space group, the Ni^{2+} cations are randomly distributed in the 16d octahedral positions.

Kim et al. examined how the cation ordering in the $LiNi_{0.5}Mn_{1.5}O_4$ structure affected high-rate discharge capabilities. The ordered structure, which exhibits poor electrochemical performance, especially at high discharge rates, is less interesting than the disordered phase.

The oxidation state of nickel in the $LiNi_{0.5}Mn_{1.5}O_4$ spinel is Ni^{2+}, and the Li^+ insertion/de-insertion occurs in two reversible steps corresponding to the Ni^{2+}/Ni^{3+} and Ni^{3+}/Ni^{4+} redox reactions. $LiNi_{0.5}Mn_{1.5}O_4$ is particularly characterized by its high cyclability at room temperature, which is remarkably higher than that of other 5 V $LiM_yMn_{2-y}O_4$ spinels. This electrode also delivers a high capacity of around 135 mAh/g. Additionally, the nickel doping prevents the tetragonal structure formation upon Li^+ insertion into $LiMn_2O_4$, allowing the spinel $LiNi_{0.5}Mn_{1.5}O_4$ to remain cubic, forming $Li_2Ni_{0.5}Mn_{1.5}O_4$. However, $LiNi_{0.5}Mn_{1.5}O_4$ suffers from severe capacity fading, which hinders its practical use when cycled at high temperatures (55 °C).

The $LiNi_{0.5}Mn_{1.5}O_4$ cathode material is unstable at high temperatures (55 °C) in the presence of non-aqueous electrolytes containing $LiPF_6$-EC-DMC. This instability leads to strong dissolution of the spinel material, caused by the generation of hydrofluoric acid (HF) through the reaction of the electrolyte at the electrode surface.

Researchers have explored various approaches to address this issue. Coating the $LiNi_{0.5}Mn_{1.5}O_4$ particles with metal oxides, such as ZnO, has demonstrated improved capacity retention at 55 °C. The amphoteric nature of ZnO reduces HF concentration, thereby decreasing transition metal dissolution and enhancing specific capacity retention during cycling.

Another approach is doping $LiNi_{0.5}Mn_{1.5}O_4$ with ions like Co, Cr, Ga, or Al. Among the doped materials, $LiCr_{0.2}Ni_{0.4}Mn_{1.4}O_4$ has shown outstanding electrochemical performance, with chromium proving to be a suitable dopant for enhancing the cycling performance of $LiNi_{0.5}Mn_{1.5}O_4$ cathodes, even at high temperatures (55 °C). This doped cathode material exhibits excellent electrochemical performance at both 25 °C and 55 °C. It delivers an experimental specific discharge capacity of 140 mAh/g when cycled at relatively slow rates, with a theoretical specific capacity of 147 mAh/g. Moreover, it maintains over 99% capacity retention after 50 cycles. Even when cycling is extended to 250 cycles, the capacity retention remains above 96%. Notably, these experimental results are obtained regardless of cycling temperature, whether at room temperature or 55 °C.

The importance of this material lies in its ability to combine high-energy density with high-power density. It exhibits an exceptional rate capability, maintaining 87% of its capacity even at a 60C charge/discharge rate, equivalent to a complete discharge in just 1 min. As a result, the $LiCr_{0.2}Ni_{0.4}Mn_{1.4}O_4$ spinel cathode is able to store an energy density of 654 Wh/kg and deliver a power density of 27.5 kW/kg at the high operating temperature of 55 °C. This material represents the highest rate capability reported to date for $LiNi_{0.5}Mn_{1.5}O_4$-based cathodes, making it one of the most promising positive electrode materials for practical high-power lithium-ion batteries.

2.1.3 Polyanion Materials

The group of materials known as Na Super Ionic Conductor (NASICON) is the most studied in this class. This includes materials with the formulas $LiMPO_4$ (where M = Fe, Co, Mn) and $LiM1M2(XO_4)_3$ (where M1 and M2 = Nb, Ti, Fe, V; and X = S, Mo, W).

The effect of the anion group XO_4 and the metal M on the discharge potential has been extensively studied [7]. For example, the Fe^{3+}/Fe^{2+} redox couple in the $Fe_2(SO_4)_3/LiFe_2(SO_4)_3$ system has a potential of 3.6 V versus Li/Li^+. However, substituting S with Mo or W reduces the discharge potential of this same redox couple to 3 V. This demonstrates the significant impact the choice of anion group XO_4 can have on the electrochemical properties.

The electronegativity of metal X influences the strength of the Fe–O bond. Substituting S with Mo or W increases the polarization of O toward Fe, making the Fe–O bond more covalent and reducing the difference between the Fe^{3+}/Fe^{2+} and Li/Li^+ couples.

A similar inductive effect applies to the role of the metal M when the XO_4 group is fixed. For the compound $Li_xX_2(PO_4)_3$ where X = V, Ti or Fe, the profile of the potential curves strongly depends on the metal ion X [8].

The NASICON structure compound of formula $Li_xFe_2(XO_4)_3$ has an approximate discharge potential of 2.8 V, with a potential difference of around 0.8 V. The $LiFePO_4$ compound has become the most promising positive electrode material in the NASICON group. It was first introduced by Padhi et al. [9].

The olivine structure of $LiFePO_4$ can be described as similar to a spinel structure. Both are characterized by the general chemical formula AB_2O_4, where A and B occupy the tetrahedral and octahedral sites of a stacked oxygen framework. The key difference is in the stacking pattern—face-centered cubic for spinel, but distorted compact hexagonal for olivine. Figure 3 is a schematic representation of the $LiFePO_4$ crystallographic structure.

Unlike the spinel structure, the olivine $LiFePO_4$ has the octahedral sites occupied by two types of cations, M^{2+} and Li^+. This allows for ordered arrangement between the two cation types [9]. The overall framework consists of MO_6 octahedral and PO_4 tetrahedral linked at edges and vertices. This 3D structure forms tunnels along the (010) and (001) directions, where the lithium ions reside in octahedral sites. The overall orthorhombic structure of $LiFePO_4$ is described in the Pnma space group.

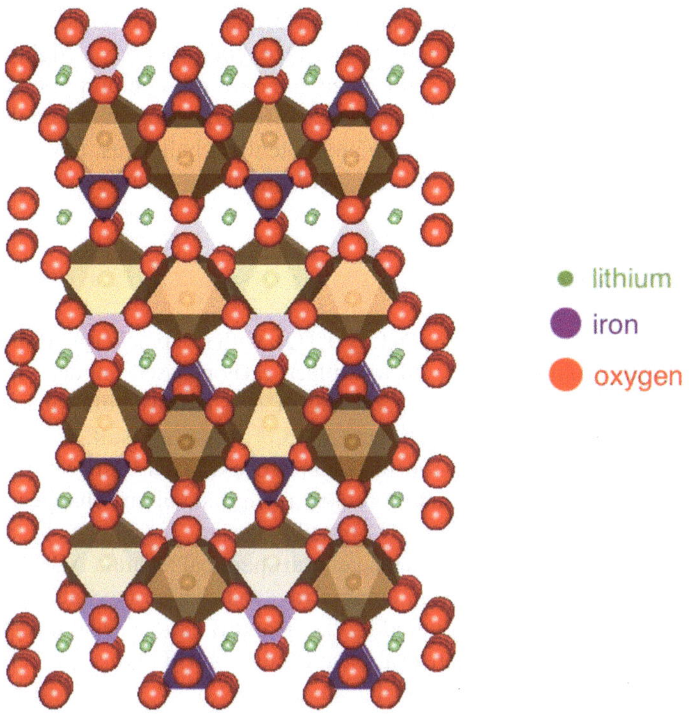

Fig. 3 Schematic representation of the olivine $LiFePO_4$ crystallographic structure (Structure designed in VESTA [6])

2.2 *Negative Electrode Materials for LIBs*

2.2.1 Carbon Graphite

The currently available commercial batteries rely on a specific type of anode material—an intercalated carbon, such as coke, graphite, or MCMB (mesocarbon microbead) carbon. Of these options, MCMB carbon is considered the best choice for commercial batteries at the moment, due to its impressive capacity of up to 350 mAh/g [10, 11].

However, this anode material does have some notable drawbacks. One key issue is the loss of a certain quantity of lithium ions that becomes inactive during the battery's initial charging cycle [12]. This irreversible capacity loss can reduce the anode's overall capacity down to around 250 mAh/g. The inactivity of these lithium ions also affects the capacity that can be delivered by the positive electrode, which is typically made of lithium cobalt oxide ($LiCoO_2$). As a result, the full battery is not simply denoted as $LiCoO_2/C_6$, but rather as $Li_{(1-X)}CoO_2/Li_XC_6$, where the variable "X" represents the amount of inactive lithium cations.

Another factor that can impact battery performance is the physical structure of the carbon anode. It is not a perfectly flat surface, but rather has a slight inclination or unevenness. This topography can influence the overall operating potential of the battery cell. Researchers have continued to explore ways to optimize anode materials and designs to minimize these capacity and potential challenges, in order to further improve the energy density and efficiency of lithium-ion batteries.

2.2.2 Spinel $Li_4Ti_5O_{12}$

The lithium titanate compound $Li_4Ti_5O_{12}$ (LTO), with its spinel crystal structure [13, 14], is the most promising material to replace carbon-based anodes. This interest stems from LTO's ability to repeatedly insert and remove lithium ions without any significant change in the cell volume [15, 16]. As a result, it enables excellent reversibility of the lithium-ion insertion process. This "zero-strain" property, where lithium insertion and extraction causes negligible volume changes, ensures an exceptionally long cycling life for LTO-based batteries.

The LTO spinel structure follows the general chemical formula AB_2O_4, where the oxygen atoms form a close-packed cubic array, and the tetrahedral and octahedral interstitial sites are partially occupied by the A and B cations. The inserted lithium ions specifically occupy the octahedral interstitial sites within this spinel framework.

The LTO material exhibits a flat plateau around 1.55 V versus Li/Li^+ and has a theoretical specific capacity of 175 $mAhg^{-1}$. Additionally, its spinel structure can accommodate three lithium ions, corresponding to the chemical formula $Li_7Ti_5O_{12}$ [17].

The spinel structure of lithium titanate oxide (LTO) provides 3D pathways for the intercalation and extraction of lithium ions. This relatively high-voltage operation also prevents the formation of lithium dendrites on the anode surface, a key safety issue in lithium-ion batteries (LIBs). Furthermore, it inhibits the formation of a potentially unsafe solid electrolyte interface (SEI) layer, which can occur at lower potentials due to electrolyte reduction. Moreover, due to its high operating potential the formation of lithium dendrite on the anode surface is prevented. Actually, because its safety characteristic, it is the only anode material commercialized into electric vehicles.

However, the high anode potential of LTO limits the overall energy density of a battery using this material. To compensate, the positive electrode should have an equally high electrochemical reaction voltage.

2.2.3 Silicon Anode

Silicon has emerged as a highly promising anode material for next-generation lithium-ion batteries due to its remarkable theoretical capacity that far exceeds that of the commonly used graphite. With a theoretical capacity of 4200 mAh/g; i.e., a staggering 10 times greater than that of the commonly used graphite anodes, silicon offers an enticing path to significantly boosting the energy density of lithium-ion cells. This remarkable capacity stems from silicon's ability to accommodate a large amount of lithium ions during the charging process, allowing for the storage of significantly more energy compared to graphite.

However, this high storage capacity comes with a significant drawback, the massive volume changes the material undergoes during the lithiation and delithiation process. As silicon absorbs and stores lithium ions, its volume can expand by over 300%, causing severe mechanical stresses that lead to pulverization of the active material, loss of electrical contact, and rapid capacity fade. Extensive research efforts have therefore focused on developing strategies to mitigate these volume changes and stabilize the silicon anode structure. A key breakthrough has been the use of silicon nanostructures, such as silicon nanowires, nanospheres, nanoparticles, and thin films. These nanoscale architectures provide ample void space to accommodate the volume changes, while also shortening the lithium ion diffusion pathways to enable rapid charging and discharging.

Additionally, silicon anodes have been combined with conductive carbon coatings, graphene, metal nanoparticles, and conductive polymers to address the inherently low electrical conductivity of silicon. These composite designs help to maintain the structural integrity of the anode and ensure efficient charge transport, leading to significant improvements in the cycle life and rate capability of silicon-based lithium-ion batteries. In fact, researchers have turned to silicon-carbon composite materials as an innovative solution. By combining the high specific capacity of silicon with the good electrical conductivity and structural stability of carbon, these composite anodes are able to effectively buffer the volume changes experienced by the silicon, resulting in significantly improved electrochemical cycling performance.

As the scientific community continues to unravel the complex failure mechanisms of silicon anodes, further innovations in nanomaterial design and composite engineering hold great promise for realizing the full potential of this transformative anode technology. The development of high-performance silicon anodes represents a crucial step forward in the quest for high-energy–density lithium-ion batteries capable of powering the next generation of electronic devices and electric vehicles.

3 LIBs Applications

Lithium-ion batteries (LIBs) have become increasingly prevalent across a wide range of applications, powering everything from renewable energy grids to portable electronics. In solar energy systems, LIBs enable the storage of excess electricity generated by solar panels for use during peak demand or cloudy periods. Their high-energy density, longevity, and fast charging make them well-suited for large-scale energy storage needs.

Lithium-ion batteries (LIBs) are indispensable for more than just solar applications. Their reliability and ability to provide consistent, on-demand power make them an essential backup solution for critical infrastructure such as hospitals and data centers, ensuring continuity of operations during blackouts or other disruptions to the main power grid. LIBs also power emergency power and uninterrupted power supply (UPS) systems, as well as advanced surveillance and alarm systems, including the cameras, sensors, and communication devices that enhance security and emergency response capabilities.

The ubiquity of LIBs extends to the field of portable electronics, powering everything from digital cameras and smartphones to smartwatches and laptops. Their compact size, lightweight construction, and high-energy density enable manufacturers to create sleek, powerful devices that can operate for extended periods without frequent recharging. This technological breakthrough has revolutionized the way we capture, store, and interact with digital content on the move.

Furthermore, LIBs became essential to the rise of electric mobility. LIB-powered electric vehicles (EVs) and hybrids are quickly gaining market share, providing eco-friendly transportation options that can match the performance of traditional internal combustion engines.

Lithium-ion batteries (LIBs) have become indispensable in the medical field, powering critical devices like pacemakers as well as a vast array of other life-saving medical equipment. Their unparalleled reliability, safety, and ability to deliver consistent power make them essential components that enable healthcare professionals to provide patients with the highest quality of care.

Overall, LIBs are an indispensable technology that revolutionized power across a wide range of applications: from renewable energy storage to portable electronics. Their versatility and capabilities have made them omnipresent, transforming how we power our world.

4 The Role of LIBs Recycling in the Environment Protection

As the world accelerates its transition to renewable energy, the role of lithium-ion battery (LIB) recycling will become increasingly crucial. LIB recycling serves as a vital link in the broader effort to create a more sustainable and self-sufficient energy ecosystem for generations to come. The growing demand for these batteries is leading to a concerning accumulation of waste, as traditional disposal often results in the loss of valuable materials that could be recovered and reused.

Recycling LIBs allows us to extract and reintroduce the precious metals, such as lithium, cobalt, and nickel, back into the manufacturing supply chain. This closed-loop approach not only reduces the environmental impact of mining new resources but also enhances the overall sustainability and circularity of the energy sector.

Moreover, effective LIB recycling can help mitigate the risks of supply chain disruptions and price volatility for these key materials, ensuring a more secure and reliable source of the components necessary to power the clean energy technologies that are essential for combating climate change.

The importance of recycling lithium-ion batteries cannot be overstated. The global demand for rechargeable batteries, especially for portable electronics and electric vehicles, is growing exponentially, leading to a significant increase in Li-ion battery waste. These batteries contain potentially toxic materials that can contaminate the environment if not properly disposed of. Improper disposal can also result in harmful chemical reactions. Recycling Li-ion batteries can reduce the demand for new materials, conserve Earth's limited natural resources, and reduce the environmental impact of mining. Ultimately, recycling Li-ion batteries helps reduce energy consumption and CO_2 emissions, minimizes environmental toxicity, and decreases the volume of waste.

4.1 Recycling Processes

4.1.1 Pretreatment

Before recycling lithium-ion battery cathode material, pretreatment is necessary to mitigate risks of battery instability and access the material of interest. Used lithium-ion batteries must be completely discharged before safe pretreatment and separation.

Two main discharge methods have emerged:

- Chemical discharge involves immersing spent batteries in a saline solution. This uses water ions as current carriers and energy conversion agents. While simple and adaptable, it can corrode the metal shell and release harmful gases.
- Physical discharge focuses on the choice of discharge medium. For example, Kang et al. used a roller press to mechanically deform and short-circuit the battery in distilled water with active cooling. This generates significant heat from exothermy, requiring more controlled discharge for high-power, high-density batteries.

The disassembly of used lithium-ion batteries (LIBs) is the second phase of the recycling process, aiming to extract the battery contents once the case has been discharged. This disassembly is typically done manually using knives and saws to separate the individual cells.

In the manufacturing of LIBs, the cathode active materials are mixed with polyvinylidene fluoride (PVDF) as a binder and then coated onto an aluminum foil current collector, forming the foil cathode. For effective recovery of valuable elements during LIB recycling, it is critical to separate the active materials from the binder and aluminum foil. The main separation methods currently used include organic solvent dissolution, alkaline solution dissolution, and heat treatment.

The organic solvent dissolution method relies on the principle of isotropic phase solubility to dissolve PVDF and separate the active materials of the cathode from the aluminum foil and PVDF. Commonly used solvents for this purpose include N,N-dimethylformamide (DMF), N,N-dimethylacetamide (DMAC), N-methylpyrrolidone (NMP), and dimethyl sulfoxide (DMSO). Using DMSO at 60 °C for 85 min is a suitable option due to its moderate cost, low toxicity, and environmental safety compared to NMP and other solvents. However, this method is limited by the high cost and toxicity of organic solvents in general.

High-energy ball milling, a key mechano-chemical process, can significantly alter the physical and chemical properties of raw materials. By integrating mechanical processing techniques, this process modifies the materials at a fundamental level, enhancing their activity and potentially inducing unexpected chemical reactions. For instance:

$$LiCoO_2 + NaCl \rightarrow LiCl + NaCoO_2 \quad (3)$$

Mechano-chemical pretreatment is frequently used to process the active cathode particles from used lithium-ion batteries. This method alters the crystalline structure of the particles, improving the efficiency of extracting the precious metals. The recovered metals can be obtained at room temperature without the need for corrosive solvents, making it an environmentally friendly approach.

Currently, there are three main processes used to recycle lithium-ion batteries: pyrometallurgy, hydrometallurgy, and the direct recovery method.

4.1.2 Pyrometallurgy

The pyrometallurgical process for recycling lithium-ion batteries (LIBs) involves high-temperature melting. In this process, the battery components, including cathodes (e.g., NMC, LMO) and organic materials like plastic and separators, are broken down. Used LIBs are directly fed into a melting furnace without any pretreatment. The high-temperature environment (around 1000 °C) allows for the reduction of carbon, transforming the metals into relatively stable alloys containing cobalt, iron, and nickel. These metal alloys can then be separated using hydrometallurgical techniques to extract the pure materials.

Interestingly, the plastics, organic solvents, and graphite in the batteries actually provide heat through combustion, reducing the overall fuel requirements for the process. Additionally, aluminum in the batteries serves as a reducing agent during the reaction. The lithium, which is a critical element for LIBs, remains in the slag fraction and can be recovered through additional processes. The advantages of pyrometallurgy lie in its operational simplicity and technological maturity, making it possible to avoid sorting and reducing materials. However, this process still faces several major challenges despite its prior use. The main disadvantages include strict requirements for post-treatment of waste gases, such as carbon dioxide, carbon monoxide, sulfur dioxide, volatile organic compounds, and dust. Pyrometallurgy is also characterized by high energy consumption. Additionally, the majority of valuable metals like lithium, aluminum, and manganese end up in slag that is generally not recovered or recycled. Finally, components like electrolytes and/or plastics are burned and lost during this process.

4.1.3 Hydrometallurgy

Hydrometallurgical recycling is used to process the fine black-mass fraction obtained by dismantling, crushing, and pretreating batteries. This aqueous chemistry-based process aims to eliminate impurities that could compromise the efficiency and selectivity of the metal recovery. It typically involves three key stages: leaching the spent cathode material, treating the leach solution to remove impurities and purify it, and finally recovering the metal salts.

The hydrometallurgy process for recovering precious metals from materials like lithium-ion batteries (LIBs) typically involves a leaching step that is considered the most significant. This leaching step aims to convert the metals present in the cathode material obtained during pretreatments into ionic solutions that can be further processed.

A variety of inorganic acids such as HCl, HNO_3, and H_2SO_4, as well as organic acids like citric acid and malic acid, are commonly used for leaching LIBs. It is also possible to conduct the leaching in an alkaline environment to improve selectivity. Sometimes the addition of H_2O_2 as a reducing agent is necessary to increase leaching efficiency. For example, a study using aspartic acid showed that increasing the H_2O_2 amount from 1 to 4% vol. boosted the leaching efficiency of cobalt and lithium from 1 to 60%, likely due to the reduction of Co^{3+} to the more easily leachable Co^{2+}. Key parameters influencing the leaching process include reactant concentration, temperature, and reaction time.

Metal impurities are typically removed through chemical precipitation, taking advantage of their distinct solubilities. In this case, a combination of chemical precipitation and pH adjustment allows for the effective removal of Al and Fe ions, while Cu ions prove more challenging due to a solubility constant similar to that of Co, Ni, and Mn.

The purified solution then undergoes metal recovery, where the metals can be extracted via chemical precipitation, solvent extraction, or electrolysis. Once the metals have been recovered, the remaining lithium-enriched solution is directed to a specialized lithium recovery section, where the lithium is extracted either by chemical precipitation using sodium carbonate or by crystallization through distillation.

Hydrometallurgical processes offer several attractive advantages, such as high metal recovery rates, high product purity, low energy consumption, and minimal gas emissions. However, these processes also have some drawbacks, including high reagent consumption, significant waste emissions, and complicated separation processes.

4.1.4 Direct Treatment

Cathodes near the end of their lifespan typically lose around 20% of their usable lithium content due to various factors. These include the formation of solid electrolyte interphase (SEI) and parasitic reactions that produce compounds like lithium fluoride

(LiF) and lithium carbonate (Li_2CO_3). This lithium loss contributes to increased cellular impedance and shifts in electrode potentials.

Direct recovery has emerged as a sustainable recycling solution, as it can correct compositional and structural defects without destroying the original crystal structure or decomposing electroactive materials. Key processes in direct recovery include lithium supplementation and crystal restoration. This method is expected to provide battery manufacturers with recycled electroactive materials as raw inputs.

Unlike energy-intensive pyrometallurgy and hydrometallurgy, direct recycling is a more environmentally friendly process that requires less energy and fewer chemicals. This direct recovery method does not generate significant air pollution or water waste. For example, recycling 1 kg of $LiNi_{1/3}Co_{1/3}Mn_{1/3}O_2$ (NMC111) cathode material via direct recycling saves up to 50% in energy usage and reduces gas emissions by 85% compared to other methods. Additionally, the direct regeneration process helps preserve the value of degraded electroactive materials. In fact, the total cost of recycling LMO batteries by pyrometallurgical, hydrometallurgical, and direct methods was estimated at 2.43$/kg, 1.30$/kg, and 0.94$/kg of processed used battery cells, respectively.

5 The Role of LIBs Recycling in the Environment Protection

After presenting the history of the emergence of lithium-ion batteries and their applications, as well as the current progress in the development of electrodes and the recycling operation, it is appropriate to analyze and discuss the role of lithium-ion batteries in promoting the energy transition.

The transition to renewable energy sources like solar and wind is crucial for a more sustainable future, but this shift cannot be fully realized without advancements in energy storage technology. Lithium-ion batteries (LIBs) have emerged as a dominant and transformative technology, offering capabilities that make renewable energy integration more feasible. These high-performance batteries boast impressive energy density, allowing them to pack substantial power into a compact form. Their long cycle life, enabling thousands of recharge cycles without significant degradation, makes them well-suited for grid-scale energy storage applications requiring reliable, long-lasting performance to help stabilize the flow of renewable electricity. This is especially important as intermittent solar and wind generation continues to grow, with LIBs acting as crucial buffers to smooth fluctuations and ensure a steady, uninterrupted supply of clean power.

The rapid charging capabilities of lithium-ion batteries (LIBs) are enabling the development of electric vehicles (EVs) that can be quickly recharged, helping to overcome a key barrier to wider EV adoption. The latest LIB-powered EVs can add hundreds of miles of range in just minutes, making them a far more practical and convenient option than gas-guzzling vehicles. This fast charging speed, combined with the inherent efficiency and environmental benefits of electric powertrains, is accelerating the transition to sustainable transportation—a crucial step in meeting emissions reduction targets. Overall, the versatility and performance advantages of LIBs are proving to be a game-changer, unlocking the potential of renewable energy and electric mobility in ways previous battery technologies could not.

The dramatic 98% decline in lithium-ion battery (LIB) prices over the past decade has been a key driver in making electric vehicles (EVs) increasingly cost-competitive compared to traditional gasoline-powered cars. Just 10 years ago, in 2011, the average LIB cost stood at a staggering 917 $/kWh. However, by 2021, that figure had plummeted to a mere 132$/kWh: an 86% reduction that has made EVs far more financially accessible for mainstream consumers. This dramatic cost decline has been fueled by the rapid scaling up of global battery manufacturing capacity. As economies of scale have taken hold and production volumes have skyrocketed, battery makers have been able to drive down per-unit costs through more efficient processes, greater supply chain optimization, and ongoing technological innovations.

Advancements in battery chemistry, cell design, and manufacturing have been instrumental, as researchers relentlessly work to boost energy density, cut material costs, and optimize production. Yet, industry experts say battery prices must drop another 30% to truly compete with traditional combustion engine vehicles. Reaching this milestone would likely spur rapid, global adoption of sustainable electric transportation.

Although the rapid decline in battery costs has slowed in recent years, largely due to supply chain issues for key raw materials like lithium, cobalt, and nickel, the battery industry continues to be a hub of innovation and technological advancement. Exciting new lithium-ion battery (LIB) designs, such as the cutting-edge "4680" cells developed through the Tesla-Panasonic partnership, are pushing the boundaries of energy density, charge rates, and cost-effectiveness. These larger, higher-capacity 4680 cells have the potential to deliver significantly improved performance and affordability compared to previous generations of LIBs.

The battery sector is witnessing not just incremental improvements to existing lithium-ion technology, but the development of more transformative next-generation solutions. For instance, silicon-based anode materials hold the promise of boosting energy density by 30–40% over conventional graphite anodes. Meanwhile, the rapid advancement of solid-state battery technologies, which offer inherent safety and energy density advantages, could ultimately displace today's liquid electrolyte lithium-ion batteries (LIBs) altogether. While still in relatively early stages, these solid-state breakthroughs are generating significant excitement and investment as they inch closer to commercial viability.

Continued innovation and cost reductions in battery technology are essential for the growth and stability of the renewable energy industry, as well as enabling a broader transition to a sustainable, low-carbon future across transportation, buildings, and the wider economy. As battery performance and affordability continue to improve, these energy storage solutions will become indispensable for supporting large-scale integration of variable renewable power, electrifying vehicle fleets, and decarbonizing various other sectors. The future of a greener, more sustainable global economy hinges heavily on the battery industry's ability to maintain its innovative trajectory.

The main role of the lithium-ion battery is the energy storage for later use in other applications, and Table 2 shows the latest contributions to this field. In addition to the literature dealing with energy storage applications to justify the efficiency of the lithium-ion battery, Table 1 shows the contributions between 2022 and 2023 using mainly artificial intelligence and control technologies for the thermal management and safety of these batteries. In this context, the Feedforward Neural Network (FNN) and the Convolutional Neural Network (CNN) have been adopted to predict the degradation of batteries by analyzing their parameters (voltage, current, temperature, etc.) [18]. In addition, safety control technologies involve the adoption of battery cooling technologies to protect the batteries against thermal runaway [19]. This protection is provided, as well as, by the diagnosis of battery short circuits using the voltage cosine similarity method [20]. Although they are widely deployed, the use of lithium-ion batteries in decentralized energy systems contributes between 30 and 60% of greenhouse gas emissions [21].

Table 2 Last contributions of the energy storage by lithium-ion batteries

The main role of Lithium-ion battery	The last contributions about this technology	Objectives
Energy storage	[18] The use of artificial neural networks in managing the state of the battery	The use of Feedforward Neural Network (FNN) to predict battery degradation with an error of 8.6% over the first 100 cycles
		The use of Convolutional Neural Network (CNN) to analyse parameters such as temperature, voltage and current to predict and estimate the degradation of the state of the battery
	[20] The use of voltage cosine similarity for diagnosing lithium-ion battery short circuits	The voltage cosine similarity method will allow diagnosis of internal short-circuit faults by constructing a two-dimensional vector between the modulus of battery current and voltage. This method measures the change in voltage in order to calculate the cosine similarity and detect the battery faults. The results showed that the method performs rapid diagnostics when the simulated internal short-circuit fault resistance is less than 5Ω
	[21] Analysis of the harmful effects of lithium-ion batteries on global warming	The use of lithium-ion batteries in decentralized systems to store electrical energy generated from intermittent sources for self-consumption contributes between 30 and 60% of greenhouse gas emissions. These emissions are generally between 30 $KgCO_2$/kWhc and 270 $KgCO_2$/kWhc for each 1 kWhc stored

Energy storage by lithium-ion batteries will certainly help power instruments and components such as electric vehicles and smart grids to satisfy energy demand while preserving nature's biodiversity. According to Table 3, the contribution of the literature, between 2017 and 2024, on the main applications of lithium-ion batteries is based on the thermal management of electric vehicle batteries, the analysis of factors affecting their life cycle and the estimation of their states of charge. R. El Kassar et al. adopted systems for heating (e.g. internal preheating by pulsed current) and cooling (e.g. assembly of a heat sink with air channels) lithium-ion batteries for electric vehicles [22]. For lithium-ion batteries, whether used in electric vehicles or other components, the factors influencing their lifetime are essentially as follows: Operating temperature, multiplier and charge/discharge cut-off voltage [23]. Zhang et al. also introduce approaches for improving electric vehicles by estimating their lifetime using neural networks and Ridge linear regression, as well as the Kalman filter and the electrochemical impedance spectroscopy method [23]. In the context of improving the technological reliability of electric vehicles, Selvaraj et al. estimated the state of charge of lithium-ion batteries in order to guarantee their safety, using the Coulomb counting method, the proportional integral observer and the fuzzy inference system [24]. In the same period, from 2017 to 2024, scientific contributions on smart grids, as a major application of lithium-ion batteries, focus on the application of optimization algorithms and the reutilization of batteries already operating in electric vehicles. Hasanien et al. use the Gazelle optimization algorithm to minimize the fitness function that depends on the squared error between the estimated and measured voltage of lithium-ion batteries for their optimal use in smart grids [25]. On the other hand, the lithium-ion batteries already used in electric vehicles can be implemented in the smart grid and the research of Chiang et al. are developing a converterless energy management system for these batteries to enable energy consumption from renewable resources during off-peak hours [26]. Overall, the development of the implementation of lithium-ion batteries in smart grids has motivated researchers and manufacturers to introduce innovative systems such as the battery converter accompanied by a ventilator to control the battery temperature and the super capacitor to control the charging and discharging of the lithium-ion battery [27].

In view of the major applications for lithium-ion batteries in powering electric vehicles and smart grids, other potential applications could expand their markets and enhance their technical and economic feasibility. The 5th and 6th generations of mobile phones, decentralized energy systems and portable computers are all possible future applications for lithium-ion batteries. Between 2018 and 2023, the contributions of 5G are focused mainly on the conception of battery monitoring systems to manage 5G stations and the real-time determination of battery parameters using event-driven sampling and the use of 5G to store the results [28]. The security of 5G wireless communication services is also among the contributions of this generation where a photovoltaic system powered by a lithium-ion battery has been developed to secure the system in the case of current drop [29]. For portable computers, the contributions between 1995 and 2014 focus mainly on the deployment of lithium-ion batteries, with a metal oxide cathode and a carbon anode in these devices [30]. In this context, a study has developed a logistic model to predict the production of waste

Table 3 Last contributions of the major applications of lithium-ion batteries: the electric vehicles and smart grid

The major applications of lithium-ion batteries	The last contributions about this technology	Objectives
Electric vehicles (reducing fossil fuel dependence)	[22] Thermal management of lithium-ion batteries to be integrated in electric vehicles	In view of their high densities and their ability to guarantee access to sustainable energy, lithium-ion batteries are being integrated into electric vehicles to reduce the carbon footprint. For this purpose, heating and cooling techniques have been adopted to improve the thermal management of the (lithium-ion) batteries in these vehicles. The objective of adopting heating systems is to increase the temperature in low-temperature conditions while ensuring low deterioration. These systems include internal preheating using pulsed current, high-frequency AC preheating, etc. For the cooling system, the assembly of a heat sink with air channels and a set of phase-change materials, as well as a sandwich configuration of the heat pipe cooling system, will enable a reduction in temperature of around 43%
	[23] Presentation of the factors affecting the lifetime of lithium-ion batteries for electric vehicles, as well as the method for estimating this lifetime	The factors affecting lithium-ion battery lifetime are as follows: – **The operating temperature**: Research has shown that increasing the temperature from −20 °C to 50 °C reduces the rate of aging of the battery, which in turn increases its lifetime. In fact, the optimum temperature for a lithium-ion battery is around 30 °C, where the lifetime exceeds 1800 cycles – **The charge and discharge multiplier**: A high number of cycles leads to a high charge multiplier and a rapid rate of degradation, which reduces the lifetime of the lithium-ion battery. According to recent research, a temperature of 20 °C with a number of cycles of less than 500 reduces the rate of decrease in battery capacity – **The charging and discharging cut-off voltage**: A high charging cut-off voltage and a low discharging cut-off voltage both reduce the battery's capacity. As a result, a battery's state of charge or discharge is linked to the appearance of micro-cracks which, when the number of cycles is high, will subsequently cause the generation of large cracks and ultimately irreversible damage. In view that 4.3 V, 4.35 V and 4.4 V are considered as cut-off voltages, 4.3 V retains the battery capacity at a rate of 93.4% and the other 2 values reduce this retention at rates of 87.4% and 80.3%, respectively

(continued)

Table 3 (continued)

The major applications of lithium-ion batteries	The last contributions about this technology	Objectives
		Improving the reliability of electric vehicles is essentially based on predicting and estimating the lifetime of the lithium-ion battery, which effectively replaces their maintenance. This estimation is essentially based on approaches that focus on data and modeling – **Approaches based on data**: It analyzes data (e.g. vehicle monitoring data, etc.) using neural networks, Ridge linear regression and support vector regression. Therefore, the Ridge linear regression model predicts the battery's state of health with a mean absolute error (MAE) of less than 5% – **Approaches based on modeling**: As the analysis of physico-chemical reactions during the charging and discharging process is only accurate at the internal molecular level, the analysis of battery failure to estimate their lifetime requires regularity and accuracy. To achieve this, the use of models such as the Kalman filter and the equivalent circuit model is recommended. In this context, a mathematical model has been developed based on the operating temperature of the batteries and their charging and discharging rates, and optimized to predict lifetime even after 2000 cycles. In addition, electrochemical impedance spectroscopy also estimates the lifetime using the equivalent circuit model with optimum accuracy. The results of the approaches of modeling used for the estimation of the lifetime of lithium-ion batteries is illustrated by the innovative application of an electric vehicle that travels 23,266 km at high power in 1800s

(continued)

Table 3 (continued)

The major applications of lithium-ion batteries	The last contributions about this technology	Objectives
	[26] The estimation of the state of charge of lithium-ion batteries in electric vehicles	The estimation of the state of charge of lithium-ion batteries used in electric vehicles is a key indicator for testing their reliability and guaranteeing their safety. There are several existing methods for estimating the state of charge of batteries: Direct methods (electrochemical impedance spectroscopy method, Couloumb counting method, etc.), battery model (Burger model, Thevenin model, etc.), observer method (proportional integral observer, Luenberger observer, etc.) and data-based method (fuzzy inference system, deep learning neural network, etc.). Below is a brief explication of some of these approaches: – **Couloumb counting method**: This method is based on the integration of the instantaneous battery current over time. The precise state of charge is determined during the thermodynamic equilibrium condition – **The integral proportional observer**: This method is characterized by rapid convergence between the measured voltage and the estimated voltage, and has an error in estimation of less than 2%. This method is characterized by rapid convergence between the measured voltage and the estimated voltage, and has an error in estimation of less than 2%. The proportional integral observer is proposed for the RC model of lithium-ion batteries and will reduce the calculation steps while offering high accuracy – **The fuzzy inference system**: This method estimates the state of charge using fuzzy techniques, by taking into consideration the battery degradation process. The fuzzy inference system adjusts the measurement noise and estimates the non-linear gain of the Kalman filter

(continued)

Table 3 (continued)

The major applications of lithium-ion batteries	The last contributions about this technology	Objectives
Smart grid	[25] The application of the Gazelle optimization algorithm to lithium-ion batteries to be used in smart grids	The use of the gazelle optimization algorithm addresses the main challenge, which is the non-linear optimization of lithium-ion batteries for their application in smart grids. The vision of this approach is based on the minimization of the fitness function, which depends on the quadratic integral error between the estimated and measured lithium-ion battery voltage. As a result, the gazelle optimization algorithm justified its convergence and minimized the fitness function while reducing data processing time compared with other algorithms
	[26] Energy management for lithium-ion batteries discarded by electric vehicles for use in smart grid	This study is intended to develop an energy management system without converter for lithium-ion batteries in the smart grid balance procedure after they have been used in electric vehicles. The objective of this energy management system is to avoid overcharging batteries while allowing energy to be consumed from renewable resources during off-peak hours. As a result, it reduces the voltage drop up to 50%, while addressing the challenge of the degradation of lithium-ion batteries already used in electric vehicles
	[27] A detailed review of patents for lithium-ion batteries connected to the grid and used in energy storage	The document analyzes the patents on lithium-ion batteries connected to the grid. In general, the patents focus on the following applications: – The storage of electrical energy generated by a wind turbine is used to satisfy energy demand in the case of a grid failure. (WO 2020/228918 A1(2020)) (JCL Bagger et al. 2020) – The introduction of a battery converter to store the energy produced from PV accompanied with a ventilator to control the battery temperature. (US 2012/0235481 A1, 2012.) (N. Takeshi et al.) The invention of a super capacitor for controlling the charging and discharging of the lithium-ion battery for an optimal grid connection. (WO2021/ 214,228 A1, 2021.) (R. Alan et al.)

from batteries based on carbon nanotubes [31]. While the contributions of 6G and decentralized energy systems regarding the application of lithium-ion batteries do not exist, except for the contribution of Gutsch and Leker [21] on the impact of the implementation of lithium-ion batteries in these decentralized systems in terms of greenhouse gas emissions reaching 60%. (Table 4) oxide cathode and a carbon anode in these devices [30]. In this context, a study has developed a logistic model to predict the production of waste from batteries based on carbon nanotubes [31]. While the contributions of 6G and decentralized energy systems regarding the application of lithium-ion batteries do not exist, except for the contribution of Gutsch and Leker [21] on the impact of the implementation of lithium-ion batteries in these decentralized systems in terms of greenhouse gas emissions reaching 60%. (Table 4) oxide cathode and a carbon anode in these devices [30]. In this context, a study has developed a logistic model to predict the production of waste from batteries based on carbon nanotubes [31]. While the contributions of 6G and decentralized energy systems regarding the application of lithium-ion batteries do not exist, except for the contribution of [21] on the impact of the implementation of lithium-ion batteries in these decentralized systems in terms of greenhouse gas emissions reaching 60%. (Table 4) oxide cathode and a carbon anode in these devices [30]. In this context, a study has developed a logistic model to predict the production of waste from batteries based on carbon nanotubes [31]. While the contributions of 6G and decentralized energy systems regarding the application of lithium-ion batteries do not exist, except for the contribution of [21] on the impact of the implementation of lithium-ion batteries in these decentralized systems in terms of greenhouse gas emissions reaching 60%. (Table 4).

In general, the applications of lithium-ion batteries are not only focused on energy storage and electric vehicles, or on smart grids, but also in other areas such as emergency power systems and advanced surveillance and alarm systems, digital cameras and smartwatches, and in health applications including pacemakers.

However, why focus in this section only on lithium-ion battery applications in electric vehicles and smart grids?

There is no doubt that the smart grid's use of the energy stored in the lithium-ion battery to power pacemakers, sensors, communication devices and all the other electrical appliances needed to cover the daily energy needs. In addition, electric vehicles contribute to mobility while offering environmentally friendly transport, helping to improve the performance of urban and industrial transport by optimizing the logistics function between suppliers and customers. For this reason, electric vehicles and smart grids are the major applications for lithium-ion batteries, which will contribute to sustainability and promote the other possible applications, such as portable computers and decentralized energy systems, as well as the 5th and 6th generations of mobile phones. Decentralized energy systems enable energy autonomy while consuming energy produced by a nearby entity, without the need to overload the power grid or transport electricity over long distances. In this way, the lithium-ion battery stores the electricity produced by a wind turbine or photovoltaic panels in the decentralized system, which can then be used to power tertiary buildings. As already mentioned, this electricity will power devices in each building, including heating and air-conditioning

Table 4 Last contributions of the other possible applications of lithium-ion batteries: 5G, 6G, the portable computers PC and the decentralized energy systems

Other applications for lithium-ion batteries	Year of publication	Topic
5G	2023	The design of a cloud monitoring system for the lithium-ion battery to manage 5G stations [28]
	2021	The determination in real time of the state of health of lithium-ion batteries by acquiring their parameters (voltage and current) using event-driven sampling and the use of 5G to record the results in the cloud [28]
	2018	The development of a photovoltaic system led by a lithium-ion battery with a control method to secure 5G wireless communication services in case of current failure [29]
6G	N/A	N/A
Portable computer PC	2014	The study analyzes the use of CNT carbon nanotubes to improve the performance of the anodes and cathodes of lithium-ion batteries for their applications in portable computers. It provides a logistic model for predicting the generation of waste in these new lithium-ion CNT batteries, as well as their effects on the environment [31]
	1995	The use of lithium-ion batteries, with a metal oxide cathode and carbon anode, in electronic applications such as mobile phones and portable computers [30]
Decentralized energy systems	N/A	N/A

systems and portables computers. In this context, the computer industry must be developed by designing more lithium-ion battery-powered computers to respect the environment and take advantage of their potential applications such as quantum computing, data management and all the potential applications contributing to the digital transition. At the end, the use of the lithium-ion battery for the development of 5G is incomplete and still requires research work, in particular the management of smart and sustainable building data, the powering of monitoring systems for stations operating on 5G and the autonomy of devices connected to 5G. This is further incentive to deploy 6G to assess network coverage and throughput, and to pilot drones and electric vehicles.

Researchers, therefore, need to point their compasses toward the further development of algorithms for optimizing lithium-ion batteries and methods for estimating their state of charge and lifetime, for their optimal application in energy storage, electric vehicles and smart grids. This research should focus on optimizing the cooling rate of lithium-ion batteries by using AgO nanofluid from 23.95% up to 50% or 70% in [24], as well as reducing the error of 8.6% occurred in the first 100 cycles when using a Feedforward Neural Network (FNN) to predict battery degradation in [18]. The reduction of battery drop to 20% and not just 50% in [26] is another perspective for the management of energy in lithium-ion batteries discarded by electric vehicles and that they will be used to balance the smart grid. Certainly, these perspectives of research must include the development of the portable computer industry through the integration of lithium-ion batteries and also the development of 5G and 6G and the enhancement of decentralized energy systems through this sustainable battery although [21] estimates their contributions worth up to 30% of greenhouse gas emissions.

The question must be posed is, are lithium-ion batteries enough to promote the energy transition?

The energy transition, i.e., the transition from energy consumption based on fossil fuels to energy consumption based on renewable resources, is a political decision aimed primarily at achieving sustainable development and sustainability. The principle of sustainable development aims to adopt approaches that preserve nature, and lithium-ion batteries have proved their performance in terms of storage and reductions in greenhouse gases. To understand the contribution of lithium-ion batteries in promoting sustainable development and subsequently the energy transition, it is important to determine their impact on climate change. In this context, the literature focuses mainly on the impact of used and recycled lithium-ion batteries on greenhouse gas emissions. Jenu et al. (Table 5) analyze the impact of battery stress on its lifetime and the impact of using these batteries to store intermittent electricity generated by PV technology on the energy mix of Bulgaria, Germany, Netherlands, Spain, and Finland [32]. Therefore, the reduction in temperature (as a stress factor) reduces the number of cycles and therefore reduces the carbon footprint. However, the study of the impact of PV on the countries mentioned shows the high carbon intensity in the electricity grid considering that their energy mix includes coal from 40 to 45%. Table 5 also presents the contribution of Li et al. on batteries recycled by pyrometallurgy, which reduce the carbon footprint by 10.76 kg of CO_2 equivalent [33]. Although the concept of fighting climate change is directly linked to the achievement of sustainable development and energy transition, it is important to look at the contributions of lithium-ion batteries in this context.

Table 6 indicates that there are no direct studies on the achievement of sustainable development by lithium-ion batteries, although there is a contribution by Stephan et al. on the presentation of the motivations for decision-makers to invest in the industrialization of lithium-ion batteries in terms of energy transition [34]. Although the studies already presented, on lithium-ion battery energy storage and its use in electric vehicles and smart grids, reflect and contribute to the implementation of the energy transition, the literature is seriously lacking in direct contributions on the role of lithium-ion batteries in promoting sustainable development and consequently the energy transition.

The question arises once again: are lithium-ion batteries enough to promote the energy transition?

Indeed, the industrialization of electric vehicles and smart grid infrastructures is part of the deployment of the energy transition. An electric motor will reduce the carbon footprint and lower the level of atmospheric pollution, as well as combating global warming. Furthermore, powering smart grids with lithium-ion batteries will

Table 5 Publications on the contribution of lithium-ion batteries to climate change (2020–2023)

The contribution on climate change	Year of publication	General ideas
[32] Analysis of the stress and life cycle of batteries to reveal their impact on climate change	2020	The study examines the impact of using lithium-ion batteries to store intermittent electricity generated by PV panels and its contribution to reducing the carbon footprint. It analyzes battery stress factors, including degradation factors such as loss of lithium stock and loss of anode and cathode materials, and their lifetime in terms of management and impact on climate change. This study leads to the following results: – The analyze of the impact of using lithium-ion battery for intermittent electricity storage in Bulgaria, Germany, Netherlands, Spain, and Finland on climate change showed that carbon has a high intensity in the electricity grid as coal represents 40–45% in the energy mix – The modifications of the batteries temperature, as stress parameter, from 25 °C to 15 °C results in a decrease in life cycle numbers from 2810 to 2420 cycles, which subsequently reduces the carbon footprint
[33] The impact of using recycled lithium-ion batteries on climate change	2023	This study is intended to analyze the impact of using recycled lithium-ion batteries on the environment via a meta-analytical approach and statistical study of research work to quantify greenhouse gas emissions. The results of the investigation showed that recycling 1 kg of used lithium-ion batteries emits up to 44.59 kg of CO_2 equivalent, with a reduction of around 10.76 kg of CO_2 equivalent. In addition, the energy consumption of this recycling process is around 3.3 MJ, with a maximum value of 154.4 MJ. The recycling process was performed using the combined hydro-pyro metallurgical process, a high-performance method that recovers up to 80% of lithium. It enables the precious metals in the lithium-ion batteries already in use to be processed by pyrometallurgy (where the metals are recovered in the form of an alloy by heating to a temperature in excess of 1000 °C) and hydrometallurgy (where the alloy is treated in an aqueous medium to separate the metals by electrolysis)

encourage the consumption and production of intermittent, environmentally friendly electricity (photovoltaic and wind power) to encourage the implementation of decentralized energy systems. As long as the literature does not contain direct contributions on the role of lithium-ion batteries in achieving sustainability, sustainable development and the energy transition, it is not sufficient to know whether batteries achieve the energy transition or whether or not they are sufficient for this implementation. These contributions should include experimental data, experience feedback and

Table 6 Publications on the contribution of lithium-ion batteries in sustainable development and the energy transition

The axis title	Year of publication	The main idea
Sustainable development	N/A	N/A
Energy transition	2021	[34] The presentation of the different factors and mechanisms enabling policy-makers and leaders to embark on the industrialization of clean technologies such as the lithium-ion battery

approaches, as well as perspectives on the potential role of lithium-ion batteries in the implementation of sustainable development and the energy transition. Without overlooking that, the energy transition could be achieved through various measures such as energy efficiency in buildings, the production of electricity from solar thermal energy, the green hydrogen. It is therefore necessary to adopt approaches for the widespread use of lithium-ion batteries in daily activities and in industry, and to provide data on the feedback from these approaches and, above all, to analyze their impact on climate, biodiversity and the carbon footprint in order to know whether they are really sufficient for the implementation of the energy transition.

Overall, developing algorithms for optimizing battery thermal management models, estimating their states of charge and determining the factors influencing their lifetimes achieve the promotion of the energy transition through lithium-ion batteries. However, the use of digital technologies for the development of lithium-ion battery technology is insufficient until their impact on sustainable development and the reduction of the carbon footprint can be analyzed. It is therefore necessary to provide data and experience feedback in order to perform analytical and statistical studies about the implementation of the energy transition using lithium-ion battery technology.

6 Conclusion and Perspectives

The present chapter illustrates the lithium-ion batteries (LIBs) technologies and gives a general review on their major role in the global energy transition. The chapter starts by presenting the history and the evolution of LIBs technology, which is linked to the earlier development of nickel–cadmium (Ni–Cd) batteries in 1889 by Swedish scientist Waldemar Jungner. Until the 90 s of the technological revolution of these batteries, in particular when Sony introduced the first commercially available lithium-ion battery. The significant advancements in LIBs came thanks to the groundbreaking work of Professor John B. Goodenough, who is later awarded the 2019 Nobel Prize in Chemistry alongside Akira Yoshino and Stan Whittingham, made, then, crucial and major improvements in LIBs. In addition, the chapter presents the current progress in cathode and anode for LIBs especially the layered materials, the spinel materials, the polyanion materials, the carbon graphite and the silicone anode. The large applications of LIBs in storing the electricity excess, generated by solar panels for later use, hospitals and data centers, portable electronics (digital cameras, smartphones, smartwatches and laptops) and even more electric and hybrid vehicles are presented. To protect the environment and conserve the biodiversity, recycling the LIBs, through several processes such as the pretreatment and pyrometallurgy is crucial to decrease the volume of wastes and minimizes the toxicity. At the end, the chapter analyzes the role of LIBs on promoting the energy transition through presenting an overview of the researcher's contributions on this field. The main key findings of this analysis is summarized as follows:

- The investigation indicates that the literature need more applications of the artificial intelligence. It use only the Feedforward Neural Network (FNN) and Convolutional Neural Network (CNN) in managing the state of the battery and apply the other conventional approaches such as the air and liquid cooling and the phase-change material for the thermal management of LIBs.
- The researchers must direct their compass toward the integration of LIBs in the 5th and 6th generations of mobile phones (5G and 6G) and in the decentralized energy systems to provide more comfort as well as achieve the goals of the energy transition.
- The researchers need to produce further research works on the contribution of LIBs of implementing the sustainable development and therefore the energy transition, in addition to the contributions on energy storage and electric vehicles, and on smart grids.
- The organisms must provide data and experience feedback in order to perform analytical and statistical studies about the impact of using LIBs on sustainable development and the reduction of the carbon footprint.

The contribution of LIBs in the global energy transition must not stop in this stage but include the following perspectives of future research works:

- The prediction of the LIBs maintenance by initiating cracks in order to analyze the thermomechanical behavior of batteries and avoid the thermal runaway.
- The development of algorithms and AI approaches to manage the state of charge of LIBs.
- The global transition to the exploitation of the renewable resources and the benefit of LIBs for storing the energy to offer more accessibility for several sectors such as industry, agriculture, mobility, and services.

References

1. Trueb LF, Rüetschi P (1998) Nickel-Cadmium-Akkumulatoren. In: Trueb LF, Rüetschi P (éds) Batterien und Akkumulatoren: Mobile Energiequellen für heute und morgen. Springer, pp 90–100. https://doi.org/10.1007/978-3-642-58741-2_11
2. Scrosati B, Garche J (2010) Lithium batteries: Status, prospects and future. J Power Sources 195(9):2419–2430. https://doi.org/10.1016/j.jpowsour.2009.11.048
3. Murphy DW, Broadhead J, Steele BCH (Éds) (1980) Materials for advanced batteries. Springer US. https://doi.org/10.1007/978-1-4684-3851-2
4. Yazami R, Munshi MZA (1995) Novel anodes for solid state batteries. In: Handbook of solid state batteries and capacitors. World Scientific, pp 425–466. https://doi.org/10.1142/9789812831828_0020
5. Sun Y-K (1999) Cycling behaviour of $LiCoO_2$ cathode materials prepared by PAA-assisted sol–gel method for rechargeable lithium batteries. J Power Sources 83(1):223–226. https://doi.org/10.1016/S0378-7753(99)00280-3
6. Momma K, Izumi F (2011) VESTA 3 for three-dimensional visualization of crystal, volumetric and morphology data. J Appl Crystallogr 44(6):1272–1276. https://doi.org/10.1107/S0021889811038970
7. Manthiram A, Goodenough JB (1989) Lithium insertion into $Fe_2(SO_4)_3$ frameworks. J Power Sources 26(3):403–408. https://doi.org/10.1016/0378-7753(89)80153-3
8. Lithium Batteries: Science and Technology | SpringerLink. (s.d.). Consulté 14 novembre 2024, à l'adresse https://link.springer.com/book/https://doi.org/10.1007/978-0-387-92675-9
9. Padhi AK, Nanjundaswamy KS, Goodenough JB (1997) Phospho-olivines as positive-electrode materials for rechargeable lithium batteries. J Electrochem Soc 144(4):1188. https://doi.org/10.1149/1.1837571
10. Dahn JR (1991) Phase diagram of LixC6. Phys Rev B, Condensed Matter 44(17):9170–9177. https://doi.org/10.1103/physrevb.44.9170
11. Takami N, Satoh A, Hara M, Ohsaki T (1995) Rechargeable lithium-ion cells using graphitized mesophase-pitch-based carbon fiber anodes. J Electrochem Soc 142(8):2564. https://doi.org/10.1149/1.2050054
12. Deng D, Lee JY (2014) Meso-oblate spheroids of thermal-stabile linker-free aggregates with size-tunable subunits for reversible lithium storage. ACS Appl Mater Interfaces 6(2):1173–1179. https://doi.org/10.1021/am404853n
13. Shin H-C, Pyun S-I (2001) Investigation of lithium transport through lithium cobalt dioxide thin film sputter-deposited by analysis of cyclic voltammogram. Electrochim Acta 46(16):2477–2485. https://doi.org/10.1016/S0013-4686(01)00457-1
14. The Structural Stability of Transition Metal Oxide Insertion Electrodes for Lithium Batteries—Handbook of Battery Materials—Wiley Online Library. (s. d.). Consulté 14 novembre 2024, à l'adresse https://onlinelibrary.wiley.com/doi/https://doi.org/10.1002/9783527611676.ch13
15. Nakahara K, Nakajima R, Matsushima T, Majima H (2003) Preparation of particulate $Li_4Ti_5O_{12}$ having excellent characteristics as an electrode active material for power storage cells. J Power Sources 117(1):131–136. https://doi.org/10.1016/S0378-7753(03)00169-1
16. Panero S, Satolli D, Salomon M, Scrosati B (2000) A new type of lithium-ion cell based on the $Li_4Ti_5O_{12}/Li_2Co_{0.4}Fe_{0.4}Mn_{3.2}O_8$ high-voltage, electrode combination. Electrochem Commun 2(11):810–813. https://doi.org/10.1016/S1388-2481(00)00127-2
17. Kubiak P, Garcia A, Womes M, Aldon L, Olivier-Fourcade J, Lippens P-E, Jumas J-C (2003) Phase transition in the spinel $Li_4Ti_5O_{12}$ induced by lithium insertion. J Power Sources 119–121:626–630. https://doi.org/10.1016/S0378-7753(03)00186-1
18. Zou Y, Lin Z, Li D, Liu Z (2023) Advancements in artificial neural networks for health management of energy storage lithium-ion batteries: a comprehensive review. J Energy Storage 73:109069. https://doi.org/10.1016/j.est.2023.109069
19. Lv Y, Geng X, Luo W, Chu T, Li H, Liu D, Cheng H, Chen J, He X, Li C (2023) Review on influence factors and prevention control technologies of lithium-ion battery energy storage safety. J Energy Storage 72:108389. https://doi.org/10.1016/j.est.2023.108389

20. Wu X, Wei Z, Wen T, Du J, Sun J, Shtang AA (2023) Research on short-circuit fault-diagnosis strategy of lithium-ion battery in an energy-storage system based on voltage cosine similarity. J Energy Storage 71:108012. https://doi.org/10.1016/j.est.2023.108012
21. Gutsch M, Leker J (2022) Global warming potential of lithium-ion battery energy storage systems: a review. J Energy Storage 52:105030. https://doi.org/10.1016/j.est.2022.105030
22. El Kassar R, Al Takash A, Faraj J, Hammoud M, Khaled M, Ramadan HS (2024) Recent advances in lithium-ion battery integration with thermal management systems for electric vehicles: a summary review. J Energy Storage 91:112061. https://doi.org/10.1016/j.est.2024.112061
23. Zhang X, Jiang S, Lin Z, Gui Q, Chen F (2023) Model construction and performance analysis for asymmetric compound parabolic concentrator with circular absorber. Energy 267:126597. https://doi.org/10.1016/j.energy.2022.126597
24. Selvaraj V, Vairavasundaram I (2023) A comprehensive review of state of charge estimation in lithium-ion batteries used in electric vehicles. J Energy Storage 72:108777. https://doi.org/10.1016/j.est.2023.108777
25. Hasanien HM, Alsaleh I, Tostado-Véliz M, Alassaf A, Alateeq A, Jurado F (2023) Optimal parameters estimation of lithium-ion battery in smart grid applications based on gazelle optimization algorithm. Energy 285:129509. https://doi.org/10.1016/j.energy.2023.129509
26. Lee C-C, Xuan C, Wang F, Wang K (2024) Path analysis of green finance on energy transition under climate change. Energy Economics 107891. https://doi.org/10.1016/j.eneco.2024.107891
27. Wali SB, Hannan MA, Ker PJ, Rahman SA, Le KN, Begum RA, Tiong SK, Mahlia TMI (2024) Grid-connected lithium-ion battery energy storage system towards sustainable energy: a patent landscape analysis and technology updates. J Energy Storage 77:109986. https://doi.org/10.1016/j.est.2023.109986
28. Qaisar SM, AlQathami M (2021) Event-Driven sampling based Li-ion batteries SoH estimation in the 5G framework. Procedia Comput Sci 182:109–114. https://doi.org/10.1016/j.procs.2021.02.015
29. Nakamura M, Takeno K (2018) Green base station using robust solar system and high performance lithium ion battery for next generation wireless network (5G) and against mega disaster. In: 2018 International power electronics conference (IPEC-Niigata 2018—ECCE Asia), pp 201–206. https://doi.org/10.23919/IPEC.2018.8507634
30. Megahed S, Ebner W (1995) Lithium-ion battery for electronic applications. J Power Sources 54(1):155–162. https://doi.org/10.1016/0378-7753(94)02059-C
31. Espinoza VS, Erbis S, Pourzahedi L, Eckelman MJ, Isaacs JA (2014) Material flow analysis of carbon nanotube lithium-ion batteries used in portable computers. ACS Sustain Chem Eng 2(7):1642–1648. https://doi.org/10.1021/sc500111y
32. Jenu S, Deviatkin I, Hentunen A, Myllysilta M, Viik S, Pihlatie M (2020) Reducing the climate change impacts of lithium-ion batteries by their cautious management through integration of stress factors and life cycle assessment. J Energy Storage 27:101023. https://doi.org/10.1016/j.est.2019.101023
33. Li J, Li L, Yang R, Jiao J (2023) Assessment of the lifecycle carbon emission and energy consumption of lithium-ion power batteries recycling: a systematic review and meta-analysis. J Energy Storage 65:107306. https://doi.org/10.1016/j.est.2023.107306
34. Stephan A, Anadon LD, Hoffmann VH (2021) How has external knowledge contributed to lithium-ion batteries for the energy transition? iScience 24(1), 101995. https://doi.org/10.1016/j.isci.2020.101995

Infusing Sustainability: Mint Tea By-Products as Thermal Insulating Mortar Additives for Energy Transition in the Construction Industry

Othmane Horma, Boutahar Laaouar, Sara El Hassani, Aboubakr El Hammouti, Mohammed Amine Moussaoui, and Ahmed Mezrhab

Abstract The use of biomass in construction materials has emerged as a promising technique to enhance thermal insulation and mitigate carbon emissions, addressing crucial energy and ecological challenges. This paper introduces novel construction materials made from by-products of mint tea, one of Morocco's most abundant and culturally significant beverages. The thermophysical properties of the developed cement-based bio-composites integrating crushed spent tea and mint stems were determined using the hot-disk method. Experimental results indicate a 20% reduction in density and significant decreases in thermal conductivity and diffusivity, each exceeding 50%. These findings highlight the potential of these materials as lightweight insulators. ANSYS Fluent simulations demonstrate that our bio-composites clearly outperform traditional mortar in wall applications, suggesting a promising direction for future sustainable building practices.

Keywords Mint stems · Spent tea · Energy · Efficient mortar · Thermophysical characterization · Thermal simulation

1 Introduction

The construction industry is a major consumer of energy and producer of CO_2 emissions globally. Traditional building materials and methods contribute significantly to environmental degradation, prompting a critical need for sustainable solutions. The

O. Horma (✉) · B. Laaouar · A. El Hammouti · M. A. Moussaoui · A. Mezrhab
Mechanics and Energy Laboratory, Mohammed First University, Oujda, Morocco
e-mail: horma.othmane@ump.ac.ma

S. El Hassani
Laboratory of Energetics, Abdelmalek Essaâdi University, Tetouan, Morocco

A. El Hammouti
Institut de Recherche de la Construction, ESTP, Cachan, France

F. Latrache et al. (eds.), *Digital Technology for an Innovative Energy Transition*, Green Energy and Technology, https://doi.org/10.1007/978-3-031-89657-6_10

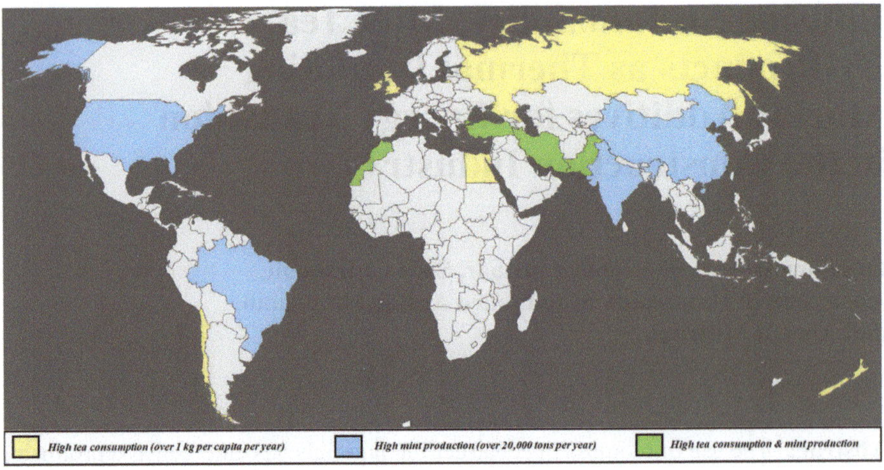

Fig. 1 Most tea consuming and mint producing countries

pursuit of such alternatives has become increasingly urgent as the sector continues to grow, with cement production alone projected to exceed five billion metric tons by 2030 [1], accounting for a substantial portion of global anthropogenic CO_2 emissions [2].

Researchers have turned their attention to the potential of industrial, municipal, and biomass by-products as part of sustainable construction materials [3–5]. These materials not only help reduce the environmental footprint but also enhance the properties of building components. For example, incorporating bio-based additives into cement has been shown to produce lightweight, energy-efficient, and eco-friendly construction elements [6, 7]. Such innovations are crucial for transitioning towards more sustainable building practices.

In Morocco, a country deeply intertwined with tea and mint in its cultural fabric, the production of these commodities generates significant quantities of bio-waste. Annually, tens of thousands of tons of spent tea and mint stems are discarded or underutilized, often only finding use in animal feed or low-value agricultural applications [8, 9]. This study proposes a novel approach to valorize these bio-wastes by integrating them into cement-based mortars. Figure 1 illustrates Morocco's significant role as one of the top tea-consuming countries and a major producer of mint, highlighting the potential for local resource utilization to foster sustainable construction practices [10, 11].

1.1 Literature Review and Research Gap

Currently, spent tea has virtually no productive use. Spent coffee grounds, on the other hand, have been thoroughly investigated for their potential in construction materials.

Both spent tea and spent coffee grounds are by-products of the beverage preparation process, making them similar in terms of extraction and availability. Likewise, many plant stems and fibers have been studied as additives, but to our knowledge, no one has investigated the use of mint stems. This similarity in research direction underscores the potential for our study to fill a significant gap in the literature.

Incorporating spent coffee grounds into construction materials has been shown to enhance various properties. For instance, Charai et al. [12] found that adding spent coffee grounds (SCG) improves thermal insulation of cement-based bio-composites. They found that the addition of 10% by weight of SCG to cement reduced the density and thermal conductivity by 20% and 62%, respectively. Lachheb et al. [13] investigated the addition of up to 6% by weight of SCG into a plaster-based binder. The findings showed that this integration could reduce the thermal conductivity by 38% and more importantly, diminish the annual CO_2-related emissions by 1500 kg. These results demonstrate the beneficial impact of this post-infusion material on the construction industry in terms of energy efficiency and environment.

Various plant stems, leaves, and fibers have been studied as additives in construction materials. Research highlights the benefits of natural fibers, such as improved thermal and acoustic insulation and increased mechanical strength. Benaniba et al. [14] studied the influence of incorporating date palm fibers (DPF) on the thermal and mechanical properties of mortar. The authors found that a 6% DPF addition by weight reduced the density and thermal conductivity by 25% and 29%, respectively. Additionally, 6% content of these fibers slightly decreased the compressive strength while improving the flexural strength. Studies on Pennisetum, straws and alfa fibers [15–17] reported improvements in thermal properties and density of conventional binders, although a decrease in mechanical strength was noted. These findings highlight the need to balance the trade-off between thermal performance gain and loss of mechanical strength.

Despite the extensive research on other bio-based additives, the integration of mint tea by-products into construction materials has not been adequately explored. This research gap presents an opportunity to investigate the unique properties of mint stems and spent tea, and their potential to enhance the sustainability and performance of cement-based mortars. By addressing this gap, our study aims to contribute valuable insights into the development of eco-friendly construction materials that leverage local bio-wastes, thereby promoting sustainable building practices.

1.2 Study Objectives and Methodological Approach

This chapter aims to explore the utilization of crushed spent tea and mint stems in the construction industry, focusing initially on their thermophysical characterization. We assess the density, thermal conductivity, and diffusivity of the resulting bio-composites using non-destructive methods such as the Hot Disk method. Subsequently, the study employs Computational Fluid Dynamics (CFD) simulations to evaluate the influence of these innovative mortars on the thermal behavior of building envelopes. By harnessing these underexploited bio-wastes, the research seeks not only to contribute to the body of knowledge in sustainable materials but also to propose practical, eco-friendly alternatives that align with Morocco's cultural practices and environmental goals.

2 Materials and Methods

2.1 Sample Preparation

Spent tea leaves were collected from local cafés after routine teapot rinsing, while mint stems were sourced from local vegetable shops. Both bio-sourced materials were subjected to sun-drying for one day to eliminate moisture and humidity, which helps preserve these plant-based materials and enables their use as additives in cement-based composites.

Locally sourced C35 cement and sand in a 3:1 ratio were used for the development of the mortars. Increments of 5% by weight of the crushed spent tea (CST) and mint stems (CMS) additives and a water-to-cement ratio of 0.5 were used to fabricate the mixtures. Figure 2 illustrates the raw materials and the end products, while Table 1 details the composition of the prepared samples.

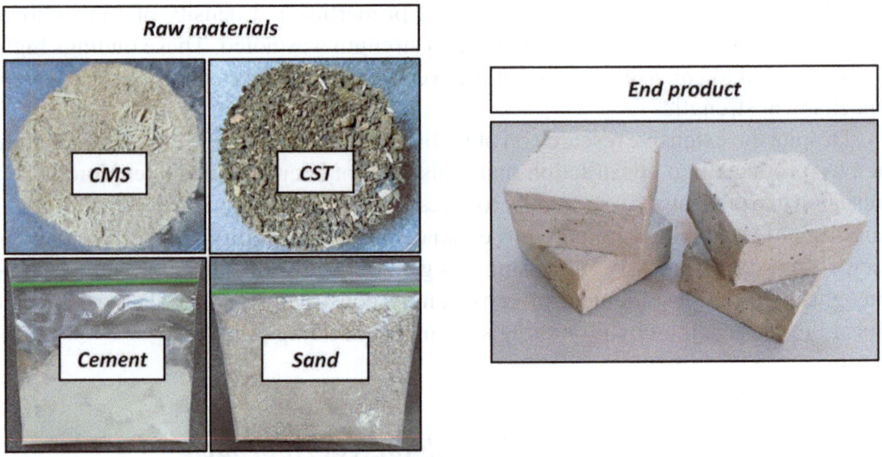

Fig. 2 Raw materials used in the studied mortars and the final product

Table 1 Composition of the studied bio-composite mortars

Specimens	CMS content (wt%)	CST content (wt%)	Cement (wt%)	Sand (wt%)	Water to cement ratio
M	0	0	100	33	0.5
CMSM5	5	0			
CSTM5	0	5			
CMSM10	10	0			
CSTM10	0	10			

2.2 *Thermophysical Characterization*

The Hot Disk method (HDM), renowned for its rapid assessment capabilities, was employed to determine the thermophysical properties of solid materials. This technique operates in the transient phase of thermal transfer, yielding properties such as thermal conductivity, diffusivity, and specific heat. The HDM setup and its components are displayed in Fig. 3.

The measurement process starts by sandwiching the HDM sensor, which acts simultaneously as a thermocouple and a heating source, between two identical samples. These samples are then heated by a constant current of 80 mW power for 80 s. Afterward, the temperature variation of the specimens is recorded by the TPS instrument. Finally, the properties are calculated using the hot-disk analyzer software version 7.1.5.

It is noteworthy to mention that the density of the mortars was calculated by dividing the mass by the volume of the specimens.

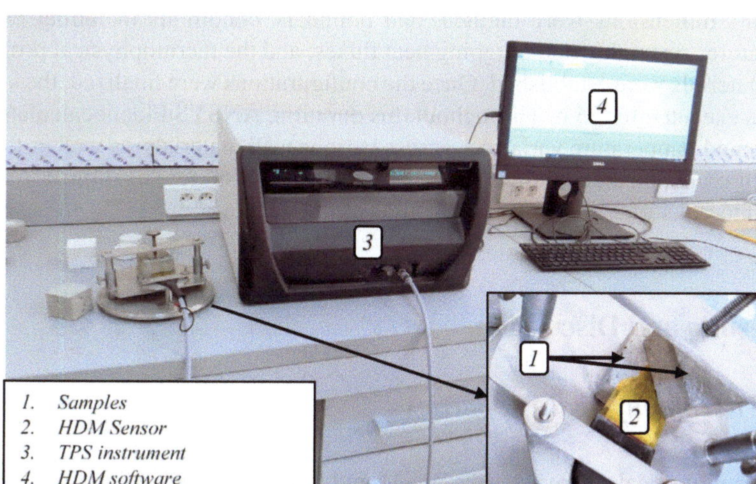

Fig. 3 Hot disk method setup

2.3 Thermal Behavior Analysis: Simulation Setup for Mortar

Following the determination of the thermophysical properties of the crushed spent tea mortar (CSTM) and mint stem mortar (CMSM), simulations were conducted using ANSYS Fluent. ANSYS Fluent is a popular software for performing numerical simulations related to fluid dynamics and thermodynamics [18]. In this study, it was used for wall modeling and simulations of heat transfer through different wall materials. These simulations assessed the influence of these innovative mortars on the thermal behavior of a typical Moroccan wall.

The simulation was performed for a 24-h period, specifically on July 23, 2023, a day that experienced significant temperature variations with maximum $\left(T_{out}^{max}\right)$ and minimal temperatures $\left(T_{out}^{min}\right)$ were 45 °C and 26 °C, respectively.

The mathematical model used to approximate real temperature variations, previously employed in several studies [19, 20], is given as follows:

$$T_{sa}(t) = \frac{T_{out}^{max} + T_{out}^{min}}{2} + \frac{T_{out}^{max} - T_{out}^{min}}{2} \sin\left(\omega t - \frac{\pi}{2}\right) \quad (1)$$

In this equation, ω represents the angular frequency, calculated as $2\pi/86400$ to account for the daily cycle, where 86,400 is the number of seconds in a day.

A simple scenario was conducted for this simulation, consisting of replacing standard mortar layers by our innovative mortars in a typical Moroccan wall [21]. Given their superior thermophysical properties both CMSM10 and CSTM10 were integrated as non-load bearing mortar coating in order to improve the thermal performance of the wall.

The simulation process started with the creation of the wall model, comprising a 200 mm concrete core sandwiched between two 20 mm mortar layers. Subsequently, the wall's dimensions were defined, and boundary conditions including external temperatures, incoming and outgoing heat fluxes, and the thermophysical properties of the materials were established. Once the configurations were finalized, the simulation was executed for 25 h. Throughout this duration, ANSYS Fluent calculated and documented temperature variations across various wall materials, as well as the heat fluxes entering and leaving the system. Finally, the results were examined, focusing on the internal surface temperature profiles of the wall.

3 Results and Discussion

3.1 Thermophysical Properties

The thermophysical properties of CMSM and CSTM composites are presented in Table 2 along with their densities. Incorporating 10 wt % CMSM or CSTM reduces bulk density by ~20 % relative to reference mortar (M). Figure 4 illustrates the

Table 2 Thermophysical properties and density of the studied mortars

Samples	λ (W/mK)	α (mm^2/s)	ρ (kg/m^3)	Specific heat equation	c_p (J/kgK)
M	0.831	0.432	1843	$c_p = \lambda/\alpha\rho$	1043
CMSM5	0.410	0.255	1639		981
CSTM5	0.362	0.230	1655		950
CMSM10	0.284	0.202	1458		964
CSTM10	0.286	0.175	1472		1110

influence of CMS and CST on the thermal conductivity (λ) and diffusivity (α) of the studied mortars.

The results indicate that incorporating CMS and CST significantly reduces both thermal conductivity and diffusivity, thereby enhancing their insulation properties. Specifically, the thermal conductivity of CMSM10 and CSTM10 is reduced by approximately 65% compared to the reference mortar. Similarly, the thermal diffusivity of CMSM10 and CSTM10 decreases by around 53% and 60%, respectively. These reductions follow an exponential decay pattern, which is attributed to the porous structure and low thermal conductivity of the bio-sourced additives. Previous studies on the incorporation of spent tea and mint stem aggregates [3, 22] corroborate the present findings, exhibiting similar trends in the evolution of thermal properties while reducing density at a slightly higher rate due to the larger size of the aggregates.

The specific heat capacity c_p variation is proportionate to the decrease in the thermal conductivity, diffusivity and density, as shown in the equation within Table 2. The addition of CMS resulted in a slight decrease of c_p, while an increase was observed for the integration of CST. This could be explained by the lower thermal diffusivity of CSTM, suggesting that CST may positively impact the thermal performance of mortar, especially during summertime. A recent study [23] on peanut shells (PS) incorporation into cement-based materials found similar results. The study showed that as PS content increased, the specific heat capacity decreased initially at 2% of PS content by weight, before increasing for the specimens integrating up to 6% PS content. This increase was due to the pores created within the cement matrix, which impeded heat flow and decreased the thermal diffusivity.

Overall, the results demonstrate the potential of CMS and CST composites as effective thermal insulating materials for sustainable construction applications.

3.2 *Numerical Analysis*

The simulation results of the heat transfer through the walls using different mortars [M, CMSM10, and CSTM10] are presented in Fig. 5. The findings indicate that even thin layers of mortars incorporating CMS and CST can significantly improve the

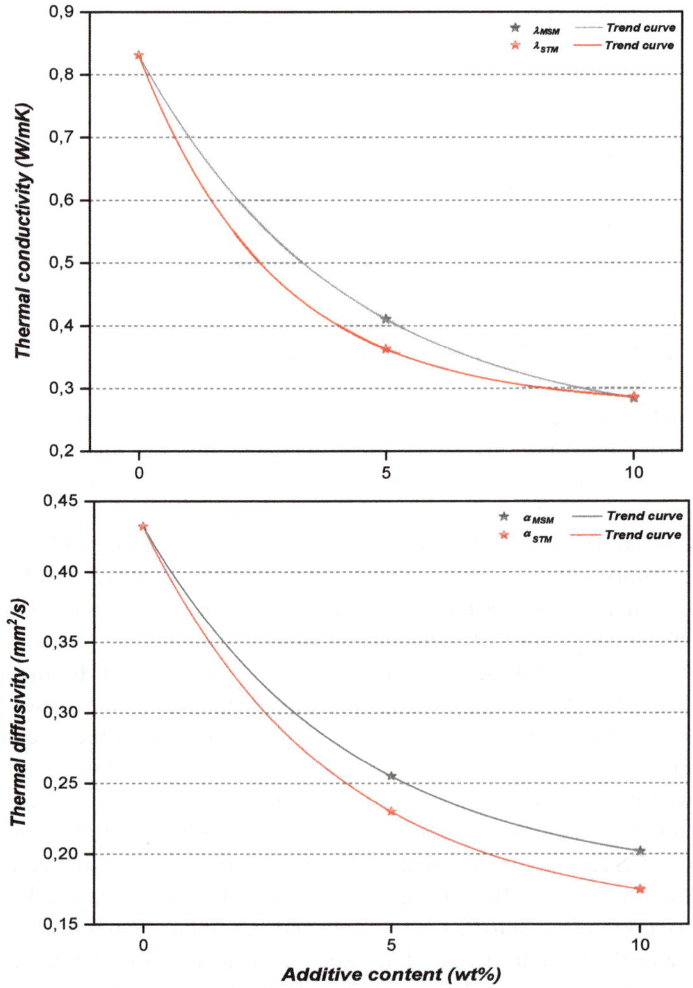

Fig. 4 Thermal conductivity (top) and diffusivity (bottom) versus CMS and CST content

thermal performance. Both CMSM10 and CSTM10 reduced the maximum temperature by 1.5 °C and increased the time lags ($\varphi_{M-\mathrm{CMSM}}$ and $\varphi_{M-\mathrm{CSTM}}$) by over an hour when compared to regular mortar (M). The slight difference (20 min) between the time lag of CSTM10 and CMSM10 can be attributed to the higher specific heat and lower thermal diffusivity of CSTM. Figure 6 displays the visualization of the thermal profiles of the simulated walls after 19 h, when the M wall reached its peak internal surface temperature at 30.5 °C. The temperature profile of the walls using CSTM10 and CMSM10 was practically identical as both reached approximately 29 °C simultaneously. Previous studies [23, 24] presented similar results, where the substitution of conventional concrete and adobe with bio-composite counterparts led to improvements in the thermal performance of the studied walls.

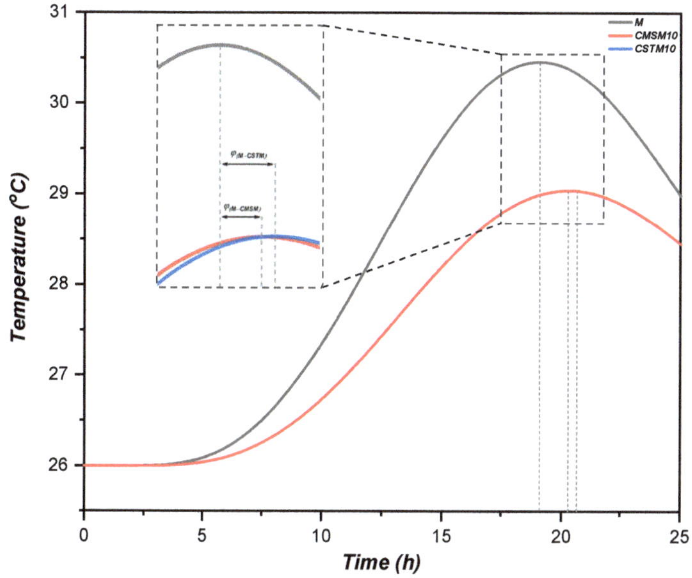

Fig. 5 Internal temperature variation of the studied wall

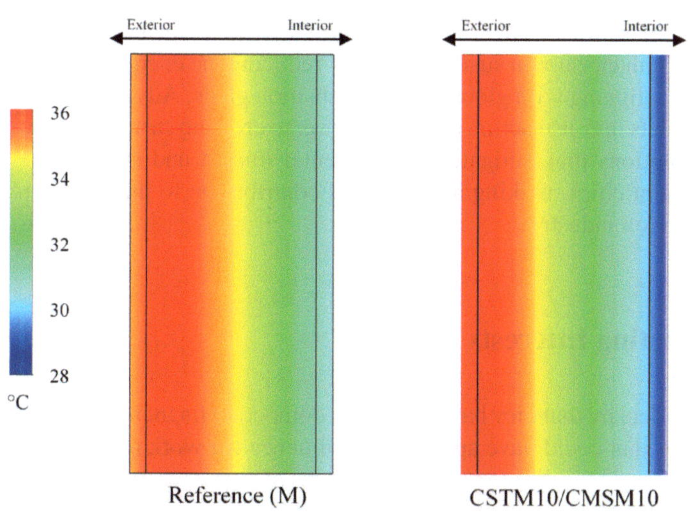

Fig. 6 Temperature profiles of the studied walls at 19 h

In summary, these numerical results demonstrate the influence of our novel bio-composites on building envelopes, showing their potential as thermal insulating building components.

4 Conclusion

This chapter investigated the integration of mint tea by-products into cement-based materials, aiming to valorize these bio-wastes and develop lightweight, energy-efficient mortars. The findings can be summarized as follows:

- The addition of 10 wt% of CMS and CST decreased the density of mortars by approximately 20%.
- The integration of CMS and CST reduced thermal conductivity by 65% compared to the reference mortar.
- Thin layers of CMSM10 and CSTM10 significantly improved the thermal performance of the wall, increasing the thermal shift by over an hour and decreasing the maximum temperature by 1.5 °C.

This work not only established the insulating potential of these mortars, but also set the stage for further research to find the right proportions to achieve an optimal balance between physical, thermal and mechanical properties. Future studies should investigate the long-term durability and hygrothermal properties of these mortars under various environmental conditions.

These innovative materials have significant potential applications in the development of smart cities. Their superior thermal properties can contribute to energy savings and enhance urban sustainability by reducing the energy demand for heating and cooling, aligning with global energy transition goals. Advancing the use of mint tea by-products in construction moves us closer to creating sustainable, cost-effective building solutions that mitigate environmental impact and support the growth of smart, resilient cities, making a substantial contribution to the energy transition in the construction industry.

5 Competing Interests

The authors declare that they have no known competing financial interests or personal relationships that could have appeared to influence the work reported in this paper.

References

1. Datis Export Group (2020) Worldwide Cement Production From 2015 to 2019. https://datis-inc.com/blog/worldwide-cement-production-from-2015-to-2019/
2. Monteiro PJM, Miller SA, Horvath A (2017) Towards sustainable concrete. Nat Mater 16(7):698–699. https://doi.org/10.1038/nmat4930
3. Horma O, Charai M, El Hassani S, El Hammouti A, Mezrhab A (2022) Thermo-physical and mechanical characterization of cement-based mortar incorporating spent tea. J Build Eng 52 (November 2021):104392. https://doi.org/10.1016/j.jobe.2022.104392
4. Horma O, Charai M, El Hassani S, El Hammouti A, Moussaoui MA, Mezrhab A (2022) Thermal performance study of a cement-based mortar incorporating EPS beads (8 May):2–11. https://doi.org/10.3389/fbuil.2022.882942
5. Zhang Z, Wong YC, Arulrajah A, Horpibulsuk S (2018) A review of studies on bricks using alternative materials and approaches. Constr Build Mater 188:1101–1118. https://doi.org/10.1016/j.conbuildmat.2018.08.152
6. Tarun J, Karade SR, Singh LP (2019) A review of the properties of hemp concrete for green building applications. J Clean Prod 239:117852. https://doi.org/10.1016/j.jclepro.2019.117852
7. Prusty JK, Patro SK (2015) Properties of fresh and hardened concrete using agro-waste as partial replacement of coarse aggregate—a review. Constr Build Mater 82:101–113. https://doi.org/10.1016/j.conbuildmat.2015.02.063
8. Celik MY, Celik F (2014) Turkish tea: the only tea washed with snow that snow falls on. Int J Basic Clin Stud 3(2):1–15
9. Moyer K (2021) The U.S. produces over 70% of the world's mint. AgHires. https://blog.aghires.com/u-s-produces-70-worlds-mint/. 28 August 2022
10. Faysse N, Rais I, Mekki AAE, Jourdain D (2017) Prospects for a certified mint supply chain in morocco based on an assessment of consumers' willingness to pay. New Medit 16(2):47–54
11. Research, Department Statista (2020) Global per capita tea consumption as of 2016, by Country. https://www.statista.com/statistics/507950/global-per-capita-tea-consumption-by-country/. 24 Jan 2020
12. Mouatassim C, Horma O, El A, Mezrhab A, Karkri M (2022) Thermophysical characteristics of cement-based mortar incorporating spent coffee grounds. Mater Today: Proc 57:867–870. https://doi.org/10.1016/j.matpr.2022.02.488
13. Lachheb A, Allouhi A, El Marhoune M, Saadani R, Kousksou T, Jamil A, Rahmoune M, Oussouaddi O (2019) Thermal insulation improvement in construction materials by adding spent coffee grounds: an experimental and simulation study. J Clean Prod 209:1411–1419. https://doi.org/10.1016/j.jclepro.2018.11.140
14. Samir B, Driss Z, Djendel M, Raouache E, Boubaaya R (2020) Thermo-Mechanical characterization of a bio-composite mortar reinforced with date palm fiber. J Eng Fibers Fabrics 15. https://doi.org/10.1177/1558925020948234
15. Ajouguim S, Talibi S, Djelal-Dantec C, Hajjou H, Waqif M, Stefanidou M, Saadi L (2019) Effect of Alfa fibers on the mechanical and thermal properties of compacted earth bricks. Mater Today: Proc 37:4049–4057. https://doi.org/10.1016/j.matpr.2020.07.539
16. Charai M, Salhi M, Horma O, Mezrhab A, Karkri M (2022) Thermal and mechanical characterization of adobes bio-sourced with pennisetum setaceum fibers and an application for modern buildings. Constr Build Mater 326(August 2021):126809. https://doi.org/10.1016/j.conbuildmat.2022.126809
17. El Hammouti A, Charai M, Channouf S, Horma O, Mezrhab A, Karkri M, Tankari MA (2023a) Application analysis and environmental impact of straw reinforced gypsum plaster for improving the energy efficiency in buildings in the six climate zones of Morocco. J Build Eng 74. https://doi.org/10.1016/j.jobe.2023.106829
18. Mishra Rajesh BK, Behera MM, Petru M (2021) Finite element modeling based thermodynamic simulation of aerogel embedded nonwoven thermal insulation material. Int J Therm Sci 164:106898. https://doi.org/10.1016/j.ijthermalsci.2021.106898

19. El hammouti A, Charai M, Channouf S, Horma O, Nasri H, Mezrhab A, Karkri M, Tankari MA (2023b) Laboratory-testing and industrial scale performance of different clays from Eastern Morocco for brick manufacturing. Constr Build Mater 370. https://doi.org/10.1016/j.conbuildmat.2023.130624
20. Sun C, Shu S, Ding G, Zhang X, Xinghua Hu (2013) Investigation of time lags and decrement factors for different building outside temperatures. Energy Build 61:1–7. https://doi.org/10.1016/j.enbuild.2013.02.003
21. Horma O, Charai M, Mezrhab A, Karkri M (2020) Thermal characterization of cement-plaster-expanded polystyrene composites. In: 2020 5th International conference on renewable energies for developing countries (REDEC), pp 1–6. https://doi.org/10.1109/REDEC49234.2020.9163883
22. Horma O, El Hammouti A, El Hassani S, Kirgiz MS, Channouf S, Moussaoui MA, Mezrhab A (2024a) A sustainable approach to mortar production using mint stem aggregates: investigating the thermal, mechanical and physical performance. Mater Civil Eng. https://doi.org/10.1061/JMCEE7/MTENG-17828
23. Horma O, Channouf S, El Hammouti A, El Hassani S, Miri H, Moussaoui MA, Mezrhab A (2024b) Enhancing concrete sustainability using crushed peanut shells: an analysis of thermophysical properties, durability, and application potential in construction. J Build Eng 90. https://doi.org/10.1016/j.jobe.2024.109434
24. El Hammouti A, Channouf S, Charai M, Horma O, Miri H, Mezrhab A, Karkri M, Tankari MA (2023c) Resource deposit, characterization and energy saving potential of olive pomace as a promising aggregate for energy efficient earth bricks in Eastern Morocco. Constr Build Mater 393. https://doi.org/10.1016/j.conbuildmat.2023.131989

Evaluating Moroccan Soils for Sustainable Building Applications: An Experimental and Computational Analyses

Boutahar Laaouar, Othmane Horma, Aboubakr El Hammouti, Sara El Hassani, Mohammed Amine Moussaoui, and Ahmed Mezrhab

Abstract The global demand for energy in the building sector has significantly increased, primarily due to the widespread use of air conditioning to maintain comfortable temperatures. This growing demand underscores the urgent need to adopt sustainable practices, with a particular focus on developing net-zero or green buildings to promote environmental responsibility. In this chapter, we explore the suitability of three soil types sourced from the Anoual region (32° 40′ 41″ N, 3° 05′ 42″ W) of Morocco for building bricks. We comprehensively analyze each soil type's compositions and microstructural features, employing granulometric analysis, scanning electron microscopy, X-ray diffraction, and X-ray fluorescence to uncover their complete physicochemical properties. Additionally, we conduct detailed investigations of the thermal properties of the studied soil samples. Using the Hot Disk method, we determine parameters such as thermal conductivity, diffusivity, and specific heat capacity, providing insights into the thermal behavior of these soils under various conditions. Finally, to assess the thermal performance of these soils in practical applications, we conducted a Computational Fluid Dynamics (CFD) analysis using ANSYS software to demonstrate their impact on improving energy efficiency in buildings.

Keywords Adobe construction · Numerical simulation · Physicochemical characterization · Soil analysis · Thermophysical properties

---

B. Laaouar (✉) · O. Horma · M. A. Moussaoui · A. Mezrhab
Mechanics and Energy Laboratory, Mohammed First University, Oujda, Morocco
e-mail: boutahar.laaouar@ump.ac.ma

A. El Hammouti
Institut de Recherche de la Construction, ESTP, Cachan, France

S. El Hassani
Laboratory of Energetics, Abdelmalek Essaâdi University, Tetouan, Morocco

F. Latrache et al. (eds.), *Digital Technology for an Innovative Energy Transition*, Green Energy and Technology, https://doi.org/10.1007/978-3-031-89657-6_11

1 Introduction

Decarbonizing the energy sector by 2050 is crucial, with the construction industry playing a significant role in this effort [1]. According to the International Energy Agency (IEA), buildings accounts for approximately 36% of total energy consumption and nearly 40% of greenhouse gas (GHG) emissions. Research indicates that raw earth materials can reduce construction energy use and improve indoor comfort. These materials, derived from natural sources such as local soils, are produced through straightforward in situ processes. Historically, raw earth-based materials have been widely used due to their excellent hygrothermal properties, abundance, and ecological benefits. However, selecting an eco-material made from raw earth requires meeting specific criteria regarding properties related to heat transfer.

A promising technique that has been extensively studied over the last two decades to improve the insulation properties of materials involves incorporating bio-sourced materials into conventional binders such as clay, cement, or gypsum [2–4]. However, determining novel deposits of these binders and their characterization is an essential step before investigating their use in the conception of energy-efficient bio-composites. About 30–50% of the world's population, particularly in parts of Africa, Asia, and Latin America, still live in earthen structures [5]. This highlights the ongoing relevance and potential of earthen materials in modern construction. Recently, there has been renewed interest in these structures due to growing concerns about the negative environmental impacts of carbon dioxide emissions from modern building materials. As an environmentally friendly material, earth can be sourced locally, ensuring sustainability in construction. Additionally, earthen structures offer excellent thermal and moisture insulation, effectively storing heat and balancing air humidity within buildings [6].

In addition to the incorporation of bio-sourced materials into clay, the building industry also supports integrating phase change materials (PCMs) into building envelopes using advanced technologies, as evidenced by successful proof-of-concept studies [7, 8]. The integration of PCMs into building walls by Anter et al. [9] improved thermal performance in hot climates, as shown by the case study in Aswan, Egypt. Various PCM types and thicknesses were tested, with RT-35HC showing the best results. Optimal placement was found to be 1.5 cm from both wall surfaces, reducing indoor temperatures and energy gain significantly. This setup achieved a 66% reduction in overall energy gain during summer. A recent review study has compiled and discussed PCM-related encapsulation methods [10]. The findings show that, although promising for thermal storage, PCM microcapsules require improvements in encapsulation efficiency, particle size reduction, and standard characterization methods.

Locally sourced building materials are increasingly replacing expensive and energy-intensive manufactured products in sustainable construction [11, 12]. Dabaieh et al. [13] demonstrated that unfired clay materials are essential for mitigating GHG emissions by comparing the life cycle carbon emissions between fired and sun-dried clay bricks. Sun-dried bricks require 0.033 MJ and emit 0.24 kg CO_2

per 1000 bricks, whereas fired bricks need 4250 MJ and emit 5502 kg CO_2. This highlights the potential for substantial reductions in CO_2 emissions and energy use by choosing sun-dried bricks for construction. Previous research highlights the impact of soil composition on brick thermal performance. A study by El Hammouti et al. [14] compares the effect of soil composition on the thermal performance of unexploited clay-based building material. Mineralogical, geotechnical, and thermal analyses were conducted on two clay soils from different sites in eastern Morocco. The thermal properties of the developed bricks were tested, and the results were used in building performance simulations to evaluate and compare the energy efficiency of a residential building in semi-arid Moroccan climates.

This study compares the mineralogical, chemical, and thermophysical properties of three types of soils used in brick manufacture. These eco-friendly materials have no negative impact on the environment, and their simple preparation process makes them cost-effective. This approach helps reduce construction costs, improve project feasibility, and provide affordable housing solutions. To further reduce energy consumption, this study represents an initial step toward developing sustainable building practices. Future research will focus on fiber-reinforced bricks, aiming to mechanically strengthen the structure and improve the thermal properties of these materials.

2 Methodology

2.1 Geotechnical and Chemical Characterizations of the Soils

The raw earth materials were sourced from the Anoual region (32° 40′ 41″ north, 3° 05′ 42″ west) of Morocco. Two methods were utilized to evaluate the particle size distribution of the soils: Sedimentation analysis for particles smaller than 0.075 mm in diameter and Sieve analysis for particles ranging from 0.075 mm to 100 mm. The findings displayed in Fig. 1, shows that the particle size distribution aligns with the recommended range stipulated by the standard for compressed earth blocks XP 13–901 [15].

According to the Atterberg limit test results, the soils' plasticity adheres to the recommended values as per the standard. The red soil, with a liquid limit of 32.66%, plastic limit of 22.66%, and plasticity index of 10, has a density of 1994 kg/m^3, making it moderately plastic and suitable for load-bearing applications. Beige soil, having a liquid limit of 25.75%, plastic limit of 18.01%, and plasticity index of 7.74, is less compact with a density of 1398 kg/m^3, suitable for lighter construction needs. Gray soil, the most plastic with a liquid limit of 36.33%, plastic limit of 23.48%, and plasticity index of 12.85, has the highest density of 2117 kg/m^3, making it ideal for high stability and robust structural support.

The physicochemical composition of the three soil samples was determined using XRD, scanning electron microscopy, and energy dispersive X-ray spectroscopy (SEM–EDS). The XRD diffractogram, given in Fig. 2, presents the mineralogical

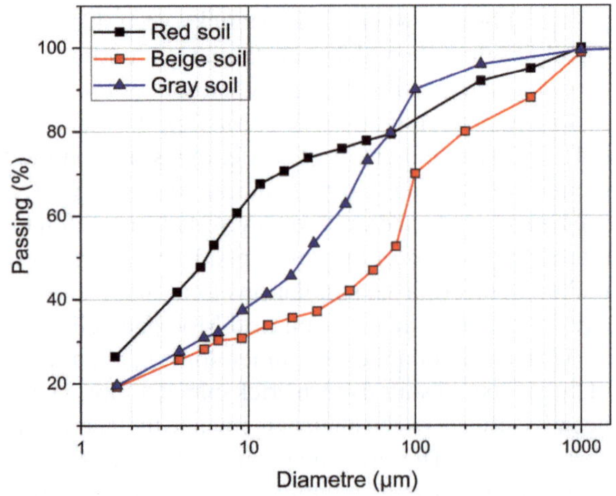

Fig. 1 Particle size distribution of clayey samples

composition of our samples. As expected from earth-based materials, quartz and calcite are the dominant components in all the studied soils.

The soil samples predominantly contain quartz (SiO_2), alongside calcite, feldspar, and chlorite. This abundance of SiO_2 suggests the clay's suitability for construction. Understanding silicon dioxide's (SiO_2) role is crucial for evaluating the clay's construction potential, impacting properties like strength, durability, and thermal insulation. Elevated SiO_2 content enhances mechanical strength, reduces water absorption, and improves thermal insulation in clay materials [16]. Figure 3 depicts the micromorphology of our clayey samples, as well as the chemical composition yielded through EDS, and confirms the XRD results as the elements Si and O are predominant in all the studied soils.

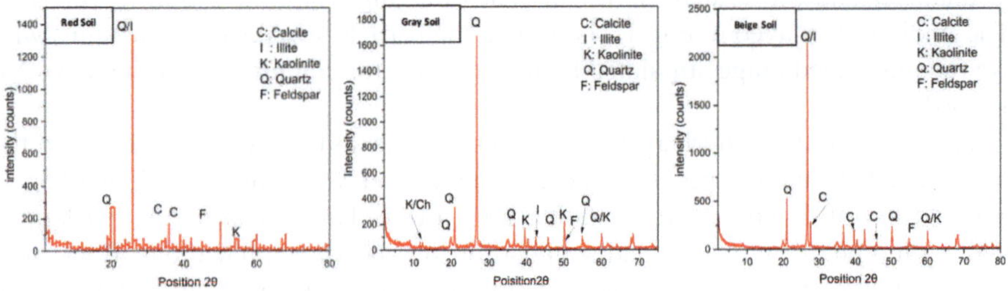

Fig. 2 X-ray diffraction of the soils

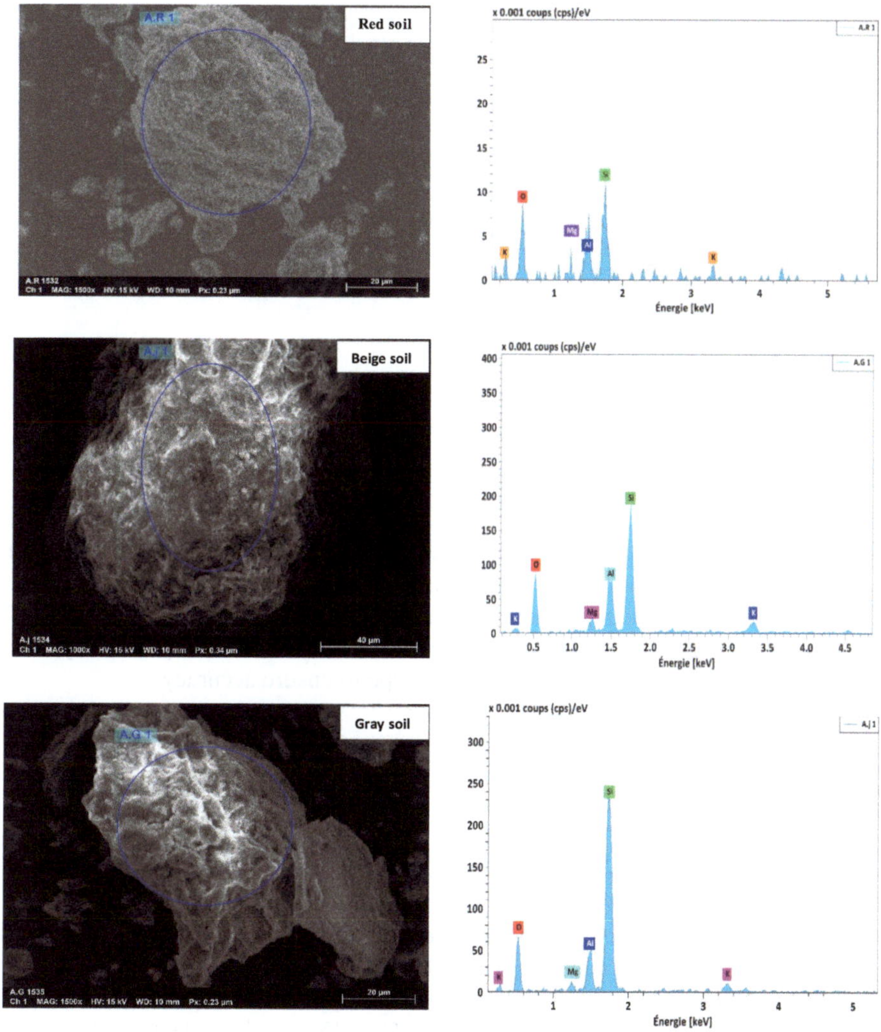

Fig. 3 SEM images and EDS of the studied soil types

2.2 *Thermal Test*

In this study, we determined the thermal properties of the produced adobes using the TPS 2200 Hot Disk device (Hot Disk, Sweden) (Fig. 4), which simultaneously evaluates thermal conductivity, thermal diffusivity, and volumetric heat capacity. Measurements involved placing a TPS double spiral Kapton sensor (ref. 5501), with a radius of 6.403 mm, between two cylindrical samples from each soil type, each measuring 2 cm in height and 6 cm in diameter. To ensure thermal stability, the TPS sensor was clamped manually between the samples. The sensor was then heated with

Fig. 4 Measurement of the thermal conductivity

a constant power of 80 to 150 mW for 80 s. The temperature increase of the samples over time was tracked using the same sensor, and the thermal transport properties were simultaneously calculated using the Hot Disk Analyzer software version 7.1.5, based on the recorded time-dependent temperature rise. To ensure repeatability, three measurements were conducted for each soil type to ensure accuracy.

2.3 Simulation of Adobe Walls

In the second phase of our study, we conducted a comparative analysis of adobe walls constructed with different soil types (Fig. 5). Using ANSYS FLUENT software, we simulated and analyzed the thermal behavior of three distinct wall types over a 24-h period [17]. This simulation was based on data from a specific day, July 23, 2023, in Oujda, Morocco, which experienced significant temperature variations, with outdoor temperatures ranging from 26 °C to 45 °C. This selection allowed us to demonstrate the effectiveness of our materials in regulating substantial thermal fluctuations. ANSYS FLUENT, a numerical simulation tool, is commonly used to solve fluid dynamics and thermodynamics equations [18]. The mathematical model and boundary condition equations used to approximate the temperature variations within the software are provided below:

- The Boundary conditions:

$$-\lambda\frac{\partial T}{\partial x} = h_{\text{ext}}(T_{sa} - T_e) \quad \text{For } x = 0 \tag{1}$$

$$-\lambda\frac{\partial T}{\partial x} = h_{\text{int}}(T_i - T_{\text{int}}) \quad \text{For } x = e \tag{2}$$

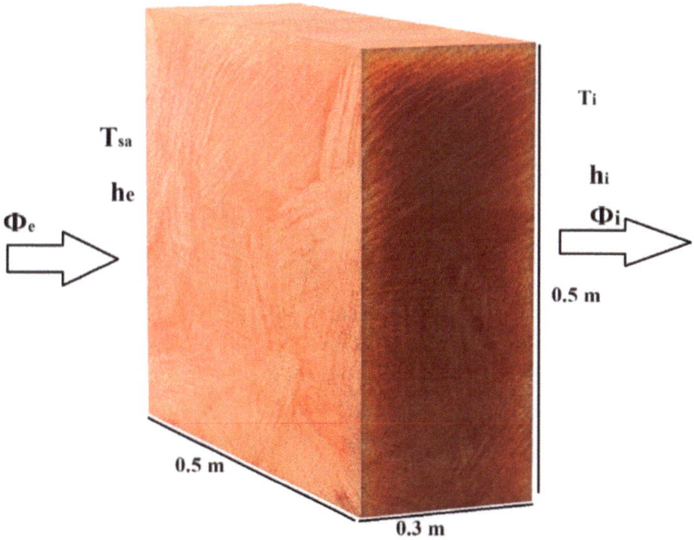

Fig. 5 Process of heat flux transfer across the studied wall

- Mathematical model:

$$T_{as}(t) = \frac{T_{\max} + T_{\min}}{2} + \frac{T_{\max} - T_{\min}}{2} \sin\left(\frac{2\pi t}{p} - \frac{\pi}{2}\right) \tag{3}$$

The convective heat transmission coefficients are denoted by h_{ext} for the exterior and h_{int} for the interior. The temperatures T_e and T_{int} represent the exterior and interior air temperatures, respectively. The outdoor temperature T_{as} is the temperature of the ground and air, which follows a sinusoidal variation with a period p of 24 h. The maximum and minimum outdoor temperatures are denoted by $T_{\max}$ and $T_{\min}$. In this study, we considered the following values: $T_{\max} = 45$ °C; $T_{\min} = 26$ °C; $T_{\text{int}} = 26$ °C; $h_{\text{ext}} = 16$ W/m^{2}K and $h_{\text{int}} = 8$ W/m^{2}K.

3 Results and Discussion

3.1 Thermophysical Properties

The three adobe soils differ markedly in their heat-transfer behavior (Table 1), a direct result of their mineralogy and particle-size distributions. Red soil exhibits the lowest thermal conductivity, reflecting its higher porosity and insulating mineral phases. Thermal diffusivity mirrors this pattern, as red soil responds more slowly to temperature changes than beige or gray soil. Previous studies [19] have found similar

Table 1 Thermophysical properties of the studied adobe soil samples

Soils	Thermal conductivity (W/m.K)	Thermal diffusivity (mm^2/s)	Density (kg/m^3)	Specific heat capacity (J/kgK)
Beige soil	1.022	0.796	1398	917.95
Gray soil	0.963	0.666	2117	681.81
Red Soil	0.845	0.743	1994	655.53

thermophysical properties in clays of eastern Moroccan origins, further supporting these findings.

By using low-conductivity soils in earth bricks, the reliance on cementitious materials, and their associated CO_2 emissions, can be substantially reduced, offering a more sustainable alternative for thermally efficient masonry.

3.2 *Simulation Results*

Prior to presenting the thermal performances of the adobe walls, it is important to mention that our simulation model was validated by aligning our findings with previous studies [3, 20], which used similar mathematical frameworks, ensuring the reliability of the reported time periods and temperatures. The variation of the sol-air temperature, which characterizes the temperature of the outside heat wave, is illustrated in Fig. 6. This heat wave is modeled using a mathematical approximation of the real-time variation of the ambient temperature over a 24-h period on the selected day.

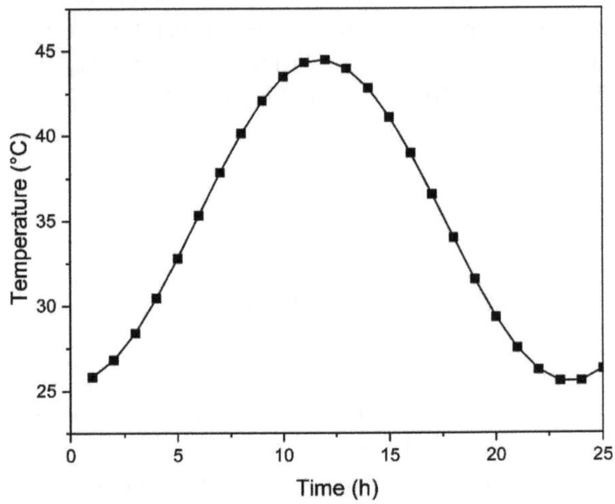

Fig. 6 Hourly variation of the sol-air temperature over a 24-hour period used for the thermal simulation

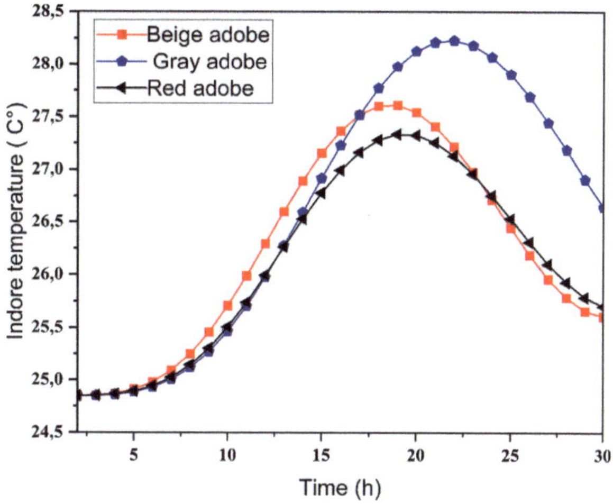

Fig. 7 Average temperature on the inner faces of the wall throughout the day for the three soil types

Figure 7 illustrates the heat distribution across the three wall configurations, highlighting the thermal behavior of each soil type. The red soil adobe wall exhibits superior insulation performance, which is attributed to its lower thermal conductivity compared to the other soil types.

For the proposed walls, we calculated the decrement factor and the time lag [20, 21]. Among the configurations, the red soil adobe wall showed superior performance with a decrement factor (f) of 0.604. This indicates that it provides lower temperature changes compared to the beige soil brick walls ($f = 0.615$) and the gray soil wall ($f = 0.623$). Additionally, the maximum temperature on the inner surface of the walls was reached by the beige and red soil bricks at around 18 h and 19 h, respectively. Even though the gray soil wall exhibits the highest time lag ($\phi \approx 9$ h), its maximum temperature is significantly higher at 28.2 °C. With a maximum temperature of 27.2 °C and a time lag of 6 h, the red adobe wall presents the best thermal performance out of the studied soil types. This indicates that this type of soil can improve occupant comfort inside the building by delaying and reducing the peak inner temperatures.

4 Conclusion

This study investigated new clay deposits from the Anoual region of Morocco, focusing on their mineralogical, chemical, and thermophysical properties. The physicochemical analysis showed high quartz content and standard plasticity, indicating the soil's strength and suitability for construction.

Experimental results revealed that red soil had the lowest thermal conductivity, suggesting its superior potential as an insulating material compared to gray and beige soils. Thermal simulations further showed that adobe walls built with red soil achieved the best performance, reducing the thermal decrement factor and extending the time lag, thereby smoothing indoor temperature swings and enhancing occupant comfort.

Future research should explore integrating biomass and fiber into the red soil to develop lightweight, energy-efficient, eco-friendly, and low-cost adobe bricks. Additionally, investigating the incorporation of phase change materials (PCMs) for latent heat storage within adobe walls could further enhance their thermal performance. PCMs offer substantial benefits due to their latent heat storage properties, which can significantly improve thermal regulation.

Overall, the red soil from the Anoual region exhibits excellent potential as a high-performance insulating material for sustainable building applications, warranting further investigation and development.

5 Competing Interests

The authors declare that they have no known competing financial interests or personal relationships that could have appeared to influence the work reported in this paper.

References

1. Jagemann C, Fursch M, Hagspiel S, Nagl S. Decarbonizing Europe's Power Sector by 2050—analyzing the implications of alternative decarbonization pathways.
2. El hammouti A, Charai M, Channouf S, Horma O, Mezrhab A, Karkri M, Tankari MA (2023) Application analysis and environmental impact of straw reinforced gypsum plaster for improving the energy efficiency in buildings in the six climate zones of Morocco. J Build Eng 74:106829. https://doi.org/10.1016/j.jobe.2023.106829
3. Horma O, Channouf S, El Hammouti A, El Hassani S, Miri H, Moussaoui MA, Mezrhab A (2024) Enhancing concrete sustainability using crushed peanut shells: an analysis of thermo-physical properties, durability, and application potential in construction. J Build Eng 90:109434. https://doi.org/10.1016/j.jobe.2024.109434
4. Horma O, Charai M, El Hassani S, El Hammouti A, Mezrhab A (2022) Thermo-physical and mechanical characterization of cement-based mortar incorporating spent tea. J Build Eng 52:104392. https://doi.org/10.1016/j.jobe.2022.104392
5. Terra literature review: an overview of research in earthen architecture conservation
6. Zak P, Ashour T, Korjenic A, Korjenic S, Wei Wu (2016) The influence of natural reinforcement fibers, gypsum and cement on compressive strength of earth bricks materials. Constr Build Mater 106:179–188. https://doi.org/10.1016/j.conbuildmat.2015.12.031
7. Aridi R, Yehya A (2022) Review on the sustainability of phase-change materials used in buildings. Energy Conver Manage X 15:100237. https://doi.org/10.1016/j.ecmx.2022.100237
8. Guermat Z, Kabar Y, Kuznik F, Boukelia TE (2024) Numerical investigation of the integration of new bio-based PCM in building envelopes during the summer in Algerian cities. J Energy Storage 79:110111. https://doi.org/10.1016/j.est.2023.110111
9. Anter AG, Sultan AA, Hegazi AA, El Bouz MA (2023) Thermal performance and energy saving using phase change materials (PCM) integrated in building walls. J Energy Storage 67:107568. https://doi.org/10.1016/j.est.2023.107568

10. Phase Change Material (PCM) Microcapsules for Thermal Energy Storage—Peng (2020) Advances in polymer technology. Wiley Online Library. https://onlinelibrary.wiley.com/doi/full/https://doi.org/10.1155/2020/9490873 (11 juin 2024)
11. Althoey F, Ansari WS, Sufian M, Deifalla AF (2023) Advancements in low-carbon concrete as a construction material for the sustainable built environment. Dev Built Environ 16:100284. https://doi.org/10.1016/j.dibe.2023.100284
12. Dsilva J, Zarmukhambetova S, Locke J (2023) Assessment of building materials in the construction sector: a case study using life cycle assessment approach to achieve the circular economy. Heliyon 9(10):e20404. https://doi.org/10.1016/j.heliyon.2023.e20404
13. Dabaieh M, Heinonen J, El-Mahdy D, Hassan DM (2020) A comparative study of life cycle carbon emissions and embodied energy between sun-dried bricks and fired clay bricks. J Clean Prod 275:122998. https://doi.org/10.1016/j.jclepro.2020.122998
14. Hammouti El, Aboubakr MC, Mezrhab A, Nasri H, Karkri M (2022) Effect of the nature of clayey soils (Eastern Morocco) on the thermophysical properties of bricks. Mater Today: Proc 60:321–327. https://doi.org/10.1016/j.matpr.2022.01.233
15. AFNOR XP (2001) Compressed earth blocks for walls and partitions: definitions-specifications-test methods-delivery acceptance conditions. Saint-Denis La Plaine Cedex (29 mai 2024)
16. Changizi F, Haddad A (2015) Strength properties of soft clay treated with mixture of nano-SiO_2 and recycled polyester fiber. J Rock Mech Geotech Eng 29. https://doi.org/10.1016/j.jrmge.2015.03.013
17. Mishra R, Behera BK, Muller M, Petru M (2021) Finite element modeling based thermodynamic simulation of aerogel embedded nonwoven thermal insulation material. Int J Therm Sci. https://doi.org/10.1016/j.ijthermalsci.2021.106898
18. El hammouti A, Charai M, Channouf S, Horma O, Nasri H, Mezrhab A, Karkri M, Tankari MA (2023) Laboratory-testing and industrial scale performance of different clays from eastern Morocco for brick manufacturing. Constr Build Mater 370:130624. https://doi.org/10.1016/j.conbuildmat.2023.130624
19. Chraibi H, El Abbassi F-E, Sakami S, Bouferra R, Kchikach A (2023) Thermal improvement of compressed earth blocks by using liquid and solid olive oil wastes in the Region of Marrakesh, Morocco. Euro-Mediterranean J Environ Integration 8(4):935–948. https://doi.org/10.1007/s41207-023-00420-z
20. Mellaikhafi A, Tilioua A, Benallel A (2022) Thermal performance assessment of a wall built with earth-based adobes and reinforced with pinnate leaves fibers. Mater Today: Proc 58:1535–1540. https://doi.org/10.1016/j.matpr.2022.03.296
21. Sun C, Shu S, Ding G, Zhang X, Xinghua Hu (2013) Investigation of time lags and decrement factors for different building outside temperatures. Energy Build 61:1–7. https://doi.org/10.1016/j.enbuild.2013.02.003

Enhancing Energy Harvesting Efficiency in Autonomous Devices Using Dual Pyro-Piezoelectric Materials

Mohammed Remaidi, Meryiem Derraz, Hassan Radoine, Mohamed Aymen Ben Achour, Chouaib Ennawaoui, and Abdelowahed Hajjaji

Abstract This chapter contributes to energy harvesting techniques, which are particularly interesting for ensuring the autonomy of various millimeter-scale and microscopic devices using current generators. It provides a concise description of pyroelectric materials, which constitute a subgroup of piezoelectric solids. It explores the origin of pyroelectric and piezoelectric conversions through a Maxwellian approach, thus offering a detailed explanation of how these two phenomena manifest in a ferroelectric crystal. Additionally, it delves deeply into the merit factors associated with these two effects. Subsequently, the chapter focuses on optimizing the power recovered by a purely resistive load, utilizing the equivalent electrical schematic of a material possessing both pyroelectric and piezoelectric properties. This allows us to model the thermomechanical losses associated with binding the electrical response to the physical characteristics of the material and the applied excitation. A numerical simulation was developed for PZT and $LiTaO_3$ to support the theoretical analysis. Finally, the obtained results are applied to identify a material capable of harvesting the maximum amount of energy.

Keywords Energy harvesting · Self-powered systems · Optimal power · Hybrid crystal · Ferroelectric material · Pyroelectricity · Piezoelectricity · Figures of merit · Thermomechanical losses · Maxwellian approach · Single crystal $LiTaO_3$ · Crystalline ceramic PZT

M. Remaidi (✉) · M. Derraz · C. Ennawaoui · A. Hajjaji
Laboratory of Engineering Sciences for Energy, National School of Applied Sciences, Chouaib Doukkali University, El Jadida, Morocco
e-mail: remaid.mohammed@ucd.ac.ma

H. Radoine
Mohammed VI Polytechnic University, Marrakesh, Morocco

M. A. B. Achour
Polytechnic University of Hauts-de-France, Valenciennes, France

F. Latrache et al. (eds.), *Digital Technology for an Innovative Energy Transition*, Green Energy and Technology, https://doi.org/10.1007/978-3-031-89657-6_12

Nomenclature

S	Area, [m^2]
e	Thickness, [m]
V	Voltage, [V] and volume, [m^3]
D	Electric displacement, [$C\cdot m^{-2}$]
σ_{ij}	Stress, [Pa]
P	Polarization, [$C\cdot m^{-2}$]
σ	Electrical conductivity, [$S\cdot m^{-1}$]
ρ	Charge volume density, [$C\cdot m^{-3}$]
C	Capacitance, [F]
i	Current, [A]
Q	Electrical charge, [C]
j	Current area density, [$A\cdot m^{-2}$]
R	Electrical resistance, [Ω]
p	Pyroelectric coefficient, [$C\cdot m^{-2}\cdot K^{-1}$]
T	Temperature, [K]
t	Time, [s]
ω	Pulsation, [$rad\cdot s^{-1}$]
ε_3	Permittivity of polar material, [$F\cdot m^{-1}$]
c_E	Volumetric heat capacity, [$J\cdot K^{-1}\cdot m^{-3}$]
d	Piezoelectric coefficient, [$C\cdot m^{-2}\cdot Pa^{-1}$

Subscript and Superscript

i	Initial
f	Final
opt	Optimal
int	Interior
c	Free
pyr	Pyroelectric
pie	Piezoelectric
n	Normal vector
T	Total

1 Introduction

For a long time, batteries have consistently played a predominant role as the primary source of power to supply low-power electronic devices, medical devices, and various embedded systems. The harnessing of renewable energy sources is emerging as a response to the challenges of the energy crisis and the growing demand for energy resources.

Energy harvesting techniques prove particularly interesting for a wide variety of self-powered systems [1], such as sensors [2, 3], monitoring devices, and biomedical devices, which require no maintenance or replacement. In recent years, the demand for energy has seen a significant increase due to the exponential growth in the number of devices. The requirements imposed by these numerous devices place a considerable burden on existing resources.

Various millimeter-scale and microscopic devices have been designed, with average power ranging from 10 to 10 mW [2–5]. Examples of microscopic energy harvesting include piezoelectric generators [6–9], which transform mechanical oscillations ($d\sigma/dt$, where σ represents mechanical stress) into electrical energy [10], thermoelectric microgenerators [11–13], photovoltaic devices (solar cells) [14], and electromagnetic microgenerators [15–17]. Pyroelectric generators are particularly suited for converting thermal energy into electrical energy compared to thermoelectric devices because they require a temporal temperature gradient (dT/dt) rather than a spatial temperature gradient (dT/dx), which is more complex to achieve on the micro/nano scale [18] (Fig. 1).

Furthermore, recent approaches have been developed to transform stationary spatial gradients into transient temperature gradients [20], facilitating the creation of energy harvesters based on pyroelectricity or radiation with pyroelectric effect [21]. Pyroelectric energy harvesting devices require little or no maintenance, unlike piezoelectric devices, as they do not involve any moving parts (Fig. 2).

Some materials with piezoelectric properties also exhibit pyroelectric characteristics, noting that ferroelectricity is a subcategory of pyroelectricity. Pyroelectricity expresses the ability of a material to change its polarization in response to a temperature variation. However, this effect is relatively limited compared to the piezoelectric effect, with an electrothermal coupling coefficient generally less than 10%, while

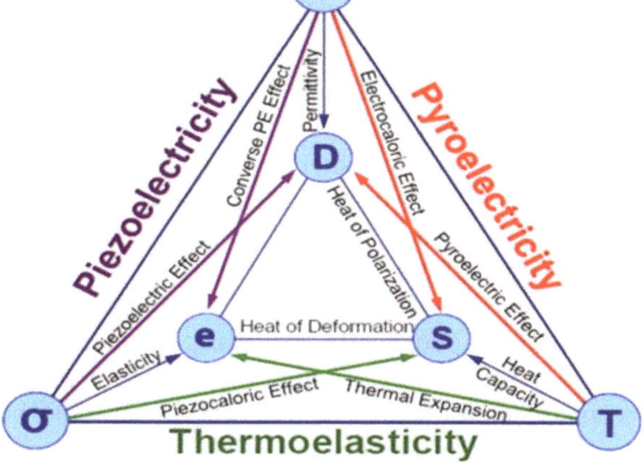

Fig. 1 Thermal (T), electrical (E), and mechanical (σ) energy states of a crystal are associated with the fundamental energy conversions: electric displacement (D), entropy (S), and strain (e) [19]

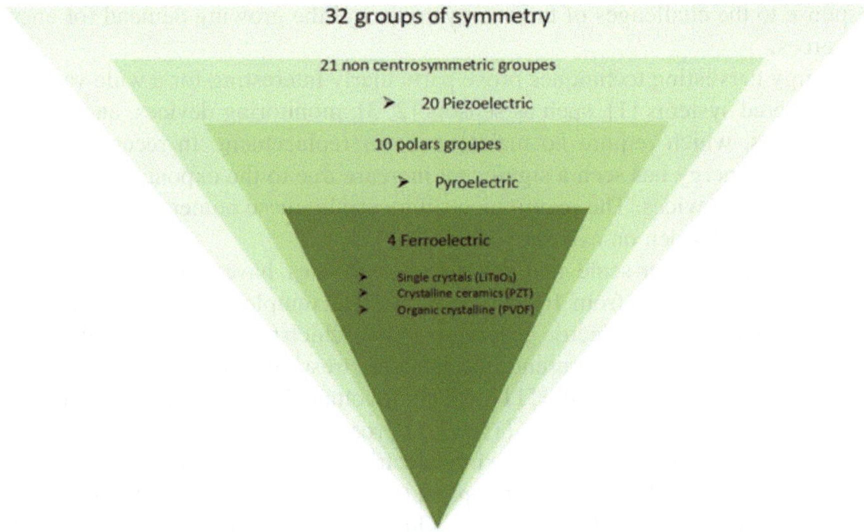

Fig. 2 Ferroelectric materials are a subgroup of pyroelectric and piezoelectric materials

piezoelectricity can reach up to 90%. It is worth noting that despite this limitation, the energies involved in thermal exchanges are often considerable compared to the mechanical energy of a vibrating system, which can have a significant impact, especially in the context of energy harvesting.

Despite the marked similarity between piezoelectricity and pyroelectricity, several distinctions exist between the two phenomena. Firstly, piezoelectricity can manifest along multiple axes (hence tensor notation), while pyroelectricity requires only matrix notation, with temperature T being a scalar. The origin of both piezoelectric and pyroelectric effects lies in the variation of spontaneous polarization under the influence of temperature or mechanical stress in certain anisotropic crystalline materials [22–24]. This effect refers to the generation of current or electric potential in non-centrosymmetric crystalline structures when subjected to a temporal temperature gradient or pressure. Pyroelectricity manifests in crystals belonging to polar classes 1, 2 m, 2 mm, 3, 3 m, 4, 4 mm, 6, and 6 mm [25, 26]. ∞,∞m [27].

In recent years, alongside implementation and experimental advances, various theories and numerical attempts have emerged to optimize piezoelectric and pyroelectric energy harvesting using conservation equations for charge, momentum, and energy [28–31]. These studies are of crucial importance in guiding future experimental efforts.

This work contributes to energy harvesting techniques by providing a concise description of pyroelectric materials, which constitute a subgroup of piezoelectric solids. It explores the origin of pyroelectric and piezoelectric conversions for non-centrosymmetric crystals, establishing a connection between microscopic and macroscopic characteristics through a Maxwellian approach. This allows for a detailed explanation and theoretical analysis of the manifestations of these two phenomena in a hybrid crystal, whether subjected to thermal or mechanical excitation, while deeply examining associated merit factors such as crystal reactivity, stored energy, and coupling coefficients. Subsequently, the chapter focuses on optimizing the power recovered by a purely resistive load, using the equivalent electrical schematic of a material possessing both pyroelectric and piezoelectric properties. Power is derived as a function of load to determine its maximum value, thereby enabling the modeling of thermomechanical losses associated with the coupling of electrical response to the material's physical characteristics and applied excitation. Furthermore, a numerical simulation is performed for PZT and $LiTaO_3$ to support the theoretical analysis. Finally, the obtained results are leveraged to identify a material capable of recovering the maximum amount of energy.

2 Hybrid Piezo-Pyroelectric Crystal Characterization: Maxwellian Approach

Although the system may be electrically neutral on a macroscopic scale, on a microscopic scale, the centrosymmetry of internal charges is not necessarily present, and electrostatic dipoles can exist (Fig. 3a), either spontaneously or induced by the application of an electric field. This characteristic arises from the nature of the chemical bonding present in certain types of crystals, which combine covalent and ionic aspects [32], thus generating particularly rich physical properties. Phenomena such as piezoelectricity and pyroelectricity are observed, for example, in materials such as $BaTiO_3$ and PZT [32]. Phenomena such as piezoelectricity and pyroelectricity are observed, for example, in materials such as $BaTiO_3$ and PZT.

The electric field outside (or inside) the ferroelectric crystal introduces free charges at the surface, and it is assumed that the polarization is uniform, meaning that P_s is constant. In the absence of free charges within the volume, it is therefore given by the following expression, where P is the area charge density:

$$\vec{E}(t) = \frac{P(t)}{\varepsilon} \cdot \vec{n} \tag{1}$$

2.1 Equilibrium

Regarding the vertical direction and thermomechanical equilibrium within the ferroelectric phase, the equilibrium state of a polar crystal exhibits a spontaneous surface charge density [24], $P_{\text{int}}(T(t_i))$, consisting of positive charges on the lower face and negative charges on the opposite face (Fig. 3a).

$$E_{T,3}(t_i) = 0 \Leftrightarrow P_{\text{int},3}(T(t_i)) = P_{c,3}(T(t_i)) \tag{2}$$

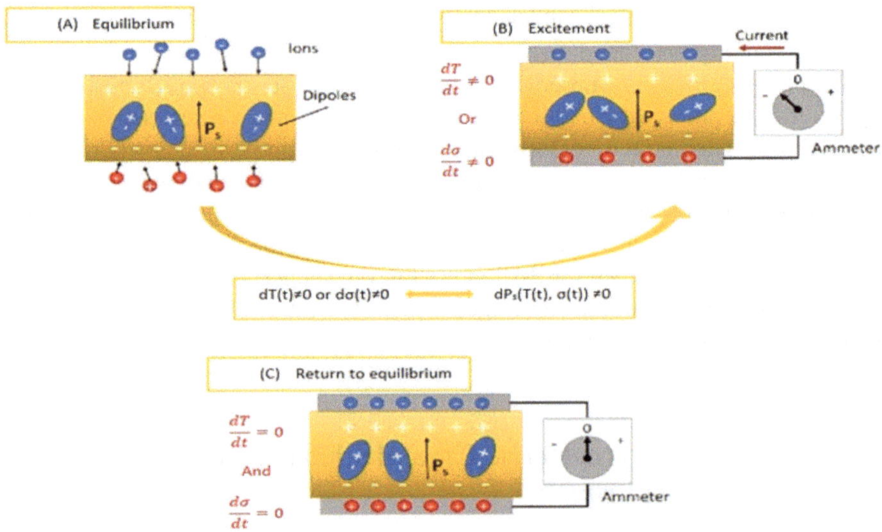

Fig. 3 Polarization changes induced by the temporal variation of temperature or pressure in a polar material

where $P_{c,z}$ is the surface density of free charges; Eq. (2) represents the condition of electrical equilibrium.

2.2 *Dynamic Regime*

The variation of thermomechanical stresses ($dT(t)/dt = 0$ and $d\sigma(t)/dt = 0$) at the material's surface leads to a reduction in the bound charges present on the surfaces [22, 24–26, 33, 34]. This charge imbalance, in the case of an open circuit (Fig. 3b), generates a transient field denoted as $E_T(t)$.

According to the electrical equilibrium condition, the total net field created inside our crystal due to the variation in pressure and temperature is expressed as follows, where p_3, d_{33} and are, respectively, the pyroelectric and piezoelectric coefficients.

$$E_{T,3}(t) = \frac{p_3 \cdot \int_{T(t_i)}^{T(t)} dT(t) + d_{33} \cdot \int_{\sigma(t_i)}^{\sigma(t)} d\sigma(t)}{\varepsilon_3} \tag{3}$$

Here, ε_{33} represents the permittivity of the polar material. This field can be neutralized either by free charges, which can be generated by an internal current [35], or by the application of an external field [36], thus ensuring the restoration of electrical balance (Fig. 3c).

The total electrical filed is uniform along the vertical direction, so the expression of the total potential as follows:

$$V_T(t) = -\frac{\left(p_3 \cdot \int_{T(t_i)}^{T(t)} \mathrm{d}T(t) + \mathrm{d}_{33} \cdot \int_{\sigma(t_i)}^{\sigma(t)} \mathrm{d}\sigma(t)\right) \cdot e}{\varepsilon_3} \quad (4)$$

where e represents the thickness. According to Eq. (4), it is observed that the total potential depends on the thickness, permittivity, as well as the coefficients p_3 and d_{33}, leading to the following expression:

$$|V_T(t)| = F_{\mathrm{pyr}} \cdot \int_{T(t_i)}^{T(t)} \mathrm{d}T(t) + F_{\mathrm{pie}} \cdot \int_{\sigma(t_i)}^{\sigma(t)} \mathrm{d}\sigma(t) \quad (5)$$

F_{pyr} and F_{pie} are, respectively, the thermal and mechanical reactivities of the crystal.

$$F_{\mathrm{pyr}} = \frac{p_3 \cdot e}{\varepsilon_3}$$

$$F_{\mathrm{pie}} = \frac{d_{33} \cdot e}{\varepsilon_3}$$

These two factors make it possible to determine the potential created by a polar material during the state of thermal or mechanical excitation [20, 37, 38].

The formulation of the charge generated by the crystal can be expressed in terms of the electric displacement vector as follows:

$$\oint \vec{D}_T \cdot \vec{n} \cdot \mathrm{d}S = Q_T(t) \quad (6)$$

where

$$Q_T(t) = \left(p_3 \cdot \int_{T(t_i)}^{T(t)} \mathrm{d}T(t) + \mathrm{d}_{33} \cdot \int_{\sigma(t_i)}^{\sigma(t)} \mathrm{d}\sigma(t)\right) \cdot S \quad (7)$$

The total current induced in the polar material under thermomechanical excitation can be described using Eq. (8). This expression aligns with the one presented in the Derraz et al. [6]:

$$i_T(t) = \left(p_3 \cdot \frac{\partial T(t)}{\partial t} + \mathrm{d}_{33} \cdot \frac{\partial \sigma(t)}{\partial t}\right) \cdot S \quad (8)$$

In an open circuit, an increase in electrical potential, considering the change in temperature or pressure, results in an increase in the associated electrical energy. Since the change in energy ($W(t)$) stored in a planar dielectric is given by 1/ $2 \cdot C_p V^2{}_T(t)$, an analogous approach was employed to derive selection criteria for piezoelectric vibration harvesting materials [39]. Therefore:

$$W_T(t) = \frac{1}{2}\left(w_{\text{pyr}} \cdot \int_{T(t_i)}^{T(t)} \mathrm{d}T(t) + w_{\text{pie}} \cdot \int_{\sigma(t_i)}^{\sigma(t)} \mathrm{d}\sigma(t) \right) \tag{9}$$

With:

$$w_{\text{pyr}} = p_3 \sqrt{\frac{S \cdot e}{\sigma_p \cdot \tau}}$$

$$w_{\text{pie}} = \mathrm{d}_{33} \sqrt{\frac{S \cdot e}{\sigma_p \cdot \tau}}$$

With: σ_p and τ being the electrical conductivity and dielectric relaxation time, respectively. These two factors allow us to relate the material's reactivities to dimensional parameters and the dielectric relaxation time, which plays a significant role in charge recovery during excitation.

Equation (9) indicates that even if the material is subjected to high mechanical or thermal excitation, this does not automatically guarantee a high-energy conversion with a pyroelectric and/or piezoelectric energy recovery system. Therefore, to enhance the material selection process, electrothermal and electromechanical coupling coefficients have been proposed. These coefficients consider the specific response of the polar material to the excitation, taking into account the thermal and mechanical parameters specific to each material.

$$k^2 = \frac{W}{E_{\text{exci}}} \tag{10}$$

The thermoelectric coupling coefficient expresses the ratio of thermal energy converted into electricity [40, 41], allowing us to incorporate the impact of thermal parameters on the pyroelectric response.

$$k_{\text{pyr}}^2 = \frac{T_{\text{hot}} \cdot p_3^2}{c_E \cdot \varepsilon_3} \tag{11}$$

Here, Thot represents the available temperature at the material's surface [29], c_E corresponds to volumetric heat capacity.

By analogy, the effect of mechanical parameters on the direct piezoelectric response is described by the mechanical–electrical coupling parameter, where s_{33} represents the mechanical compliance.

$$k_{\text{pie}}^2 = \frac{\mathrm{d}_{33}^2}{s_{33} \cdot \varepsilon_3} \tag{12}$$

Table 1 presents the numerical values of the dielectric constant, Curie temperature, as well as the pyroelectric and piezoelectric coefficients of two polar materials that exhibit both piezoelectric and pyroelectric properties.

Table 1 Material properties [24, 42–44]

Material\Propriety	$p(\mu\text{C}\cdot\text{K}^{-1}\cdot\text{m}^{2})$	$d_{33}(\mu\text{C}\cdot\text{MPa}^{-1}\cdot\text{m}^{-2})$	ε_r	T_c(°C)
PZT (∞m)	$(0, 0, 570)^T$	593	1116	240
$LiTaO_3$(3 m)	$(0, 0, 190)^T$	8	47	618

3 Harvested Power

Subsequently, the power conservation principle is applied to determine the expression of the power supplied by a material that is both pyroelectric and piezoelectric. However, to simplify our study, we assume that thermal and mechanical fluctuations are applied successively (Fig. 4).

3.1 Perfect Material

The instantaneous power dissipated by Joule effect across the resistor R, through which a current intensity $I(t)$ flows, is formulated as follows:

$$P_T(t) = R \cdot I(t) \tag{13}$$

If we neglect thermal and mechanical losses, and assume that the crystal does not store a portion of the received energy, then the recovered intensity is given by Eq. (8), resulting in the new expression for the recovered power [45]:

$$P_T(t) = R \cdot \left(\left(p_3 \cdot \frac{\partial T(t)}{\partial t} \right)^2 + \left(\mathrm{d}_{33} \cdot \frac{\partial \sigma(t)}{\partial t})^2 \right) \right) \cdot S^2 \tag{14}$$

This formulation highlights that, with constant resistance and area, the instantaneous recovered power depends directly on the surface charge density generated by a

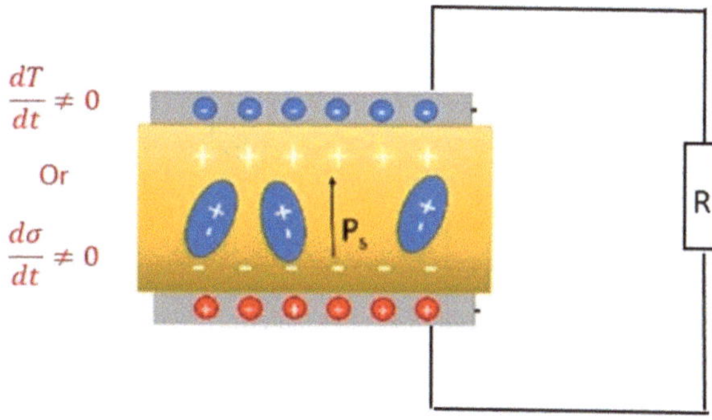

Fig. 4 Analyzed electrical schematic

thermal or mechanical excitation. The dominant factor in increasing energy recovery is the rate of variation of this excitation.

3.2 Real Material

In reality, a polar material, like all other energy conversion materials, exhibits losses. It is then a matter of finding an equivalent electrical circuit that allows us to model the lost and stored parts of the received energy.

The electrical balance between external free charges and internal bound charges in an open circuit can be expressed by the following equation:

$$\sigma \cdot E_c + \frac{\partial D_T(t)}{\partial t} = j_0 = cte \tag{15}$$

This balance is possible only if $j_0 = 0$, resulting in a new expression for the induced total current.

$$\left(p_3 \cdot \frac{\partial T(t)}{\partial t}\right) + \left(\mathrm{d}_{33} \cdot \frac{\partial \sigma(t)}{\partial t}\right) = \frac{1}{S}\left(C_p \frac{\partial V_T(t)}{\partial t} + \frac{V_T(t)}{R}\right) \tag{16}$$

Which is equivalent to:

$$i_T(t) = i_{R_T}(t) + i_{C_T}(t) \tag{17}$$

Equation (17)) explains why the polar generators (Fig. 1) are equivalent to two current generators i_{pyr} and i_{pie} (Fig. 5) in parallel with a capacitance C_T and a resistance R_T.

The power recovered across a purely ohmic resistor denoted as R, is expressed [46] as:

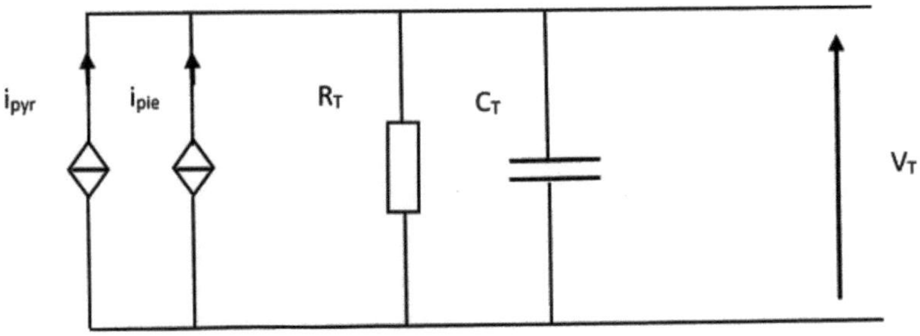

Fig. 5 Equivalent circuit of a polar material

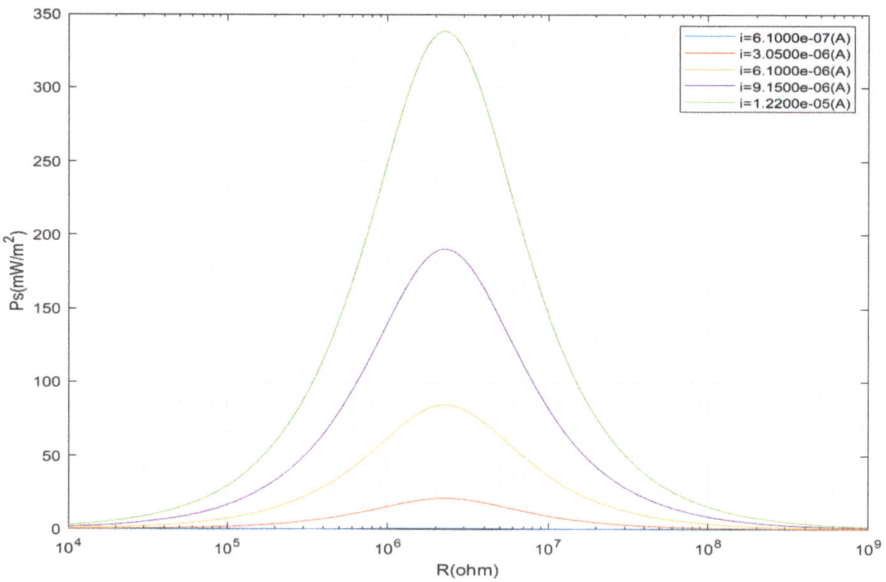

Fig. 6 Surface power density dissipated by a resistive load R for various values of $i_T\ (t)$

$$P_T(t) = R \cdot \left(\frac{R_T^2}{(R + R_T)^2 + (C_T \cdot \omega \cdot R_T \cdot R)^2} \right) \cdot i_T^2(t) \tag{18}$$

where ω is the pulsation (Fig. 6).

The maximum recovered power is given by:

$$\frac{\partial P_T(t)}{\partial R} = 0 \tag{19}$$

Since $C_T{\cdot}\omega{\cdot}R_T{>>>}1$

$$P_T(t) = \frac{1}{C_T \cdot \omega} \cdot \left(\frac{S^2}{\left(\frac{1}{C_T \cdot \omega \cdot R_T} + 1 \right)^2 + 1} \right) \cdot \left(p_3 \cdot \frac{\partial T(t)}{\partial t} \right)^2 + \left(\mathrm{d}_{33} \cdot \frac{\partial \sigma(t)}{\partial t} \right)^2 \tag{20}$$

To compare this expression with Eq. (14), one can write:

$$P_T(t) = R_{\text{opt}} \cdot F_p \cdot S^2 \cdot \left(p_3 . \frac{\partial T(t)}{\partial t} \right)^2 + \left(\mathrm{d}_{33} \cdot \frac{\partial \sigma(t)}{\partial t} \right)^2 \tag{21}$$

With:

$$F_p = \frac{1}{\left(\frac{1}{C_T \cdot \omega \cdot R_T} + 1\right)^2 + 1} \approx 0.5 \tag{22}$$

where F_p is a loss factor, which relates the optimal charge to the thermomechanical losses of a polar material.

4 Hybrid Piezo-Pyroelectric Crystal Simulation

The current section presents a dynamic regime simulation to demonstrate the usefulness of merit factors in selecting the most reliable hybrid material for thermal and mechanical energy recovery. Figure 7 displays the dimensions of the disk used in this simulation, while Table 2 provides the numerical values of the merit factors for the ferroelectric ceramic PZT and the ferroelectric crystal $LiTaO_3$.

4.1 Pyroelectric Response over Time

Figure 8 illustrates the instantaneous variation of temperature (on the left axis) and the corresponding pyroelectric response (on the right axis) over time, resulting from successive heating and cooling of PZT (Fig. 8a) and $LiTaO_3$ (Fig. 8b).

The increase in temperature from 295.15 to 335.15 K for PZT (Fig. 8a) results in a pyroelectric response where the electrical potential rises from 0 to 4.5 V. Similarly,

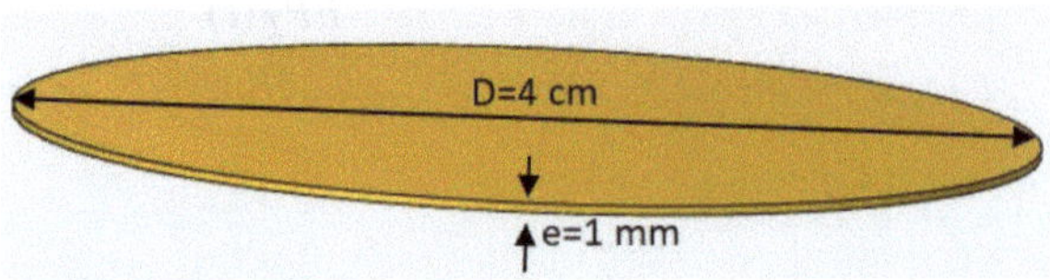

Fig. 7 Dimensions and geometry of the material

Table 2 Figures of merit for open circuit and optimal electric charge for PZT and $LiTaO_3$

Merit Fs	PZT(∞m)	$LiTaO_3$(3 m)
$F_{pyr}(V/K)$	67.68	456.57
$F_{pie}(V/\text{MPa})$	60.01	19.22
$w_{pyr} \cdot 10^{-3}\left(J^{\frac{1}{2}}/K\right)$	6.43	10.44
$w_{pyr} \cdot 10^{-3}\left(J^{\frac{1}{2}}/\text{MPa}\right)$	6.69	0.44
$k^2_{pyr}(\%)$	3.13	8.12
$k^2_{pie}(\%)$	65	32
R_{opt}	30 (MΩ)	1.5 (GΩ)

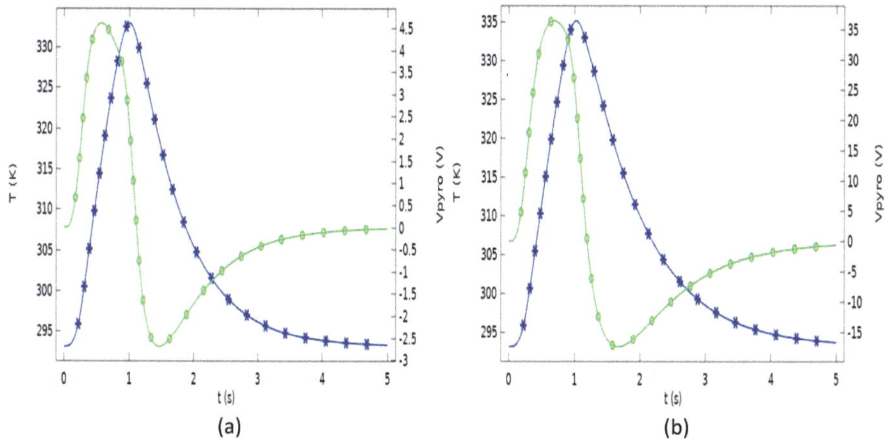

Fig. 8 Instantaneous variation of temperature and electrical potential for PZT and $LiTaO_3$ associated with optimal resistance (R_{opt})

a decrease in temperature generates a negative potential that reaches −2.6 V. During heating, the instantaneous temperature variation of $LiTaO_3$ (Fig. 8b) is similar to that observed for PZT. However, the pyroelectric response reaches a value of 35 V, and then decreases to −16 V when cooling the crystal from 335.15 to 295.15 K. These results indicate that $LiTaO_3$ exhibits higher reactivity F_{pyr} to thermal fluctuations (456.57 V/K, Table 2) compared to PZT (67.68 V/K, Table 2), even though PZT has a higher pyroelectric coefficient than $LiTaO_3$, as confirmed by the works presented in Sebald et al. [29], Kaupinmäki et al. [47].

4.2 *Piezoelectric Response over Time*

The instantaneous variation of the applied stress (left axis) on the PZT and $LiTaO_3$, along with the piezoelectric response (right axis), is illustrated in Fig. 9.

A variation in mechanical stress from 0 to 1 MPa induces a piezoelectric response in PZT (Fig. 9a), where the electrical potential reaches a maximum value of 140 V and decreases to −100 V under negative pressure. In contrast, the response of $LiTaO_3$ (Fig. 9b) to an instantaneous change in applied mechanical stress reaches a maximum value of 0.25 V and a minimum value of −0.2 V. The piezoelectric reactivity F_{pie} of PZT (60.01, Table 2) to variations in mechanical stress is thus demonstrated by these results, showing that the potential created by PZT under the same mechanical excitation is higher than that created by $LiTaO_3$. These numerical results are consistent with those presented in Dileena et al. [48], Wang et al. [49], Abbas et al. [50].

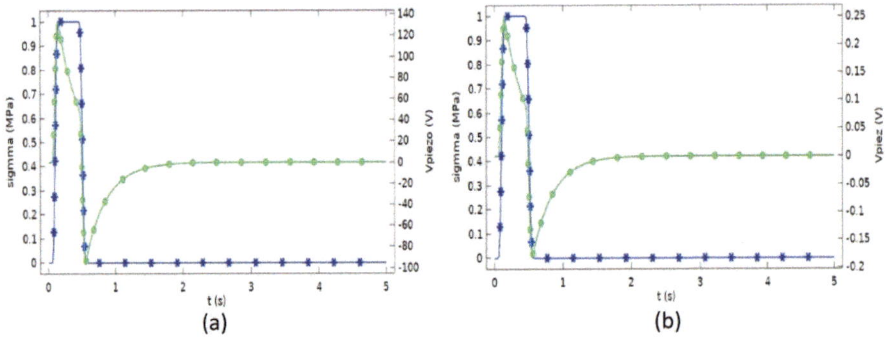

Fig. 9 Variation of mechanical stress and electrical potential for an optimal electrical charge (R_{opt}) over time for PZT and $LiTaO_3$

4.3 Power Recovered over Time

The instantaneous value of surface power density is depicted in Fig. 10 for optimal charges via pyroelectric effect (left axis) and piezoelectric effect (right axis).

The surface density of optimal electrical power created by the pyroelectric effect (left axis), generated by $LiTaO_3$, reaches a maximum value of 125 mW/m^2, while

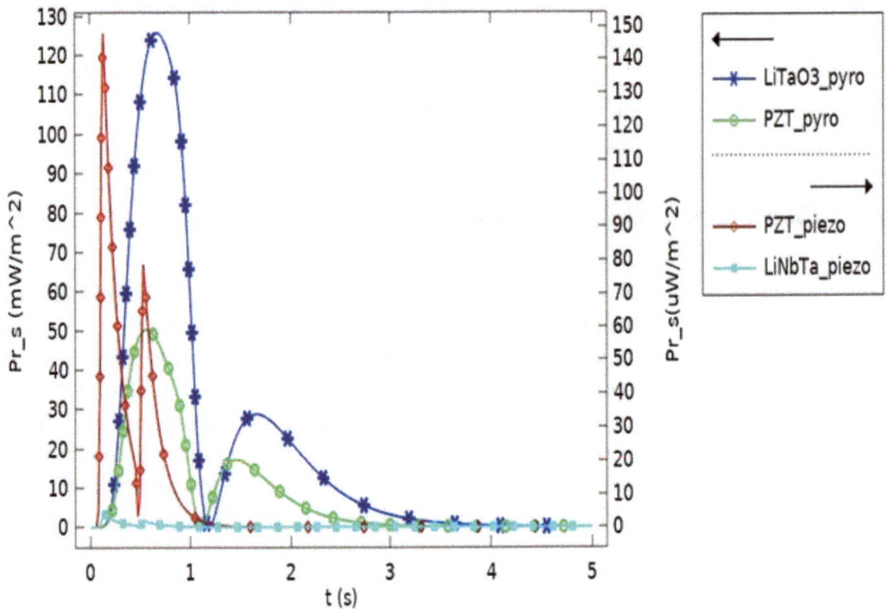

Fig. 10 Surface density of optimal electrical power created by pyroelectric and piezoelectric effects for PZT and $LiTaO_3$

PZT only achieves a maximum of 50 mW/m^2. It is worth noting that thermal fluctuations at the surface of both materials are similar, demonstrating the role of the thermomechanical coupling factor k_{pyr}^2 (Table 2), which explains why the electrical energy created by $LiTaO_3$ is greater than that created by PZT, even though the pyroelectric coefficient of PZT is larger than that of $LiTaO_3$. Concerning the piezoelectric response (Fig. 10, right axis), for the same mechanical excitation, the surface density of optimal power for PZT reaches a maximum value of 125 uW/m^2, while the electrical response of $LiTaO_3$ reaches a maximum value of 0.4 $\mu W/m^2$. This value is very low compared to the maximum electrical response of PZT. This is because the coefficient value of the mechanical–electrical coupling k_{pie}^2 (Table 2) of PZT is higher than that of $LiTaO_3$.

5 Conclusion

This study examines the relationship between pyroelectric materials, which constitute a subgroup of piezoelectric solids. It explores the origin of pyroelectric and piezoelectric conversions through a Maxwellian approach, providing a detailed explanation of how these two phenomena manifest in a ferroelectric crystal. Additionally, it analyzes in-depth the merit factors associated with these two effects.

Subsequently, the chapter focuses on optimizing the power recovered by a purely resistive load, using the equivalent electrical schematic of a material possessing both pyroelectric and piezoelectric properties. This allows for the modeling of thermo-mechanical losses associated with binding the electrical response to the physical characteristics of the material and the applied excitation.

The obtained results are applied through simulation to identify the material, between PZT and $LiTaO_3$, capable of recovering the maximum amount of energy. Furthermore, this study demonstrates that PZT can recover more mechanical energy than $LiTaO_3$, while the electrical energy created by a thermal excitation of $LiTaO_3$ is greater than that created by PZT. These findings are consistent with the values of the merit factors. Taking into account the material cost, PZT is found to be less expensive than $LiTaO_3$, making its use more economical for energy recovery.

Finally, to simplify our study, we assume that thermal and mechanical fluctuations are applied successively. This assumption can be revised to consider the correlation between the piezoelectric and pyroelectric effects, and thus determine its impact on the response of the hybrid material. Similarly, in this chapter, we focus only on the direct piezoelectric and pyroelectric effects. It would be interesting to study the thermal and mechanical behaviors when the material is subjected to an electric field, that is, the inverse piezoelectric and pyroelectric effects, for a hybrid material.

References

1. Ennawaoui C, Lifi H, Hajjaji A, Elballouti A, Laasri S, Azim A (2019) Engineering Solid Mechanics 7(1):47–58
2. Ennawaoui C, Hajjaji A, Samuel C, Sabani E, Rjafallah A, Najihi I, Azim A et al (2021) Piezoelectric and electromechanical characteristics of porous poly (ethylene-covinyl acetate) copolymer films for smart sensors and mechanical energy harvesting applications. Applied System Innovation 4(3):57
3. Ennawaoui C, Hajjaji A, Samuel C, Sabani E, Rjafallah A, Najihi I, Azim A (2021) Applied System Innovation 4(3):57
4. Lu C, Raghunathan V, Roy K (2010) Micro-scale energy harvesting: a system de-sign perspective. In: 15th Asia and South pacific design automation conference (Asp-Dac 2010), Taipei, IEEE Press, Piscataway, NJ, pp 87–92
5. Roundy S, Leland ES, Baker J, Carleton E, Reilly E, Lai E, Otis B, Rabaey JM, Wright PK, Sundararajan V (2005) Improving power output for vibration-based energy scavengers. IEEE Pervasive Comput 4:28–36
6. Derraz M, Ennawaoui C, Mastouri H, El Hmamssy Y, Abouricha N, Rjafallah A, Hajjaji A (2024) Mathematical modeling for predicting electrical energy harvested using piezoelectric composite materials for smart system applications. Micro Nano Eng 23:100253
7. Qian D, Liu WK, Zheng QJ (2008) Concurrent quantum/continuum coupling analysis of nanostructures. Comp Methods Appl Mech Eng 197:3291–3323
8. Fang H-B, Liu J-Q, Xu Z-Y, Dong L, Wang L, Chen D, Cai B-C, Liu Y (2006) Fabrication and performance of MEMS-based piezoelectric power generator for vibration energy harvesting. Microelectron J 37:1280–1284
9. Khaligh P, Zeng, Zheng C (2010) Kinetic energy harvesting using piezoelectric and electromagnetic technologies—state of the art. IEEE Trans Indus Electron 57: 850–860
10. Derraz M, Sabani E, Ennawaoui C, Balhamri A, Hajjaji A, Azim AE (2022) Morphological and ferroelectric characterizations of porous poly (ethylene-covinyl acetate) copolymer films prepared by coextrusion and pressing methods for pseudopiezoelectric effect. Mater Today: Proc 66:196–201
11. Carmo JP, Gonçalves LM, Correia JH (2010) Thermoelectric microconverter for energy harvesting systems. IEEE Trans Indus Electron 57:861–867
12. Markowski P, Pinczakowski W, Straszewski L, Dziedzic A (2007) Thick-Film thermoelectric microgenerators based on nickel-, silver-and PdAg-based compositions. In: 30th International spring seminar on electronics technology, Cluj-Napoca, pp 223–228
13. Wang W, Jia F, Huang Q, Zhang J (2005) A new type of low power thermoelectric microgenerator fabricated by nanowire array thermoelectric material. Microelectron Eng 77:223–229
14. Grätzel M (2005) Solar energy conversion by dye-sensitized photovoltaic cells. Inorganic Chem 44:6841–6851
15. Sari I, Balkan T, Kulah H (2008) An electromagnetic micro power generator for wideband environmental vibrations. Sens Actuators A: Phys 145:405–413
16. Glynne-Jones P, Tudor M, Beeby S, White N (2004) An electromagnetic, vibrationpowered generator for intelligent sensor systems. Sens Actuators A: Phys 110:344–349
17. Beeby SP, Torah R, Tudor M, Glynne-Jones P, O'Donnell T, Saha C, Roy S (2007) A micro electromagnetic generator for vibration energy harvesting. J Micromech Microeng 17:1257
18. Sebald G, Guyomar D, Agbossou A (2009) On thermoelectric and pyroelectric energy harvesting. Smart Mater Struct 18:125006
19. Nye JF (1984) Physical properties of crystals: their representation by tensors and matrices: 1st (first) Edition. Oxford Science Publications
20. Ravindran SKT, Huesgen T, Kroener M, Woias P (2011) A self-sustaining micro thermomechanic-pyroelectric generator. App Phys Lett 99:104102
21. Fang J, Frederich H, Pilon L (2010) Harvesting nanoscale thermal radiation using pyroelectric materials. J Heat Transfer 132:092701

22. Vasileska D, Goodnick SM (2011) Nano-Electronic devices. Springer, Berlin
23. Derraz M, Ennawaoui C, Achour MAB, Hajjaji A, Laadissi EM, Chhiti Y, Boulghalat M et al (2023) Morphological, piezoelectric and dielectric properties of piezoelectric lead-free ceramic Bi_4 Ti_3 O_{12} for energy harvesting applications: milling process effects. Links 111:5–6
24. Lang SB (2005) Pyroelectricity: from ancient curiosity to modern imaging tool. Phys Today 58:31–36
25. Srinivasan M (1984) Pyroelectric materials. Bull Mater Sci 6:317–325
26. Munn RW (2010) Theory of piezoelectricity, electrostriction, and pyroelectricity in molecular crystals. J Chem Phys 132:104512
27. Lubomirsky I, Stafsudd O (2012) Invited review article: practical guide for pyroelectric measurements. Rev Sci Instrum 83:051101
28. Vanderpool D, Yoon JH, Pilon L (2008) Simulations of a prototypical device using pyroelectric materials for harvesting waste heat. Int J Heat Mass Transfer 51:5052–5062
29. Sebald G, Lefeuvre E, Guyomar D (2008) Pyroelectric energy conversion: optimization principles. IEEE Trans Ultrasonics Ferroelectrics Freq Control 55(3):538–551
30. Navid D, Vanderpool A, Bah, Pilon L (2010) Towards optimization of a pyroelectric energy converter for harvesting waste heat. Int J Heat Mass Transfer 53:4060–4070
31. Moreno RC, James BA, Navid A, Pilon L (2012) Pyroelectric energy converter for harvesting waste heat: simulations versus experiments. Int J Heat Mass Transfer 55:4301–4311
32. Waghmare UV (2014) First-principles theory, coarse-grained models, and simulations of ferroelectrics. Acc Chem Res 47(11):3242–3249
33. Ye C-P, Tamagawa T, Polla DL (1991) Experimental studies on primary and secondary pyroelectric effects in $Pb(Zr_xTi_{1-x})O_3$, $PbTiO_3$, and ZnO thin films. J Appl Phys 70:5538–5543
34. Stokowski S (1976) Thermal modulation transfer function analysis of pyroelectric device characteristics. Appl Optics 15:1767–1774
35. Gribkov SP, Zal'tsberg VS, Nosova VI, Risin VE (1991) Sov Phys Solid State 33:368
36. Bourim M, Moon C, Lee S, Sidorkin V, Yoo K (2006) Pyroelectric electron emission from z face polar surface of lithium niobate monodomain single crystal. J Electroceram 17:479
37. Mangalam RVK, Agar JC, Damodaran AR, Karthik J, Martin LW (2013) Improved pyroelectric figures of merit in compositionally graded $PbZr_{1x}Ti_xO_3$ thin films. ACS Appl Mater Interfaces 5:13235–13241
38. Navid A, Pilon L (2011) Pyroelectric energy harvesting using Olsen cycles in purified and porous poly(vinylidene fluoride-trifluoroethylene) [P(VDF-TrFE)] thin films. Smart Mater Struct 20:025012
39. Priya S, (2010) Criterion for material selection in design of bulk piezoelectric energy harvesters. IEEE Trans Ultrason Ferroelectr Frequency Control 57(12):2610–2612
40. Feldhoff A (2020) Power conversion and its efficiency in thermoelectric materials. Entropy 22(8):803
41. He ZZ (2020) A coupled electrical-thermal impedance matching model for design optimization of thermoelectric generator. Appl Energy 269:115037
42. Beeby SP, Tudor MJ, White NM (2006) Energy harvesting vibration sources for microsystems applications. Meas Sci Technol 17:R175-R195
43. Newnham RE (2005) Properties of materials: anisotropy, symmetry, structure. Oxford University Press, Oxford
44. Safari A, Koray A (2008) Piezoelectric and acoustic materials for transducer applications. Springer
45. Ennawaoui C, Hajjaji A, Azim A, Boughaleb Y (2016) Theoretical modeling of power harvested by piezo-cellular polymers. Mol Cryst Liq Cryst 628(1):49–54
46. Lifi H, Ennawaoui C, Hajjaji A, Touhtouh S, Laasri S, Yessari M, Benjelloun M (2019) Sensors and energy harvesters based on (1–x) PMN-xPT piezoelectric ceramics. European Phys J Appl Phys 88(1):10901
47. Kaupinmäki S, Cox B, Arridge S, Baker C, Sinden D, Zeqiri B (2020) Pyroelectric ultrasound sensor model: directional response. Meas Sci Technol 32(3):035106

48. Dileena L, Sreeja SB, Sreekala CO (2021) A comparative study on piezoelectric and piezoresistive pressure sensor using COMSOL simulation. Mater Today: Proc 46:3121–3126
49. Wang L, Zhao L, Jiang Z, Luo G, Yang P, Han X, Maeda R (2019) High accuracy comsol simulation method of bimorph cantilever for piezoelectric vibration energy harvesting. AIP Advances 9(9)
50. Abbas S, Yuan Y, Hassan A, Zhou J, Ahmed A, Yang L, Bisengimana E (2023) Effect of the concentration ratio on the thermal performance of a conical cavity tube receiver for a solar parabolic dish concentrator system. Appl Therm Eng 227:120403. https://doi.org/10.1016/j.applthermaleng.2023.120403

Opportunities and Perspectives of the Implementation of Smart Energy Transition by 2050

Firyal Latrache, Najwa Jbira, Mohamed Saidi Hassani Alaoui, Fatima Zahra Ouariach, Soufiane Ouariach, Benaissa Bellach, and Hicham Kaddouri

Abstract Before 25 years ago, political organizations were encouraging citizens to exploit and use inexhaustible renewable resources to reduce the degree of global warming, where the average annual temperature did not exceed 1.1 °C. With the organization of international United Nations conferences on climate change and the rise in the earth's surface temperature, as well as a succession of natural disasters, awareness of clean electricity consumption and the development of green technologies is beginning to grow. The most notable measure taken during the twenty-first century to mitigate climate change is the Paris Agreement, which was signed at COP'21 on December 12, 2015 and implemented on November 4, 2016. Currently, the climate change issue aims to achieve decarbonization and sustainability. On the other hand, with this global wave of digital transformation, the objective is no longer only to achieve decarbonization, but also to provide energy comfort while implementing smart energy by 2050. This chapter analyzes some presented opportunities for implementing the energy transition and proposes some perspectives for realizing a smart energy by 2050. The approach of this chapter is to apply the smart energy

F. Latrache (✉) · B. Bellach
Mechanics, Energetics, Systems and Signals Team, Laboratory of Modelization and Scientific Calculus, National School of Applied Sciences, University Mohammed I, Oujda, Morocco
e-mail: l.firyal@ump.ac.ma

N. Jbira
Multidisciplinary Laboratory of Research and Innovation (LAMRI), Energy, Materials, Atomics and Information Fusion Team, Department of Physics, Polydisciplinary Faculty of Khouribga, Sultan Moulay Slimane University of Beni Mellal, Khouribga, Morocco

M. S. H. Alaoui · H. Kaddouri
Experimentation and Modelling Team in Mechanics and Energy Systems, National School of Applied Sciences, Abdelmalek Essaadi University, Al Hoceima, Morocco

F. Z. Ouariach
Research Team in Educational Science, Language Didactics, Human and Social Sciences, Higher Normal School of Tetouan, Abdelmalek Essaadi University, Tetouan, Morocco

S. Ouariach
Research team in Computer Science and University Pedagogical Engineering (S2IPU), Higher Normal School of Tetouan, Abdelmalek Essaadi University, Morocco, Tetouan, Morocco

F. Latrache et al. (eds.), *Digital Technology for an Innovative Energy Transition*, Green Energy and Technology, https://doi.org/10.1007/978-3-031-89657-6_13

perspective to the proposed opportunities for achieving a smart energy transition by 2050 to simultaneously satisfy energy demand, protect the environment and achieve a smart energy. At last, this chapter motivates researchers and energy experts, as well as political organizations, to direct their compasses toward investing their research in the introduction of new perspectives for achieving smart energy, and also to launch strategies and policies that will organize the realization of this smart energy along the future 25 years.

Keywords Energy transition · Digital transformation · Smart energy · Artificial intelligence · Blockchain.

1 Introduction

No one believed that the invention of the first car with an internal combustion engine in 1886 by Karl Benz would lead the world in the twenty-first century to look for tools and approaches to achieve sustainability and energy transition [1]. The comfort felt in the Benz tricycle, followed by the Ford T of 1914 and Chrysler Airflow of 1935 instead of horse-drawn carriages, right up to Dacia Logan and Hyundai, has never been thought to be a major factor in the rise in average temperature by 1.5 °C in June 2024 [2, 3]. It is not only the automobile sector that has caused climate change, but the steam engine of the Industrial Revolution and the industries of the twentieth century that came into being with the widespread use of fossil fuels, as well as the remnants of the 2 world wars, have all influenced the climate and the ecosystem vitality [4]. In the context of fossil fuels, the industrial growth and the need to evaluate the lifestyle by satisfying the energy needs motivated the world to produce electricity from coal, phosphate and natural gas while totally neglecting the harmful effects of these resources on the increase of the carbon footprint [5]. As the world has become increasingly conscious of these harmful effects, political organizations have launched strategies on sustainable development and energy efficiency, and have also been committed to increasing the proportion of renewable energies in the energy mix [6]. The Paris Agreement, which was signed on December 12, 2015 during COP'21 and became applicable on November 4, 2016, is one of the measures adopted to limit the degree of global warming by reducing the average temperature by up to 1.5 °C through the reduction of greenhouse gas emissions [7]. These emissions can of course be reduced by moving from consumption of fossil resources (coal, wood, oil, natural gas, etc.) to sustainable and renewable resources enabling the achievement of carbon neutrality (solar, wind, hydraulic and geothermal). In other terms, it signifies a commitment to a political and strategic objective, such as the energy transition, in order to achieve sustainability. According to the availability of figures and statistics, Figs. 1, 2 and 3 schematize the world electricity consumption in 2019 and 2022 and the energy consumption in U.S. in 2023. Figures 1 and 2 indicates that the industrial sector consumes, between 2019 and 2022, more electricity than other sectors (residential, commercial and public services, transport and other) and

Fig. 3 justifies this, where the petroleum is the most fossil source in transport followed by the natural gas in the residential and industrial sectors. These 3 figures justify the necessity of the energy transition and the sustainability to reduce the fossil resources share in the energetic mix [8–10].

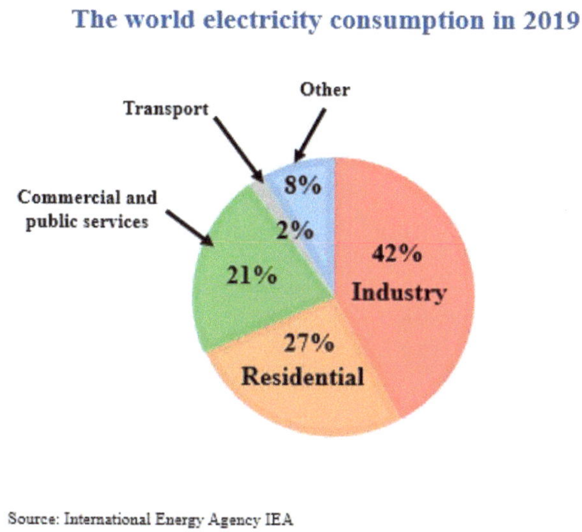

Fig. 1 World electricity consumption in 2019 by sector according to IEA [8]

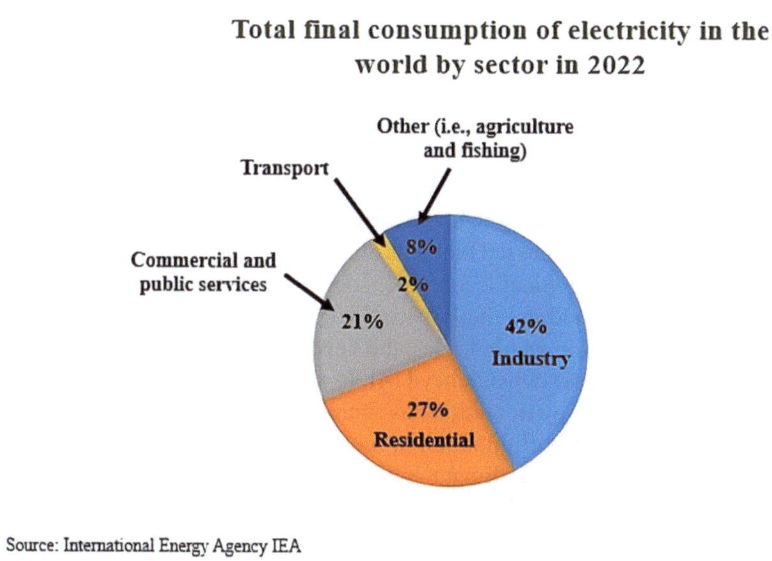

Fig. 2 World electricity consumption in 2022 by sector according to IEA [10]

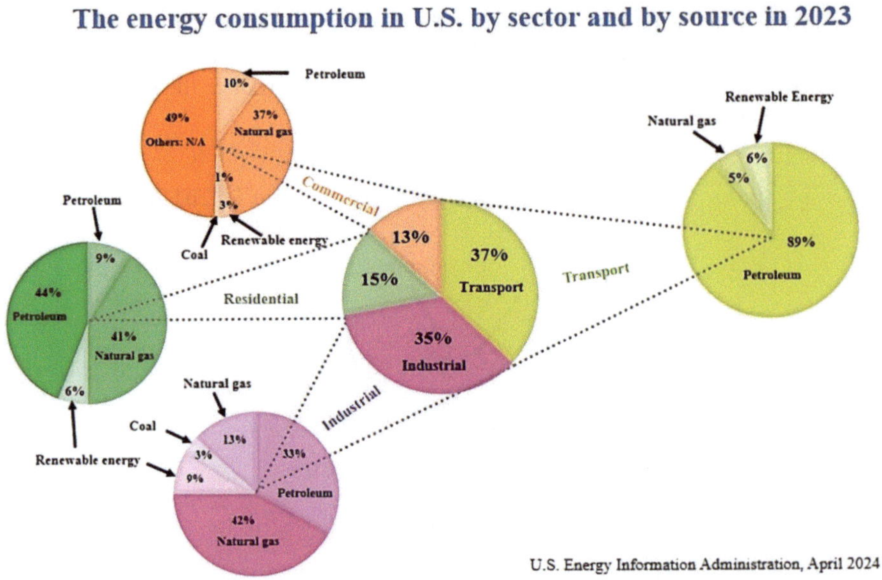

Fig. 3 Energy consumption in U.S. by sector and source in 2023 according to U. S. Energy Information Administration [9]

The main objectives of this energy transition are to reduce the level of pollution, conserve natural resources deposit (oil, coal, etc.) from extinction, eliminate conflicts and wars over natural resources and reduce the economic risks of energy crises such as the 1973-oil crisis [11, 12]. The energy transition covers a variety of sectors, including buildings, transport and waste management, to reduce our carbon footprint and transform the energy mix to 100% renewable by 2050. Y. Ghorbani et al. recommend strategies for a green energy transition based on decarbonization, focusing on raising awareness of this transition and encouraging civil society and local communities for participative collaboration, as well as inviting the general public, including scientists, to join in the decision-making process via their expertise [13]. While J. Sun et al. discussed the risks of achieving a low-carbon economic transition, which are technological, economic, social, institutional and regulatory [14]. It outlines strategies to address these risks: the transparency in decision-making, the support for risk mitigation technologies, the development of energy policies and the reduction of energy export dependency [14]. In addition to its objective of achieving decarbonization, the energy transition has socio-economic and ecological effects, as Y. Yang et al. who cited the elimination of fuel poverty, the contribution to economic growth and the mitigation of climate change [15]. Y. Yang et al. have highlighted that energy poverty will be alleviated by exploiting renewable resources, in particular intermittent resources (solar and wind power) to satisfy electricity needs in areas not connected to the grid [15]. By exploiting the solar potential, photovoltaic technology has well justified its major role in implementing the energy transition, as indicated by M. A. Kabir et al. in his study of the techno-economic feasibility of a photovoltaic system that satisfies the energy needs of universities while reducing CO_2 emissions by 2026.5 tones [16]. Without forgetting that the electricity generated from photovoltaic panels is stored in batteries for later use, and in this context, lithium-ion batteries often supply electric vehicles to contribute for a clean urban mobility without air pollution. A.R.Q. Condoretty et al. highlighted the impact of lithium-ion batteries in implementing the energy transition through their use in electric vehicles, while guaranteeing sustainability for people and clean and resilient cities [17]. For the buildings sector, the energy transition enables the achievement of the zero-energy concept while limiting the demand for peak power by consuming electricity produced from renewable sources rather than from the main grid [18]. N. Shivanaganna et al. showed a 13.30% to 50% reduction in absorbed power from the main grid in a residential area in southern India [18]. This chapter uses the concept of climate change, PV technology, buildings, lithium-ion batteries and political strategies as opportunities for achieving the energy transition.

With this wave of digital transformation, it has become essential to adopt the concept of smart energy by 2050 alongside the implementation of the energy transition. This concept of digital transformation is being adopted firstly as a circular economy perspective for sustainable building design [19] and for the development of the oil and gas sector [20], as well as a factor influencing the resistance of multinational companies to geopolitical and pandemic risks [21]. J. Reis et al. demonstrated that digital transformation is no longer used just for business, but also for human resources, smart cities and to guarantee sustainability [22]. In this context, the digital transformation reflects the United Nations' 2030 Agenda, which aims to achieve sustainable development and assess the quality of citizens' lives [22]. As long as digital transformation encourages the achievement of sustainable development by 2030, it will therefore promote the implementation of smart energy by 2050. The literature suggests approaches to energy development by 2050 aimed at achieving decarbonization [23], harnessing green hydrogen to achieve carbon neutrality [24] and assessing energy infrastructure [25]. These approaches likewise focus on achieving the energy transition by 2050 [26]. It is therefore obvious that the problematic resides in the exploitation of digital transformation to achieve smart energy by 2050, and rather the contribution of the energy transition to the implementation of this smart energy. The contributions of implementing smart energy by 2050 also focus on decarbonization [23] and the adoption of a smart energy transition based on the installation of floating wind turbines in the hydraulic reservoirs [27]. Without forgetting that the digital transformation of the marketing field adopts artificial intelligence to identify the objectives and expectations of consumers and E. G. Andrzejak et al. analyze the application of AI algorithms, such as chatbot and Google Dialogflow, to evaluate online orders in a Polish market [28]. According to M. Khalid et al., the digital transformation embodied in AI is applied in the energy sector to provide for decentralized, low-carbon electricity networks [29]. On the other hand, digital transformation adopts blockchain as a technological tool to facilitate data sharing and G. Liyanaarachchi et al. introduce the concept of "modified blockchain" for digital privacy and confronting consumers' digital fragility [30].

Despite its widespread use as a powerful tool, O. O. Egunjobi et al. showed after an analysis of 156 studies that the application of Blockchain is still limited for the energy sector [31]. Nevertheless, the literature indicates some applications of blockchain in smart vehicles [32] and in energy supply for buildings in smart communities [33]. Overall, the chapter uses artificial intelligence, soft computing and blockchain as perspectives and analyzes their roles in achieving smart energy by 2050.

The chapter analyzes the opportunities (climate change, PV technology, buildings, lithium-ion batteries and political strategies) for implementing the energy transition and the perspectives (artificial intelligence, soft computing and blockchain) for achieving smart energy by 2050. As long as the literature focuses only on achieving decarbonization (as an objective of sustainability and energy transition) in 2050 and smart energy in 2050 and not on the contribution of decarbonisation to achieving smart energy, the chapter analyzes the prospects and opportunities for achieving a smart energy transition. With the global wave of digital transformation and the need to provide sustainable energy to increase comfort and preserve the ecosystem, this chapter introduces the concept of smart energy transition by applying perspectives on opportunities and seeing whether or not environmental protection will contribute to achieving smart energy by 2050. In general, the chapter contributes to the development of intelligent energy systems while reducing the carbon footprint by 2050, and also helps to establish energy comfort and reduce energy dependency.

This chapter is organized as follows: The first section analyzes the opportunities for implementing the energy transition and the second section presents the perspectives for achieving smart energy by 2050. The final section demonstrates the impact of the opportunities provided the energy transition on the realization of smart energy by 2050.

2 Opportunities of Adopting the Energy Transition

2.1 The Dealing with the Climate Change Issue

The continuous use of fossil fuels, in particular gas, oil and coal, increases the emission of greenhouse gases into the atmosphere, which leads to a rise in temperature, and therefore to global warming. The statistics indicated that the average global temperature has risen to 1.15 °C by 2022 and that the period 2015–2022 was the warmest on record. It is evident that the increase in greenhouse gases (carbon dioxide CO_2, methane CH_4, nitrous oxide N_2O) in the atmosphere is the major factor in climate change. These greenhouse gases are present naturally in the atmosphere to guarantee life on earth. However, human and industrial activities are increasing the proportion of these gases in the atmosphere, which is damaging the ecosystem. These industrial activities also use fluorinated gases such as Chlorofluorocarbons CFCs, Hydro chlorofluorocarbons HCFCs, Hydrofluorocarbons HFCs and others in cooling systems,

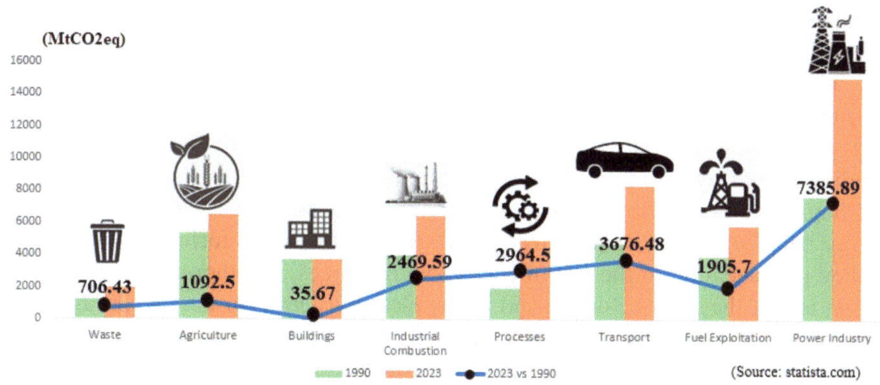

Fig. 4 World greenhouse gas emissions from 1990 to 2023 [36]

while nitrogen trifluoride NF_3 and sulfur hexafluoride SF_6 are used in the manufacture of LCDs and photovoltaic cells. These gases contribute in the other side in the increase of degree of climate change. The increase in the rate of drought has posed a challenge for hydroelectricity, which has increased demand for fossil fuels to satisfy energy needs. As a result, CO_2 emissions have risen to 490 million tons by 2022 (International Energy Agency IEA). As long as the energy transition is a modification in energy consumption and production by transitioning from an energy system based on non-renewable resources to an energy mix comprising mainly renewable resources, climate change is addressed by implementing the energy transition in energy systems. In fact, the energy transition process will be facilitated and implemented by political organizations by adopting carbon capture strategies and by motivating green financing strategies. In this context, addressing climate change through the implementation of the energy transition involves innovation to create technologies that do not pollute the environment, as well as the construction of infrastructures and efficient energy systems. In this way, these measures will promote an effective energy transition and therefore reduce the impact of climate change. These measures will reduce greenhouse gas emissions and, therefore, reduce the other harmful effects of climate change: ecosystem imbalance, rising sea and ocean levels, an increase in natural disasters (drought, floods, storms, etc.), lower agricultural productivity, etc. [3, 4, 34, 35]. Figure 4 illustrates the world greenhouse gas emissions from 1990 to 2023, where the power industry is the major contributor followed by transport and the other sector such as industrial combustion, buildings, agriculture and others [36].

2.2 *The Sustainable Building*

Sustainable building refers to the adoption of actions that enable buildings to be constructed in accordance with environmental and energy standards. According to statistics for 2022 from the International Energy Agency, the building sector represents 30% of primary energy consumption and emits 37% of global CO_2 emissions. The adoption of sustainable building reduces the carbon footprint and decreases the impact of climate change, and subsequently contributes to the adoption of the energy transition. The latest research has shown that the construction of a sustainable building comes by adopting an innovative solution, which consists of constructing buildings with zero-energy consumption. These are buildings with a zero annual energy balance: they consume (using photovoltaic and wind technology) as much as they produce, due to the use of renewable resources. Another advantage of this design is that they can be connected to the electricity grid for energy storage. The concept of zero-energy buildings is not the only way to achieve sustainable buildings. Optimizing the building materials industry will also help to achieve sustainability. Building with sustainable cement, bamboo, clay, wood and bio composites conserves biodiversity and helps to reduce the negative effects of CO_2 emissions. In the context of the environmentally building materials, the manufacture of bricks by mixing sands with other bio components makes an excellent contribution to the thermal insulation of buildings and, subsequently, to sustainability. Reducing carbon emissions in the building sector will also be achieved by incorporating phase change materials into construction materials. By storing internal energy (cold or heat) and releasing it at a later time, a building material (bricks) filled with PCM capsules makes an effective contribution to the passive cooling of a building without the need to consume active energy, which could further increase GHG emissions [18, 37, 38].

2.3 *The Optimization of Solar Photovoltaic Technology*

Photovoltaic solar energy is a technology that enables electricity to be generated using photovoltaic cells due to the photovoltaic effect. There are many types of photovoltaic panels: The stand-alone PV (with or without battery), the hybrid PV (with other systems such as hydro and wind turbines) and the grid connected PV. The literature indicates that there are several solar photovoltaic technologies materials such as the Amorphous Silicon (a-Si), the Copper Indium Gallium Diselenide (CIGS), the Cadmium Telluride (CdTe) and the Perovskite, which reach 26% of efficiency when it is used only and 30% when it is employed as tandem cell with silicon. The Perovskite materials show its significant potential for increasing efficiency and reducing costs. This photovoltaic technology generates clean electricity, reducing the impact of climate change and contributing to sustainable building and therefore in energy transition. In view of the large-scale R&D projects for photovoltaic solar power plants being installed everywhere, optimizing this technology has become a crucial choice. This optimization consists of maximizing energy production and minimizing losses and costs knowing that the price of electricity produced using photovoltaic technology has already dropped between 2010 and 2017 [39, 40].

The adoption of a sun tracker is one of the most widely used approaches for optimizing electricity production from photovoltaic technology, by determining the optimum orientation of the panels based on the angle of inclination, which is in turn determined by the geographical location. There are 2 types of sun tracker: a single-axis sun tracker (vertical and horizontal) and a 2-axis sun tracker. Another method of optimizing the efficiency of electricity production using this technology is to maintain the panels by cleaning them of dust and salt, as well as the fine particles produced by pollution. It is recommended to clean photovoltaic solar panels with lukewarm water to avoid thermal shock. There are also automated systems for cleaning PV panels: the F1A Robot from SOLARCLEANO (1000 m^2/h), the IG502 Robot from InHand Networks, the INDORS Robot, the B30M2 Robot from Xiamen, etc. There is also a technique that involves a self-cleaning mechanism, through the action of water droplets, by applying a coating made up of thin layers of titanium photo catalyst to a PV panel. This technique slows the rise in temperature from 1 to 5 °C. With the adoption of a sun tracker and a PV cleaning system, cooling the panels contributes perfectly to optimizing this technology. PV cooling techniques are divided into 3 categories: Passive cooling (based on gas, liquid and phase change material PCM), active cooling (based on gas, liquid and thermoelectric) and hybrid cooling (based on passive and active techniques) [41–45].

This section has presented the most widely used approaches and techniques for optimizing PV technology with the aim of promoting an innovative energy transition. This field of optimization is very vast, and it is recommended that the following parameters be analyzed and developed in order to improve electricity production, in other research work and in other R&D projects: Solar radiation, temperature, humidity, wind speed and wind speed [45].

2.4 The Use of Li-Ion Batteries

The role of batteries is to store and accumulate electrical energy for later use in everyday applications such as electric cars and telephones, and to satisfy self-consumption and electricity needs in areas not connected to the electricity grid. The lead battery, the nickel–cadmium battery, the nickel-metal hydride battery and the lithium-ion battery are the existing types of battery, and current research is focusing on the lithium-ion battery because of its long life and high energy density. In a liquid electrolyte and between an anode (graphite) and a cathode (metal oxide), the lithium-ion battery releases electrical energy by reversible exchange of the lithium-ion. The lithium-ion battery has a minimum voltage of 3.6 V (for older Li-ion batteries) or 3.7 V (for newer Li-ion batteries). The disadvantages of this technology are that there is a risk of thermal runaway, where heat accumulates in the battery caused by damage to the battery, resulting in an uncontrolled rise in temperature. Furthermore, the current development of lithium-ion battery technology requires the adoption of a battery management system that involves predicting thermal runaway by detecting the release of quantities of volatile organic compounds (VOCs). The UPS battery management system developed by SOCOMEC: Innovative Power Solutions guarantees the reliability of the lithium-ion battery and prevents damage and thermal runaway [46–48].

As mentioned above, electric vehicles are among the main uses for lithium-ion batteries, given that the transport sector is the main source of CO_2 emissions. According to European Environment Agency statistics for 2022, road transport contributes 71.7% of greenhouse gas emissions compared with other sectors. In reality, the contribution of the transport sector to greenhouse gas emissions is not only reflected in the emissions linked to fuel consumption, but also in the production of this fuel and the materials used in the manufacture of cars. Electric vehicle technology first appeared in 1834, and went through various stages with the appearance of nickel batteries in 1899 and lithium-ion batteries in 1977. During this period, hybrid vehicles (electric motor-internal combustion engine) were developed, until Tesla produced the first autonomous electric vehicle (250 miles on a on a single charge). Current research is based on increasing the proportion of electric vehicles in the market by 31% by 2050, given that their engines are more environmentally friendly than internal combustion engines, and in view of the increase in the portion of renewable energies in the energy mix. In general, the lithium-ion battery's contribution to the energy transition does not necessarily mean that it should be used in electric vehicles, but because electric vehicles are a major contributor to CO_2 emissions and therefore to the adverse effects of climate change. The development of lithium-ion batteries must go alongside green hydrogen, which is a powerful vector for reducing the carbon footprint [46, 49, 50].

2.5 The Role of Politics in Achieving an Innovative Energy Transition

Improving sustainability in a nation is one of the functions of political organisms. In fact, sustainability and the energy transition are inextricably linked: They enable the production and consumption of energy to satisfy the needs of habitats by protecting the environment. As long as the harmful effects of climate change continue to appear day by day, in the form of drought and shortages of water resources, and the subsequent deterioration in biodiversity, the public authorities must take regulatory, economic and financial measures [6, 15].

In this context, the policy of the European Investment Bank in collaboration with the European Union aims to adopt a policy of loans in the renewable energy and energy efficiency sectors, amounting to more than 5.7 billion euros and 10 billion euros, respectively. This collaboration has supported innovative and pioneering projects in wind energy (floating offshore wind turbines) and in the production of green hydrogen, as well as improving energy infrastructures, in particular energy storage devices. According to the United Nations Conference on Trade and Development UNCTAD report published in November 2021, 798 renewable energy policies have been adopted, covering 192 economies. Overall, 70% of these policies involve private investment in renewable energy projects. However, the report launched by UNCTAD shows that there is an investment gap of around \$2,000 billion and \$4,000 billion, respectively, to achieve the objectives of the energy transition and sustainable development. To achieve this targets, the report proposes actions to fill the investment gap, by focusing on strengthening international and regional partnerships and cooperation, and opening up to sustainable markets and financing [35, 51, 52].

As an example, Morocco, with an annual irradiation of over 2000 $kWh \cdot m^{-2}$, has adopted financial strategies to guarantee the energy transition and to commit to the environment on an international scale. In this context, Morocco has introduced financial incentives for environmental protection, such as the national environment and sustainable development fund, the industrial depollution fund and the liquid sanitation and wastewater treatment fund. As a result of the financial incentives, Morocco, as one of the organizers of COP22 in 2016, published a directive on the management of financial risks linked to climate change in March 2021. In addition, the Moroccan banking sector has motivated between 20 and 40% of green investment projects that respect the environment. At the international level, the total official development assistance given to Morocco for the environment and climate will reach 757.9 million dollars in 2019. The overall challenges of green finance in Morocco lie mainly in banning polluting and energy-intensive projects and developing investment in renewable energy and energy efficiency. Finally, Morocco has not only focused its efforts on implementing green finance, but also on environmental protection programs such as the national air program for the period 2018/2030 and the climate program, which aims to prevent and reduce atmospheric pollution. It has also elaborated laws and legal texts such as law n° 12–99 on the protection of the environment, law n° 81–12 on the littoral and law n° 36–15 on water [42, 52–55]. Figure 5 illustrates policies and targets of 2 countries (United States and Morocco) and 1 region (The European Union), which they are a major contributor in the achievement of the energy transition. In this context, The White House published on August 16, 2022 The Inflation Reduction Act aiming to mitigate climate change and drive a green American economy. In addition, Morocco launch on March 11, 2024 its offer about the green hydrogen to generate 9 millions of tons of this clean energy by 2050. For the European Union, the Green Deal, published on aims to end the greenhouse gas emissions by 2050 [56–58].

2.6 Analysis and Discussions About the Opportunities of Adopting the Energy Transition

The adoption of the energy transition policy is undoubtedly a matter of urgency, in view of the catastrophic climate situation. The transition to renewable energy means not only increasing the share of renewable resources in the energy mix (wind, solar, hydraulic and geothermal), but also dealing with the climate change

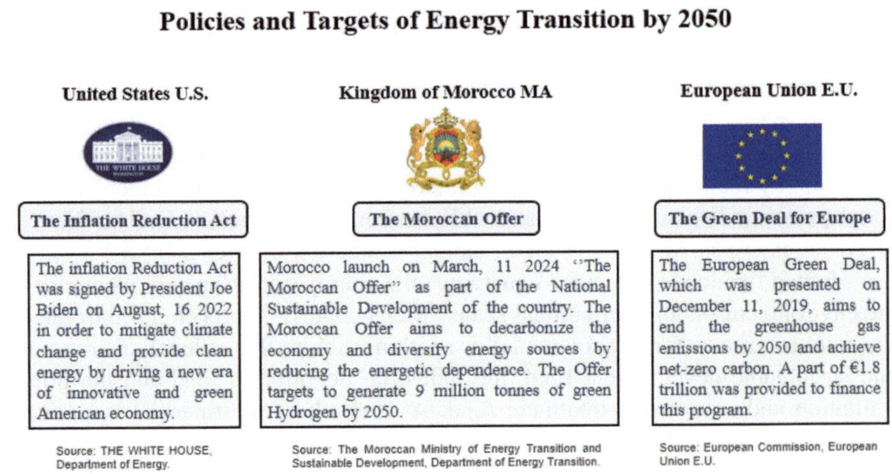

Fig. 5 Policies and Targets of the United States, the Kingdom of Morocco and the European Union to achieve the energy transition by 2050 [56–58]

issue, constructing sustainable buildings, optimizing photovoltaic solar panels, using lithium-ion batteries and contributing to the implementation of policies and strategies to achieve sustainable development. Overall, the dealing with the climate change issue consists of adopting some measures to reduce greenhouse gas emissions such as the creation of technologies that do not pollute the environment and the construction of efficient infrastructures. Then, the construction of sustainable buildings in accordance with environmental and energy standards by adopting the concept of zero-energy buildings and optimizing the buildings materials through using the phase change material for example. With regard to the production of electricity through intermittent photovoltaic source, its development to maximize production is essentially based on the adoption of sun tracker and the optimum angle of inclination of the PV panels, the setting of automated system for cleaning PV and the development and the design of the cooling the PV panels. The electrical energy produced by this technology then needs to be stored for later use, and the lithium-ion battery provides this option as its long life and high energy density. Therefore, the electric vehicles uses this technology of lithium-ion battery to reduce the greenhouse gas emissions as well as for the contribution of increasing their share, of these vehicles, in the markets by 31% in 2050. Not forgetting, of course, the major role played by the political organisms in the implementation of sustainability and energy transition by adopting financial strategies and financial incentives, as well as environmental protection programs and laws and legal texts. In this context, Fig. 6 illustrates the power generation and the installed capacities of renewable energies by 2050: solar photovoltaics PV, wind energy, hydropower, biomass, solar thermal and geothermal. The statistics indicates that the solar PV will generate more than of 50% of electricity by 2050 [8].

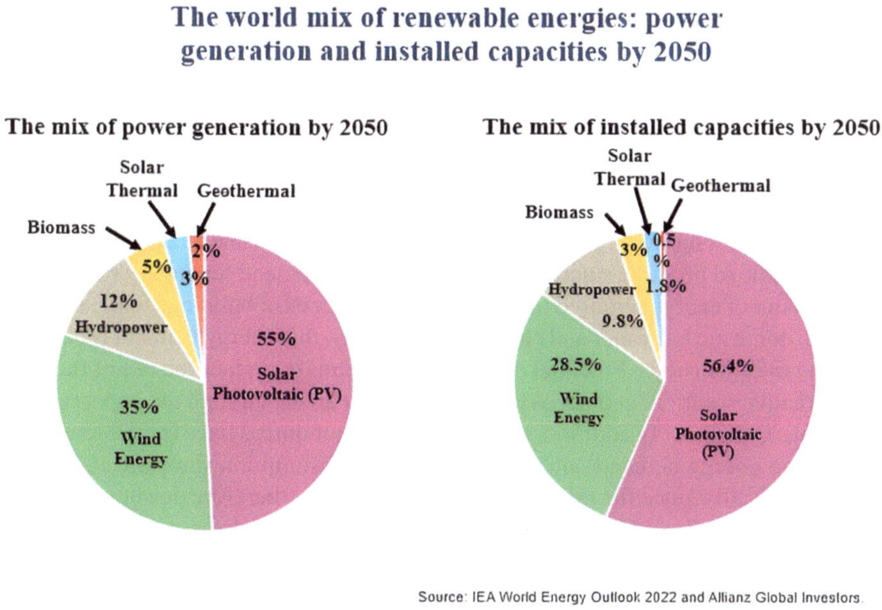

Fig. 6 World mix of renewable energies: power generation and installed capacities by 2050 from IEA [8]

There is no doubt that these opportunities complement each other: Programs and strategies founded by political organizations encourage the use of sustainable building materials and the use of intermittent sources (photovoltaic and wind power) to meet daily energy needs, such as powering electric vehicles through lithium-ion batteries. All these measures have been adopted in order to protect the environment and reduce the impact of climate change. It's true that these opportunities contribute to the adoption of the energy transition policy alongside other opportunities, such as the production of green hydrogen, but how will they have an impact on the adoption of the vision of smart energy by 2050?

To solve this issue, we have to identify the intelligent tools needed to manage the chain between producer and consumer in order to satisfy energy demand while preserving nature's biodiversity. Digitization and digitalization are making life easier, as well as facilitating the management of energy demand, energy storage and energy transport through artificial intelligence algorithms and soft computing programs, to satisfy energy needs in an environmentally friendly way. For this purpose, the following paragraph shows the approaches used by artificial intelligence to manage energy demand, storage and transport. These measures are seen as prospects for achieving the vision of smart energy by 2050.

3 Perspectives of Promoting a Smart Energy by 2050

3.1 The Perspectives of Smart Energy

3.1.1 AI for Energy Demand and Storage Energy Management

The use of artificial intelligence to manage energy demand and storage has become indispensable to promote smart energy by 2050. The demand for energy refers to the total amount of energy required to cover the energy needs of buildings for heating and cooling, domestic hot water and electricity. In reality, this energy demand is based on a variety of dimensions: Potential demand, which aims to replace the use of fuel with an alternative solution, and derived demand, which arises through the appearance of a new energy source. In addition, researchers and economists have been interested in controlling energy demand and planning energy consumption since the 1970s, and more specifically since the oil crisis in 1973. They have therefore developed models for evaluating and predicting energy consumption. Some of these models include econometric models that develop correlations between energy consumption and other variables such as energy prices, population and technologies used in strategic sectors, as well as in industry and agriculture. As part of the adoption of the energy transition policy, renewable energies must satisfy this energy demand with the aim of reducing the carbon footprint and subsequently limiting GHG emissions. For this purpose, energy storage is considered as an effective solution for covering the intermittent nature of renewable energies, in particular photovoltaic solar energy and wind power. It will therefore enable energy demand to be satisfied and will contribute to energy self-sufficiency through storage systems such as: Hydraulic storage by pump, thermal storage and battery storage, in particular the lithium-ion battery (already presented in the previous paragraph). It is obvious, then, that energy demand management and energy storage are the perfect way of ensuring a balance between energy production and final energy consumption, as well as fighting climate change by adopting the energy transition policy toward environmentally friendly renewable resources. So how can intelligent management of energy demand and storage contribute to energy development by 2050? [11, 59].

Artificial intelligence is playing a major role in the adoption of the energy transition, and is considered one of the prospects for achieving smart energy by 2050. Its integration enables energy production to be optimized by algorithmic optimization and by real-time adaptation of this production according to climatic conditions. According to the International Energy Agency, digitalization has become a necessity to improve the efficiency, security and accessibility of energy sources. The smart management of the electricity grid is another benefit of artificial intelligence, where the use of a smart electricity grid will effectively balance between supply and demand. The adoption of digitalization is not just about improving energy production, but also about developing businesses involved in the energy sector. The adoption of digitization is not just about improving energy production, but also about developing businesses operating in the energy sector by predicting energy demand

and managing resources to find the optimum price for the customer. In addition, artificial intelligence contributes to the digitalization of businesses, with the treatment, storage and management of data to ensure their competitiveness in the market. Without forgetting, through the optimization and optimal management of energy storage technologies, to ensure a constant balance between supply and demand. In the literature, energy demand management systems are based on techniques and approaches aimed primarily at controlling energy consumption and optimizing the use of electrical resources, as well as reducing costs. In this context, Zhang's model manages energy demand for intelligent buildings using a fuzzy expert framework. In addition, the oil and gas industry uses machine learning algorithms (Iqbal's model), and Baliyan's model uses the Maximum Power Point Tracking MPPT technique to optimize residential loads. The literature also shows that energy demand is managed by soft computing, such as Anthony's model, which manages residential energy with the aim of achieving energy savings of 1.25% and reducing costs by 0.5%. This approach also manages energy for the operation of microgrids using Philipo's model, with the aim of saving 3.7 kW each month. The method used for these applications is fuzzy logic control and there is another method based on the genetic algorithm, which aims to manage energy consumption, by load transfer (El Makroum model) in order to achieve energy savings of 63.48%. Despite the existence of effective solutions for the digitization of energy demand management, it suffers from a number of obstacles, mainly related to the lack of information technology, which reflects all aspects of real-time data analysis and communication via a robust communications network [13, 25, 45, 60–62].

In fact, the adoption of digitalization not only includes energy demand management, but also energy storage. Digital management of energy storage promotes the efficient and reliable use of intermittent resources (photovoltaic and wind power) to satisfy the energy demand. This worldwide wave of digitalization means that digital technologies (big data, the Internet of Things IoT, industrial computing, etc.) can be integrated into the smart management of energy storage to guarantee decentralized production. According to the literature, the digital twin is an approach used to predict the maintenance of storage systems (already mentioned: Battery storage, fuel cell storage, pumped hydroelectric energy storage, thermal energy storage and supercapacitor storage), as well as to determine their efficiency and lifetime by representing it virtually according to its data in real time. This digital twin incorporates artificial intelligence, machine learning and data analysis. In view of the difficulty of managing battery energy storage, digital twin technology offers the possibility of intelligent control, especially for lithium-ion batteries, which help to reduce greenhouse gas emissions. Among the applications of digital twin technology for digital storage management of lithium-ion batteries are solving problems of limited power output, estimating maximum discharge capacity and optimizing the power of the battery cell formation (cathode, anode, electrolyte, an aluminum plate and a copper plate). For fuel cells, digital twin technology provides failure diagnosis in terms of dynamic models in order to find the operational solutions required. Furthermore, the intelligent construction of pumped hydro stations is another application of digital twin technology for pumped hydro energy storage systems. In the case of super capacitors, the digital twin is used to predict the phenomenon of relaxation, since it has an effect on their efficiency and state. In general, the use of digital twin technology provides a solution for controlling temperature in the case of the lithium-ion battery, monitoring and optimizing other energy storage systems and finally improving their operation. However, the energy data supplied by the digital twin needs to be secured by a tool to guarantee its integrity in other contexts, and the blockchain facilitates this procedure. Therefore, blockchain has great potential in the energy sector, particularly in network management, to guarantee the transport and traceability of energy between the producer and the consumer [45, 63–65].

3.1.2 Blockchain for Energy Distribution

Blockchain is a digital technology that enables transactions to be stored and data to be transmitted by the nodes of a network. Thus, a transaction is a transfer of data between two nodes, which in turn are the entities of a blockchain and which behave like devices sharing information with each other. The advantages of blockchain are mainly transparency, decentralization and security: With transparency, transactions will be accessible to all members and will also be recorded and shared in a collaborative manner, so the data will be stored on several computers, since blockchain is decentralized. The data will then be cryptographically secured and verified by the blockchain members. Blockchain, an innovative technology that has attracted the attention of the financial world and facilitated the complexity of financial transactions, has become ambitious for the energy sector, to contribute to the adoption of the energy transition policy and subsequently to the implementation of smart energy by 2050. As energy demand and energy storage are managed by artificial intelligence, the management of energy consumption must be managed by digitization to guarantee the transportation of energy to consumers. To do this, the blockchain performs real-time energy consumption diagnostics to determine the quantity needed for consumption, as well as the energy-consuming systems that would not overload the chain between the 2 nodes (the producer and the consumer). Indeed, blockchain is motivating local authorities to obtain a permanent overview of energy infrastructures, with a view to dimensioning the network and building power stations to satisfy energy needs. In this context, a smart grid enables consumption to be monitored in real time, controlled according to energy demand, and allows intermittent renewable energies to be integrated into the grid. Once again, local and regional authorities can create energy markets using online blockchain platforms, bringing together offers from energy producers and enabling citizens to choose the best package to fight fuel poverty. Statistics from the International Energy Agency (IEA) in 2018 showed that 789 million people do not have access to electricity, and these statistics highlighted the problem of fuel poverty. To solve this problem, the adoption of decentralized energy systems through blockchain will not only reduce environmental impacts and improve sustainability, but will also generate grid-independent energy (especially allowing intermittent sources to contribute to the energy mix) and be able to contribute to regional electricity markets. There are some examples of projects involving the application of blockchain technology to the improvement and distribution of energy: France's TECSOL project, Spain's Pylon Network project and Switzerland's Power ID project. Then, France's TECSOL project plans to build charging stations for electric vehicles using photovoltaic solar energy, and also to power residential buildings; Spain's Pylon Network project aims to design low-energy consumption servers; and Switzerland's Power ID project will enable decentralized energy to be exchanged via a small network between producers and consumers, and distributed throughout the country [33, 66, 67].

3.2 *Analysis and Discussions of the Proposed Axis of Promoting a Smart Energy by 2050*

The perspectives proposed for achieving the vision of smart energy by 2050 are essentially based on the use of artificial intelligence algorithms to manage energy demand and storage, and as well as blockchain to distribute this energy. In view of the global wave of digitalization, the smart management of energy has become a necessity to improve efficiency of energetic systems and therefore combating fuel poverty. The application of artificial intelligence in managing energy demand is essentially based on the use of fuzzy expert framework, machine learning algorithms, maximum power point tracking MPPT and soft computing. The use of these artificial intelligence algorithms aims to optimize the use of electrical resources, reduce costs and control energy consumption. In fact, the use of artificial intelligence not only includes energy demand management, but is also intended to the treatment, storage and management of data of businesses involved in the energy sector. In addition, the development of energy storage technologies to help the implementation of the vision of smart energy by 2050 is encouraging the use of the digital twins, which incorporates artificial intelligence, machine learning and data analysis. The advantages of digital twin is providing intelligent control, especially of li-ion batteries, and providing failure diagnostics. All that is for enabling intermittent sources to contribute to the energy mix. On the other hand, what about energy distribution? In fact, blockchain will enable transactions to be carried out between producers and consumers, with the aim of managing energy consumption, satisfying needs and fighting fuel poverty. It is true that this technology has penetrated the world of finance and business, but has become ambitious as well as for the energy sector by performing real-time energy consumption diagnostics to determine the quantity needed for energy to be distributed.

The perspectives presented will undoubtedly contribute to the implementation of the vision of smart energy by 2050, given that the triple demand-storage-consumption is necessary for each community to meet its energy needs and combat fuel poverty. Subsequently, the adoption of these perspectives for the implementation of smart energy must be performed alongside the opportunities for achieving an innovative energy transition. Is there a link between the realization of the innovative energy transition policy and the implementation of the vision of smart energy by 2050? It is true that artificial intelligence approaches have been presented and blockchain communication networks have been identified to guarantee transactions between producer and consumer, but how these opportunities and perspectives really interlink to achieve an innovative energy transition while taking advantage of global digitalization to contribute to the vision of smart energy by 2050? In the context of positioning the opportunities of the energy transition policy and the prospects for implementing the vision of smart energy by 2050, the timing of the adoption of each strategy must be taken under consideration. This means that we need to understand the right time to adopt each approach and whether or not it is possible to implement them all at once, and whether or not adopting the energy transition will also lead to the implementation of the smart energy vision. Another issue that needs to be addressed is whether the adoption of the energy transition is temporary or permanent, i.e., whether it only contributes to the implementation of the smart energy vision or whether the 2 approaches are interdependent. We therefore need to analyze the influence of the opportunities of the energy transition on the perspectives of the smart energy vision, in order to track the global wave of digitalization while preserving the environment.

4 From the Opportunities of Energy Transition to Perspectives of Adopting Smart Energy by 2050

The previous two paragraphs discuss the opportunities for achieving the energy transition: dealing with climate change, sustainable building, optimizing solar PV technology, the use of Li-ion batteries and politics. They likewise propose perspectives for achieving smart energy by 2050, including artificial intelligence, soft computing and blockchain, which are currently the necessary tools for the development of digital technologies. This section discusses the role of the innovative energy transition in achieving smart energy by 2050: It analyzes the application of the perspectives on opportunities (Table 1) and presents the work highlighting the results of this application in terms of smart energy (Table 2). The purpose is to demonstrate the application of perspectives on opportunities to highlight that these opportunities of the innovative energy transition lead perfectly to achieving smart energy. The reasoning in this section is similar to the reasoning by contrapositive: The only difference is that if the demonstration of the application of perspectives to opportunities is true, then opportunities can lead to the realization of perspectives.

Table 1 presents the contributions between 2016 and 2024 on using artificial intelligence, soft computing and blockchain to mitigate climate change and contribute to sustainable building, as well as optimization of photovoltaic technology and the use of lithium-ion batteries. Without forgetting, of course, the role of proposed strategies, which both contribute to the energy transition and organize the opportunities offered within a legislative and political framework. S. Mehryar et al. apply the machines

Table 1 Scientific contributions of the application of smart energy perspectives on the opportunities of the innovative energy transition

Opportunities of innovative energy transition	Perspectives of smart energy vision by 2050	The application of perspectives on opportunities of innovative energy transition
The dealing with the climate change issue	Artificial intelligence, soft computing and blockchain	Artificial intelligence [68] The use of artificial intelligence to identify climate risks and analyzes climate resilience policies. The approaches used in this context are as follows: – About identifying climate risks: The machine learning approach, including artificial neural networks, is often used to forecast meteorological parameters, assess deterioration caused by natural disasters and analyze the resistance of infrastructures and communities against climate risks – About analysis of climate resilience policies: Machine learning techniques are also being used to assess the impact of policies implemented to reduce the degree of climate change. In addition, agent-based models have been adopted to analyze policy and participatory approaches to climate resilience prior to their implementation
		Soft computing [69] The application of non-linear machine learning methods, including support vector regression SVR and the relevance vector machine RVM, to improve the accuracy of the impact of climate change on the availability of water resources
		Blockchain [70] The use of blockchain to combat climate change, through the monitoring of greenhouse gas emissions and the deployment of green finance. One of the most significant examples of blockchain is the adoption of smart contracts, by applying Article 6 of the 2015 Paris Agreement, for the development of renewable energy markets. The marketing of decentralized electricity (from photovoltaic and wind technologies) is another advantage of blockchain, where a Swiss application called 'Quartierstrom' can be used to manage transactions in a neighborhood

(continued)

Table 1 (continued)

Opportunities of innovative energy transition	Perspectives of smart energy vision by 2050	The application of perspectives on opportunities of innovative energy transition
The sustainable building		Artificial intelligence [71] The design of an intelligent AI-EMM system for managing energy consumption in sustainable buildings to increase energy efficiency and predict the energy consumption of residents. In this context, LSTM long-term memory models predicted energy consumption. Consequently, the IA-EMM model justified its performance with an accuracy rate of 94.3%
		Soft computing (N/A)
		Blockchain [78] The use of blockchain for the development of sustainable buildings aims to monitor and verify greenhouse gas emissions in order to reduce the carbon footprint. The use of blockchain for these buildings involves the following stages: The phase of production and transport of building materials and the phase of construction of these buildings. In this context, V. Marinakis et al. have developed an approach for real-time analysis of energy consumption data in buildings [72].
The optimization of solar photovoltaic technology		Artificial intelligence [73] The presentation of artificial intelligence algorithms for optimizing the performance of photovoltaic cells while reducing their cost: the sun tracking for controlling the PV systems, the inverter switches for diagnostic and the algorithms Adaptive Neuro-Fuzzy Inference Systems ANFIS and Particle Swarm Optimization PSO for dimensioning

(continued)

Table 1 (continued)

Opportunities of innovative energy transition	Perspectives of smart energy vision by 2050	The application of perspectives on opportunities of innovative energy transition
		Soft computing [74] The use of Maximum Power Point Tracking MPPT to extract the maximum energy from renewable sources to reduce the proportion of fossil fuels in the energy mix. The different existing MPPT methods include the P&O method for analyzing solar irradiance, the neuro-fuzzy method for satisfying network needs and the artificial neural network for adapting deviations so that the PV system remains in the maximum point
		Blockchain (N/A)
The use of Li-ion batteries		Artificial intelligence [75] Estimation of the state of charge of lithium-ion batteries in electric vehicles using artificial intelligence techniques such as the RC equivalent circuit model, where the parameters (voltage, current and temperature) are determined by the method of least squares. In addition, the SVM support vector machine is used to manipulate the above parameters as independent variables and the fusion between the Nonlinear Autoregressive Exogenous Model (NARX) and the Long Short Term Memory (LSTM) to estimate the batteries' state of charge
		Soft computing [76] The combination of nanofluids and soft computing for thermal management in order to cooling lithium-ion batteries. In this context, deep learning, the ANN artificial neural network, and the NSGA II genetic al-gorithm for non-dominated sorting combine with air to cool a large number of lithium-ion batteries (10, 5 and 24 respectively for each method) in the form of a honeycomb. As a result, the temperature dropped from the 25–70 °C range down to 19 °C.
		Blockchain (N/A)

(continued)

Table 1 (continued)

Opportunities of innovative energy transition	Perspectives of smart energy vision by 2050	The application of perspectives on opportunities of innovative energy transition
The role of politics in achieving an innovative energy transition		Artificial intelligence [59] The study analyzes the contribution of artificial intelligence to uncover the role of policies, in particular the Paris agreement, in implementing the energy transition. To perform this analysis, the article exploits the Q uantile Vector Autoregressive QVAR method and the CQ cross-quantilogram approach to test the capacity of policies and their alignment with green technologies and markets. Although the Paris Agreement promotes the energy transition by mitigating the harmful effects of climate change, geopolitical risks have impacts on the implementation of the energy transition and green investments. Therefore, the study recommends short-, medium- and long-term policies for implementing the energy transition: reducing greenhouse gas emissions by encouraging governments to launch programs to reduce energy consumption while promoting energy efficiency. In addition, businesses operating in the field of green technologies will be encouraged to exploit artificial intelligence algorithms, as well as energy storage techniques to guarantee access to intermittent resources (solar and wind). Finally, governments and political decision-makers need to launch environmental regulations to reduce greenhouse gas emissions, focus on implementing the articles of the Paris Agreement and eliminate geopolitical risks in order to promote a sustainable energy transition

(continued)

Table 1 (continued)

Opportunities of innovative energy transition	Perspectives of smart energy vision by 2050	The application of perspectives on opportunities of innovative energy transition
		Soft computing (N/A)
		Blockchain [77] The article highlights the exploitation of blockchain technology by politicians to achieve sustainability. It sets out the various measures that need to be taken to achieve the goals of sustainable development: education, health, water availability and clean energy. In this context, the article presents the motivation for the exploitation of blockchain by the political organizations discussed during the international conference of the United Nations in 2023 COP'27: Estimating the quantity of cryptographic electricity consumed by the intelligent infrastructure Zumo to reduce the carbon footprint in the transport sector. In addition, COP'27 discussed the collaboration between 3 organizations, AgriLedger for food, SourceTrace for agriculture and ESIH (the Higher School of Infotronics in Haiti), to launch a commercial platform based on blockchain to detect the required information for the customer to be confident about the product

Table 2 Scientific contributions of the results of applying the perspectives on opportunities: Achieving smart energy by 2050

Opportunities of innovative energy	Results of the approach: application of perspectives on opportunities
The dealing with the climate change issue	Greenhouse gases [83] The use of the life cycle assessment (LCA) method to estimate greenhouse gas emissions in building construction and then to manage waste. This analysis showed that concrete is the main contributor to waste, with a rate of around 38.62 kg eq of CO_2/m^2. In this context, the management of concrete waste is based on extracting waste from untreated concrete and transporting it to waste collection areas. This analysis showed that concrete is the main contributor to waste, with a rate of around 38.62 kg eq of CO_2/m^2. In this context, the management of concrete waste is based on extracting the waste from untreated concrete and transporting it to waste collection areas. This management procedure has reduced greenhouse gas emissions by 20.3%
	Energy transition [26] The contribution of digitalization to achieving the energy transition by 2050 in Switzerland. This contribution is based on the application of the SEED-STEM model, the socio-economic energy model for digitalization SEED and the energy systems model STEM, in Switzerland to reduce carbon dioxide emissions. The strategies and measures adopted include the energy saving of intelligent systems in industry, the efficient use of digital technologies and the widespread use of ICT in households, businesses and transport. These measures therefore aim to reduce the share of transport by 12% by 2050, followed by a reduction in the use of electricity in buildings and a further 53% reduction in final energy consumption
	Climate change [79] The presentation of the approach for optimizing the electricity sector to mitigate the adverse effects of climate change from 2021 to 2050. This approach is based on the adoption of the MIBP system, which aims to reduce CO_2 emissions and decrease electricity costs. The results of this model can be seen essentially in the reduction of 10.2% in CO_2 emissions and the decrease of 38.8% of coal's share of the energy mix by 2050
	Smart Climate or/and Smart energy [23] The adoption of the EnergyPLAN tool by Aalborg University to define energy use actions with zero net carbon dioxide emissions to contribute to decarbonization by 2050. This tool has been adapted for countries and territories in Latin America and the Caribbean to introduce the 2015–2030 national energy plan for Costa Rica, increase the share of renewable energy in electricity production to 60% by 2040 and to 100% by 2050 in Puerto Rico, produce up to 10,000 MW from renewable resources, etc.

(continued)

Table 2 (continued)

Opportunities of innovative energy	Results of the approach: application of perspectives on opportunities
The sustainable building	Smart building [38] Presentation of current research aimed to optimize the intelligent management of buildings in the European Union, based mainly on the intelligent management (through reinforcement learning and multiple techniques) of energy demand and the prediction of the production (through deep neural networks) of solar and wind electricity, as well as the use of BIM and digital twins in the construction sector. The study highlights the challenges facing the deployment of intelligent energy management in buildings: The lack of awareness among the general public about integrating digitalization into buildings, the lack of digital building registers that can be designed by combining digital twins and big data and the lack of intelligent tools for energy management. The recommendations presented on the widespread use of intelligent energy management in buildings include: the use of direct-propagation neural networks for thermal comfort and multiple techniques for estimating indoor temperature; the application of adaptive load control to check network flexibility and satisfy energy demand; and the use of BIM for building diagnosis and upgrading
The optimization of solar photovoltaic technology	PV in smart cities [80] Presentation of the various uses of renewable resources in smart cities, including photovoltaic solar energy: exploiting photovoltaic technology for heating and cooling applications for fragile communities, as well as electrifying them and helping them transform into smart cities. In addition, it is essential to respect the smart cities norms in order to install more efficient PV systems and satisfy the energy demand
	Decarbonization through PV by 2050 [65] The contribution of photovoltaic technology in decarbonizing and achieving zero emissions by 2050 to achieve the articles of the Paris Agreement. The study outlines the various applications of photovoltaic technology: floating technology, photovoltaic panels installed on buildings and photovoltaic systems powering electric vehicles. Overall, the objective of photovoltaic technology is to cover electricity demand by 25% and reduce greenhouse gas emissions over 21% by 2050

(continued)

Table 2 (continued)

Opportunities of innovative energy	Results of the approach: application of perspectives on opportunities
The use of Li-ion batteries	Electric vehicles by 2050 [81] This study presents the contribution of electric vehicles to decarbonization in Taiwan by 2050. Based on the demand for electricity to power electric vehicles, which is expected to reach 507 GWh in 2050, the study indicates that the estimated greenhouse gas emissions will decrease from 0.433 kg CO_2/kWh in 2025 to 0.139 kg CO_2/kWh in 2050
	Drones [84] The use of a geometrically cylindrical lithium-ion battery to power drones to reduce the carbon footprint. The advantage of this application is the dispersal of heat generation within the drone, where this heat can make a significant contribution to global warming
The role of politics in achieving an innovative energy transition	Policies of smart energy by 2050 N/A
	Energy policies of decarbonization by 2050 [82] The use of the LEAP modeling tool to analyze demand and distribution of energy to enable policies for promoting low emissions by 2050 in the Northern Territory of Australia. This country has already put in place strategies to reduce emissions, particularly in the transport sector: the National Hydrogen Strategy, the National Electric Vehicle Strategy, the Climate Action Plan and the deep decarbonization strategy for 2050. The study indicated that the LEAP modeling tool has adopted 3 scenarios for the implementation of policies to reduce greenhouse gas emissions: The first is BAU scenario, which stipulates that the transition from gas to solar energy is driven by markets. The second scenario is consumers and investors influence the high industrial growth scenario basing on that this transition. Finally the last scenario, which is focused on the export of renewable energy (including solar energy). The results of this study have identified the obstacles to the implementation of low-carbon policies, which are essentially political (e.g., lack of political initiatives), economic (e.g., lack of investment), social (e.g., lifestyle impact) and technological (e.g., limited technological expertise). The article recommends the adoption of sustainable technologies while reducing costs. In this context, solar energy and hydrogen have been proposed for electric vehicles, as transport is the first factor to be targeted by policies to reduce carbon emissions by 2050

learning techniques to identify climate risks and analyze the proposed policies about the mitigation of climate change effects [68]. In this context, A. Sarhadi et al. examined the impact of climate change on water resources by applying support vector regression SVR and relevance vector machine RVM [69]. Based on Paris agreement articles, the blockchain technology can contribute in combating climate change by the adoption of smart contracts and development of renewable energy markets [70]. Furthermore, Y. Xiang et al. and V. Marinakis et al. had developed approaches for managing energy consumption in buildings: The design of an intelligent platform "AI-EMM" [71] and the development of an approach for real-time analysis of data [72]. Then, the Adaptive Neuro-Fuzzy Inference Systems ANFIS develop the performance of PV technology [73], as opportunity of innovative energy transition, and Maximum Power Point Tracking MPPT extracts their maximum energy [74]. Lithium-ion batteries mostly perform the storage of intermittent electricity, where A. Shah et al. estimate their state of charge by combining the Nonlinear Autoregressive Exogenous Model (NARX) and the Long Short-Term Memory(LSTM) [75]. To perform their thermal management, A. Peut et al. combine nanofluids and soft computing for cooling lithium-ion batteries [76]. Concerning the major role of strategies in achieving an innovative energy transition, the artificial intelligence can test the politics capacities by exploiting the Quantile Vector Autoregressive QVAR method and the CQ cross-quantilogram approach [59]. C. Mulligan et al. exploit blockchain technology political organizations to achieve sustainability, especially what was discussed in COP'27: e.g. the launching a commercial platform to detect the required information about the product for the customer [77]. Through this analysis, the soft computing is needed, as perspectives, in building and policies to achieve smart energy by 2050. Idem about blockchain for PV technology and lithium-ion batteries. In addition, the perspectives of smart energy have all demonstrated their abilities in achieving innovative energy transition throughout the dealing with climate change issue and the achieving of sustainable building and even more PV technology, lithium-ion batteries and policies. Overall, this section demonstrates that the proposed perspectives has been well adopted on the opportunities, which highlights the achieving of the following chain "from opportunities of innovative energy transition to perspectives of smart energy by 2050." The next section indicates the results of this approach by presenting the contributions of smart energy by 2050.

The results of the application of perspectives on opportunities must be appeared in the recently published contributions concerning smart energy by 2050. Table 2 presents these contributions, which they are published between 2019 and 2024 regarding the development of the energy transition opportunities by perspectives to achieve the smart energy objective. Y. Ma et al. adopt the interval bi-level programming MIBP approach to reduce CO_2 emissions mitigate the adverse effects of climate change from 2021 to 2050 [79]. The socio-economic energy model for digitalization SEED and the energy systems model STEM introduced by L. Stermieri et al. reduce as well as carbon dioxide emissions and contribute to decrease the share of transport by 12% by 2050 [26]. In building, JL. Hernandez et al. indicate the benefits of using BIM and digital twins in the construction sector in order to contribute on the intelligent management of energy demand and production [38]. As the smart cities concept is a major element of achieving smart cities by 2050, the energy production must be performed by renewable energies where P. Pandiyan et al. exploits the PV technology to electrify these cities [80]. This renewable resource contribute in decarbonization and achievement of zero emissions by 2050 in order to perform the articles of the Paris Agreement [65]. The smart energy can be accomplished by using lithium-ion batteries in drones and electric vehicles and Y. S. Shen et al. present their contribution in decreasing decarbonization in Taiwan, which the electric vehicles will reduce greenhouse gas emissions from 0.433 kg CO_2/kWh in 2025 to 0.139 kg CO_2/kWh in 2050 [81]. The decarbonization is among the outlined issues by political organization, where A. S. Hosein et al. use the LEAP modeling tool for analyzing demand and distribution of energy to enable policies in promoting low emissions by 2050 in the Northern Territory of Australia [82]. In general, the previous analysis indicates that smart climate is required to mitigate climate change while achieving smart energy and political organizations must launch more strategies regarding the management of the smart energy transition by 2050.

To detect otherwise the impact of the digital technologies as perspectives on achieving the energy transition, Fig. 7 indicates an extracted statistics from the IRENA Agency about the power of future renewable energy projects installed in the world by 2050. The figure shows that PV energy generates up to 615 GW followed by the other renewable sources. Furthermore, Fig. 7 indicates the major role of the electrolysis on transforming the renewable energies to the hydrogen as it is considered a clean energy source producing therefore 523 mega tones by year by 2050 [10].

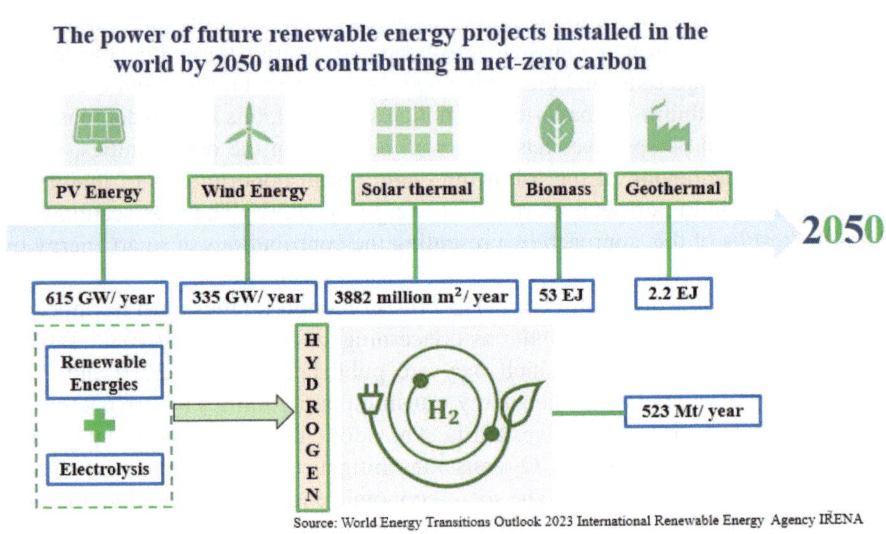

Fig. 7 Power of future renewable energy projects installed in the world by 2050 [10]

Overall, the achievement of an innovative energy transition and smart energy is directed to the same goal: the decarbonization. In fact, there are other factors affecting the smart energy transition such as social impact, conscience and flexibility and in this case, the sustainability will be performed. Therefore, decarbonization and sustainability are both goals of smart energy transition by 2050. However, what is the limitations of this process? In addition, is there recommendations to overcome it?

The limitations of smart energy transition are mainly related to geopolitical conflicts and lack of political initiatives regarding green investment and the efficient implementation in real time of environmental regulatory despite of the great effort of the political organizations. The lack of consciousness about artificial intelligence is another aspect of limitations smart energy transition achievement. Even if the recent contributions focus on all the required approaches of smart energy by 2050, the energy sector needs more intelligent tools and digitals technologies to mitigate the climate issue and satisfy the energy demand by increasing the share or renewable resources in the energy mix. Blokchain and soft computing are well recommended to analyze energy consumption and develops algorithms in order to achieve smart energy by 2050 while conserving the environmental biodiversity. According to this approach, Fig. 8 illustrated that the renewable resources will be the most consumed by 2050 with 247 exajoules, which lead to the reinforcement of the smart energy perspectives [85].

The researchers and the scientists are all invited to contribute with the technological tool of blockchain to optimize the energy production through PV technology. This blockchain can be used in analyzing the energy consumption for a country to

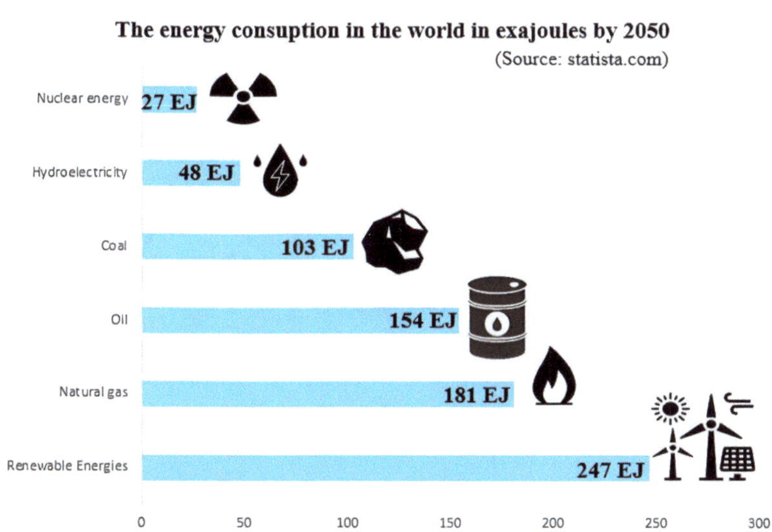

Fig. 8 Energy consumption in the world by 2050 [86]

identify the required measures for their sustainability. In this context, political organizations are invited to contribute with scientists through using intelligent tool, such as blockchain, to develop the energy sector and achieve decarbonization in the regions. The blockchain, as a numerical technology with strong capacities, can be applied for lithium-ion batteries to their supply chain and their recycling. The most important that the actors in energy sector and researchers must collaborate and exploits data and statistics to provide political organizations with required approaches to launch policies of smart energy by 2050.

At last, the achievement of innovative energy transition and smart energy is combined in one objective: the smart energy transition. For this purpose, mitigating climate change, optimizing PV energy, developing building sector, using lithium-ion batteries and adopting strategies contribute in smart energy by 2050. Then, artificial intelligence and digital technologies and also blockchain help to implement energy transition with developed technological tools. It is clear that the way for achieving smart energy by 2050 requires the promotion of energy transition, which in turn requires intelligent tools to be implemented.

TFigure 9 illustrates the global approaches of the opportunities and perspectives for achieving the smart energy transition by 2050 and some statistics about the state of green energies and technologies by 2050. This figure is considered as reference and resume about our approach on achieving a smart energy transition by 2050 [10].

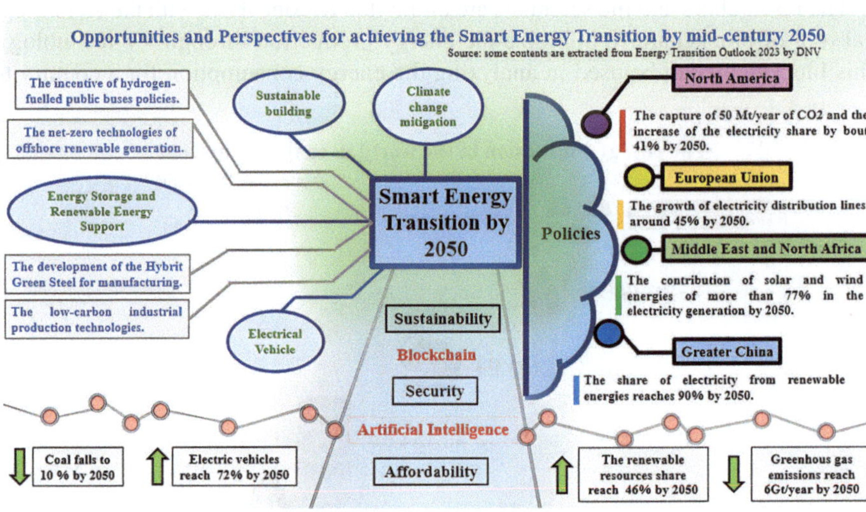

Fig. 9 Graph on the opportunities and perspectives for achieving the Smart Energy Transition by 2050 [10]

5 Conclusion and Perspectives

The chapter motivates the implementation of a smart energy transition by 2050 by proposing perspectives and opportunities for using renewable resources, reducing the carbon footprint and exploiting the global wave of digital transformation to guarantee energy comfort. The approach of this chapter consists of applying perspectives (artificial intelligence, soft computing and blockchain) on opportunities (climate change, PV technology, buildings, lithium-ion batteries and political strategies) to enable the energy transition for the achievement of smart energy. The main outcomes of this chapter stipulates that:

- Energy transition is seen as a key factor in achieving smart energy by 2050. This approach will make it possible to deal with the climate, satisfy energy needs, reduce energy dependency and open up smart energy systems.
- Artificial intelligence algorithms are widely used to confront climate change, develop buildings with zero greenhouse gas emissions, optimize PV technology, expand the use of lithium-ion batteries and benefit from government policies to ensure sustainable development.
- The application of blockchain is still limited to the development of PV technology and the use of lithium-ion batteries.
- There is a need to introduce the concept of smart climate to manage measures for confronting climate change and achieving decarbonization by 2050.
- Although the literature presents policy strategies on managing energy consumption to achieve decarbonization by 2050, the scientific community still needs policies of smart energy to be implemented by 2050.
- Geopolitical risks sometimes pose some limitations to the implementation of energy transition and the smart energy concept.

In addition, the researchers were invited to further develop the concept of smart energy transition by:

- Focusing on other opportunities, such as solar thermal, green hydrogen and smart cities, and measures to implement the energy transition.
- Introducing blockchain for the development of PV energy and lithium-ion batteries. As well as introducing other perspectives in order to achieve smart energy transition.
- Developing the concept of smart climate.
- Encouraging governments to launch policies to achieve smart energy by 2050.

6 Competing Interests

The authors and editors highly appreciate the major role played by Professor Mohamed Saidi Hassani Alaoui. Professor Mohamed Saidi Hassani Alaoui has devoted his considerable expertise to reviewing the opportunities and prospects for achieving a smart energy transition by 2050.

References

1. Palmås K (2023) Design in Marketization: the invention of car safety in automobile markets. She Ji: J Des, Econ Innov 9(1):5–20. https://doi.org/10.1016/j.sheji.2023.04.001
2. Alizon F, Shooter SB, Simpson TW (2009) Henry Ford and the Model T: lessons for product platforming and mass customization. Des Stud 30(5):588–605. https://doi.org/10.1016/j.destud.2009.03.003
3. van Daalen KR, Tonne C, Semenza JC, Rocklöv J, Markandya A, Dasandi N, Jankin S, Achebak H, Ballester J, Bechara H, Beck TM, Callaghan MW, Carvalho BM, Chambers J, Pradas MC, Courtenay O, Dasgupta S, Eckelman MJ, Farooq Z, Lowe R et al (2024) The 2024 Europe report of the Lancet countdown on health and climate change: unprecedented warming demands unprecedented action. The Lancet Public Health 9(7):e495–e522. https://doi.org/10.1016/S2468-2667(24)00055-0
4. Neokye EO, Wang X, Thakur KK, Quijon P, Nawaz RA, Basheer S (2024) Climate change impacts on oyster aquaculture—Part I: identification of key factors. Environ Res 251:118561. https://doi.org/10.1016/j.envres.2024.118561
5. Kharissova AB, Kharissova OV, Kharisov BI, Méndez YP (2024) Carbon negative footprint materials: a review. Nano-Struct Nano-Objects 37:101100. https://doi.org/10.1016/j.nanoso.2024.101100
6. Bailey D (2024) The comparative political economy of sustainability transitions: varying obstacles, accelerants and power in national capitalisms. Environ Innov Soc Trans 51:100853. https://doi.org/10.1016/j.eist.2024.100853
7. Delbecq S, Fontane J, Gourdain N, Planès T, Simatos F (2023) Sustainable aviation in the context of the Paris Agreement: a review of prospective scenarios and their technological mitigation levers. Prog Aerosp Sci 141:100920. https://doi.org/10.1016/j.paerosci.2023.100920
8. Key World Energy Statistics 2021—Analysis (2021) IEA. https://www.iea.org/reports/key-world-energy-statistics-2021
9. U.S. energy facts explained—Consumption and production—U.S. Energy Information Administration (EIA) (s. d.) Consulté 12 septembre 2024, à l'adresse https://www.eia.gov/energyexplained/us-energy-facts/
10. World Energy Transitions Outlook 2023 (s. d.) Consulté 14 septembre 2024, à l'adresse https://www.irena.org/Digital-Report/World-Energy-Transitions-Outlook-2023
11. Alpanda S, Peralta-Alva A (2010) Oil crisis, energy-saving technological change and the stock market crash of 1973–74. Rev Econ Dyn 13(4):824–842. https://doi.org/10.1016/j.red.2010.04.003
12. Lee C-C, Xuan C, Wang F, Wang K (2024) Path analysis of green finance on energy transition under climate change. Energy Economics 107891. https://doi.org/10.1016/j.eneco.2024.107891
13. Ghorbani Y, Zhang SE, Nwaila GT, Bourdeau JE, Rose DH (2024) Embracing a diverse approach to a globally inclusive green energy transition: moving beyond decarbonisation and recognising realistic carbon reduction strategies. J Clean Prod 434:140414. https://doi.org/10.1016/j.jclepro.2023.140414

14. Sun J, Yang Y, Zhou P (2024) Low-carbon transition risks in the energy sector: a systematic review. Sustainable Production and Consumption 50:115–127. https://doi.org/10.1016/j.spc.2024.07.025
15. Yang Y, Xia S, Huang P, Qian J (2024) Energy transition: connotations, mechanisms and effects. Energ Strat Rev 52:101320. https://doi.org/10.1016/j.esr.2024.101320
16. Ahsan Kabir M, Hasan MdM, Hossain T, Ahnaf A, Monir H (2024) Sustainable energy transition in Bangladeshi academic buildings: a techno-economic analysis of photovoltaic-based net zero energy systems. Energy and Buildings 312:114205. https://doi.org/10.1016/j.enbuild.2024.114205
17. Quinteros-Condoretty AR, Golroudbary SR, Albareda L, Barbiellini B, Soyer A (2021) Impact of circular design of lithium-ion batteries on supply of lithium for electric cars towards a sustainable mobility and energy transition. Procedia CIRP 100:73–78. https://doi.org/10.1016/j.procir.2021.05.012
18. Shivanaganna N, Shivamurthy KP, Boddapati V (2024) Optimal strategy for transition into nearly zero energy residential buildings: a case study. Energy 307:132742. https://doi.org/10.1016/j.energy.2024.132742
19. Banihashemi S, Meskin S, Sheikhkhoshkar M, Mohandes SR, Hajirasouli A, LeNguyen K (2024) Circular economy in construction: the digital transformation perspective. Clean Eng Technol 18:100715. https://doi.org/10.1016/j.clet.2023.100715
20. Bist N, Panchal S, Gupta R, Soni A, Sircar A (2024) Digital transformation and trends for tapping connectivity in the oil and gas sector. Hybrid Advances 6:100256. https://doi.org/10.1016/j.hybadv.2024.100256
21. George G, Schillebeeckx SJD (2022) Digital transformation, sustainability, and purpose in the multinational enterprise. J World Bus 57(3):101326. https://doi.org/10.1016/j.jwb.2022.101326
22. Reis J, Melão N (2023) Digital transformation: a meta-review and guidelines for future research. Heliyon 9(1):e12834. https://doi.org/10.1016/j.heliyon.2023.e12834
23. Icaza-Alvarez D, Jurado F, Tostado-Véliz M (2024) Smart energy planning for the decarbonization of Latin America and the Caribbean in 2050. Energy Rep 11:6160–6185. https://doi.org/10.1016/j.egyr.2024.05.067
24. Lee H, Ahn J, Choi DG, Park SY (2024) Analysis of the role of hydrogen energy in achieving carbon neutrality by 2050: a case study of the Republic of Korea. Energy 304:132023. https://doi.org/10.1016/j.energy.2024.132023
25. Hassan Q, Algburi S, Jaszczur M, Viktor P, Iqbal A, Al-Jiboory AK, Hanoon TM, Barakat M, Sameen AZ (2024) Implement and evaluate resilient energy infrastructures capable of withstanding spatial, temporal, and annual weather fluctuations in Saudi Arabia by 2050. Sustainable Futures 7:100182. https://doi.org/10.1016/j.sftr.2024.100182
26. Stermieri L, Kober T, McKenna R, Schmidt TJ, Panos E (2024) The role of digital social practices and technologies in the Swiss energy transition towards net-zero carbon dioxide emissions in 2050. Energy Policy 193:114203. https://doi.org/10.1016/j.enpol.2024.114203
27. Icaza-Alvarez D, Jurado F, Tostado-Véliz M (2023) Smart energy transition with the inclusion of floating wind energy in existing hydroelectric reservoirs with a view to 2050. Ecuadorian case study. Energy Rep 10:2804–2816. https://doi.org/10.1016/j.egyr.2023.09.066
28. Gołąb-Andrzejak E (2023) AI-powered digital transformation: tools, benefits and challenges for marketers—case study of LPP. Procedia Comput Sci 219:397–404. https://doi.org/10.1016/j.procs.2023.01.305
29. Khalid M (2024) Energy 4.0: AI-enabled digital transformation for sustainable power networks. Comput Ind Eng 193, 110253. https://doi.org/10.1016/j.cie.2024.110253
30. Liyanaarachchi G, Viglia G, Kurtaliqi F (2024) Addressing challenges of digital transformation with modified blockchain. Technol Forecast Soc Chang 201:123254. https://doi.org/10.1016/j.techfore.2024.123254
31. Egunjobi OO, Gomes A, Egwim CN, Morais H (2024) A systematic review of blockchain for energy applications. e-Prime Adv Electr Eng Electron Energy, 100751. https://doi.org/10.1016/j.prime.2024.100751

32. Lakhan A, Mohammed MA, Abdulkareem KH, Deveci M, Marhoon HA, Memon S, Nedoma J, Martinek R (2024) BEDS: blockchain energy efficient IoE sensors data scheduling for smart home and vehicle applications. Appl Energy 369:123535. https://doi.org/10.1016/j.apenergy.2024.123535
33. Energy Blockchain in Smart Communities: Towards Affordable Clean Energy Supply for the Built Environment (2024) Energy Engineering 121(8):2313-2330. https://doi.org/10.32604/ee.2024.048261
34. Ali KJ, Ehsan S, Tran A, Haugstetter M, Singh H (2024) Diagnostic excellence in the context of climate change: a review. Am J Med. https://doi.org/10.1016/j.amjmed.2024.06.010
35. Guesmi K, Si Mohammed K, Tiwari S (2024) Green horizons: enabling the energy transition through climate change policies. Int Rev Econ Financ 94:103409. https://doi.org/10.1016/j.iref.2024.103409
36. European Commission. Joint Research Centre. (2023). GHG emissions of all world countries: 2023. Publications Office. https://data.europa.eu/doi/https://doi.org/10.2760/953322
37. Bilardo M, Kämpf JH, Fabrizio E (2024) From zero energy to zero power buildings: a new paradigm for a sustainable transition of the building stock. Sustain Cities Soc 101:105136. https://doi.org/10.1016/j.scs.2023.105136
38. Hernández JL, de Miguel I, Vélez F, Vasallo A (2024) Challenges and opportunities in European smart buildings energy management: a critical review. Renew Sustain Energy Rev 199:114472. https://doi.org/10.1016/j.rser.2024.114472
39. Bisu AA, Ahmed TG, Ahmad US, Maiwada AD (2024) A SWOT analysis approach for the development of photovoltaic (PV) energy in Northern Nigeria. Clean Energy Syst 9:100128. https://doi.org/10.1016/j.cles.2024.100128
40. Hao D, Qi L, Tairab AM, Ahmed A, Azam A, Luo D, Pan Y, Zhang Z, Yan J (2022) Solar energy harvesting technologies for PV self-powered applications: a comprehensive review. Renewable Energy 188:678–697. https://doi.org/10.1016/j.renene.2022.02.066
41. admin (2023) Robots de Nettoyage Photovoltaïque. InHand Networks. https://www.inhand.com/fr/cases/photovoltaic-cleaning-robots/
42. Allouhi A, Rehman S, Krarti M (2021) Role of energy efficiency measures and hybrid PV/biomass power generation in designing 100% electric rural houses: a case study in Morocco. Energy and Buildings 236:110770. https://doi.org/10.1016/j.enbuild.2021.110770
43. Brosse automatique pour le nettoyage des panneaux photovoltaïques (s. d.) Indors. Consulté 20 juillet 2024, à l'adresse https://indors.fr/robot-de-nettoyage-automatique-de-panneaux-photovoltaques/2394-sky-series.html
44. Découvrez le robot de nettoyage solaire F1A sur le site distrimesure (s. d.) Distrimesure. Consulté 20 juillet 2024, à l'adresse https://www.distrimesure.com/fr/solarcleano/6032-f1a-robot-de-nettoyage-autonome-pour-panneaux-solaires-1000-mh-a-sec-solarcleano.html
45. El Hammoumi A, Chtita S, Motahhir S, El Ghzizal A (2022) Solar PV energy: from material to use, and the most commonly used techniques to maximize the power output of PV systems: a focus on solar trackers and floating solar panels. Energy Rep 8:11992–12010. https://doi.org/10.1016/j.egyr.2022.09.054
46. Bos V, Marie F, Gunzburger Y (2024) Lithium-based energy transition through Chilean and Australian miningscapes. The Extractive Industries and Society 17:101384. https://doi.org/10.1016/j.exis.2023.101384
47. Gérer efficacement les batteries lithium-ion (s. d.) Consulté 21 juillet 2024, à l'adresse https://www.socomec.fr/fr/solutions/par-domaines-dactivite/data-center/gerer-efficacement-les-batteries-lithium-ion
48. Ghorbani Y, Zhang SE, Bourdeau JE, Chipangamate NS, Rose DH, Valodia I, Nwaila GT (2024) The strategic role of lithium in the green energy transition: towards an OPEC-style framework for green energy-mineral exporting countries (GEMEC). Resour Policy 90:104737. https://doi.org/10.1016/j.resourpol.2024.104737
49. Maisel F, Neef C, Marscheider-Weidemann F, Nissen NF (2023) A forecast on future raw material demand and recycling potential of lithium-ion batteries in electric vehicles. Resour Conserv Recycl 192:106920. https://doi.org/10.1016/j.resconrec.2023.106920

50. Zhang J, Huang H, Zhang G, Dai Z, Wen Y, Jiang L (2024) Cycle life studies of lithium-ion power batteries for electric vehicles: a review. J Energy Storage 93:112231. https://doi.org/10.1016/j.est.2024.112231
51. Investir dans la transition énergétique: Les pays ont besoin de politiques plus équilibrées | CNUCED. (2023) https://unctad.org/fr/news/investir-dans-la-transition-energetique-les-pays-ont-besoin-de-politiques-plus-equilibrees
52. World Bank Group—International Development, Poverty, & Sustainability (s. d.) [Text/HTML]. World Bank. Consulté 10 juin 2024, à l'adresse https://www.worldbank.org/en/home
53. Adil K, Jamila E-Z (2018) *Quelle gouvernance pour la mise en oeuvre du développement durable ? : cas des energies renouvelables au Maroc* 1
54. Azouzoute A, El Ydrissi M, Elmaazouzi Z, Benhaddou M, Salihi M, Hajjaj C, Garoum M (2021) Thermal production and heat cost analysis of the potential of solar concentrators for industrial process applications: a case study in six sites in Morocco. Scientific African 12:e00765. https://doi.org/10.1016/j.sciaf.2021.e00765
55. Boulakhbar M, Lebrouhi B, Kousksou T, Smouh S, Jamil A, Maaroufi M, Zazi M (2020) Towards a large-scale integration of renewable energies in Morocco. J Energy Storage 32:101806. https://doi.org/10.1016/j.est.2020.101806
56. bazartagency (2024) Offre Maroc pour l'hydrogène vert: Les points clés. Unimagec. https://unimagec.ma/offre-maroc-pour-lhydrogene-vert-les-points-cles/
57. Inflation Reduction Act Guidebook | Clean Energy (s. d.) The White House. Consulté 14 septembre 2024, à l'adresse https://www.whitehouse.gov/cleanenergy/inflation-reduction-act-guidebook/
58. The European Green Deal—European Commission (2021) https://commission.europa.eu/strategy-and-policy/priorities-2019-2024/european-green-deal_fr
59. Chishti MZ, Xia X, Dogan E (2024) Understanding the effects of artificial intelligence on energy transition: the moderating role of Paris Agreement. Energy Economics 131:107388. https://doi.org/10.1016/j.eneco.2024.107388
60. Nebey AH (2024) Recent advancement in demand side energy management system for optimal energy utilization. Energy Rep 11:5422–5435. https://doi.org/10.1016/j.egyr.2024.05.028
61. Pinthurat W, Surinkaew T, Hredzak B (2024) An overview of reinforcement learning-based approaches for smart home energy management systems with energy storages. Renew Sustain Energy Rev 202:114648. https://doi.org/10.1016/j.rser.2024.114648
62. Seshie YM, N'Tsoukpoe KE, Neveu P, Coulibaly Y, Azoumah YK (2018) Small scale concentrating solar plants for rural electrification. Renew Sustain Energy Rev 90:195–209. https://doi.org/10.1016/j.rser.2018.03.036
63. Guo Y, Tang Q, Darkwa J, Wang H, Su W, Tang D, Mu J (2024) Multi-objective integrated optimization of geothermal heating system with energy storage using digital twin technology. Appl Therm Eng 252:123685. https://doi.org/10.1016/j.applthermaleng.2024.123685
64. Gutierrez-Rojas D, Narayanan A, Santos Nunes Almeida CR, Almeida GM, Pfau D, Tian Y, Yang X, Jung A, Nardelli PHJ (2023) A perspective on the enabling technologies of explainable AI-based industrial packetized energy management. iScience 26(12), 108415. https://doi.org/10.1016/j.isci.2023.108415
65. Maka AOM, Ghalut T, Elsaye E (2024) The pathway towards decarbonisation and net-zero emissions by 2050: the role of solar energy technology. Green Technol Sustain 2(3):100107. https://doi.org/10.1016/j.grets.2024.100107
66. Khare V, Bhatia M (2024) Renewable energy trading: assessment by blockchain. Cleaner Energy Systems 8:100119. https://doi.org/10.1016/j.cles.2024.100119
67. Ullah N, Al-Rahmi WM, Alblehai F, Fernando Y, Alharbi ZH, Zhanbayev R, Al-Adwan AS, Habes M (2024) Blockchain-powered grids: paving the way for a sustainable and efficient future. Heliyon 10(10):e31592. https://doi.org/10.1016/j.heliyon.2024.e31592
68. Mehryar S, Yazdanpanah V, Tong J (2024) AI and climate resilience governance. iScience 27(6):109812. https://doi.org/10.1016/j.isci.2024.109812
69. Sarhadi A, Burn DH, Johnson F, Mehrotra R, Sharma A (2016) Water resources climate change projections using supervised nonlinear and multivariate soft computing techniques. J Hydrol 536:119–132. https://doi.org/10.1016/j.jhydrol.2016.02.040

70. Dorfleitner G, Muck F, Scheckenbach I (2021) Blockchain applications for climate protection: a global empirical investigation. Renew Sustain Energy Rev 149:111378. https://doi.org/10.1016/j.rser.2021.111378
71. Xiang Y, Chen Y, Xu J, Chen Z (2022) Research on sustainability evaluation of green building engineering based on artificial intelligence and energy consumption. Energy Rep 8:11378–11391. https://doi.org/10.1016/j.egyr.2022.08.266
72. Marinakis V, Karakosta C, Doukas H, Androulaki S, Psarras J (2013) A building automation and control tool for remote and real time monitoring of energy consumption. Sustain Cities Soc 6:11–15. https://doi.org/10.1016/j.scs.2012.06.003
73. Jathar LD, Nikam K, Awasarmol UV, Gurav R, Patil JD, Shahapurkar K, Soudagar MEM, Khan TMY, Kalam MA, Hnydiuk-Stefan A, Gürel AE, Hoang AT, Ağbulut Ü (2024) A comprehensive analysis of the emerging modern trends in research on photovoltaic systems and desalination in the era of artificial intelligence and machine learning. Heliyon 10(3):e25407. https://doi.org/10.1016/j.heliyon.2024.e25407
74. Chausali N, Saxena J, Prasad R (2023) Nanotechnology as a sustainable approach for combating the environmental effects of climate change. J Agri Food Res 12:100541. https://doi.org/10.1016/j.jafr.2023.100541
75. Shah A, Shah K, Shah C, Shah M (2022) State of charge, remaining useful life and knee point estimation based on artificial intelligence and Machine learning in lithium-ion EV batteries: a comprehensive review. Renewable Energy Focus 42:146–164. https://doi.org/10.1016/j.ref.2022.06.001
76. Derbès S (2022) Comment la blockchain peut faciliter l'insertion des ENR dans les réseaux électriques. The Agility Effect. https://www.theagilityeffect.com/fr/article/comment-la-blockchain-peut-faciliter-linsertion-des-enr-dans-les-reseaux-electriques/
77. Mulligan C, Morsfield S, Cheikosman E (2024) Blockchain for sustainability: a systematic literature review for policy impact. Telecommunications Policy 48(2):102676. https://doi.org/10.1016/j.telpol.2023.102676
78. Yang Z, Zhu C, Zhu Y, Li X (2023) Blockchain technology in building environmental sustainability: a systematic literature review and future perspectives. Build Environ 245:110970. https://doi.org/10.1016/j.buildenv.2023.110970
79. Ma Y, Li YP, Mei H, Nie S, Huang GH, Li YF, Suo C (2024) Potential way to plan China's power system (2021–2050) for climate change mitigation. Renewable Energy 225:120257. https://doi.org/10.1016/j.renene.2024.120257
80. Pandiyan P, Saravanan S, Usha K, Kannadasan R, Alsharif MH, Kim M-K (2023) Technological advancements toward smart energy management in smart cities. Energy Rep 10:648–677. https://doi.org/10.1016/j.egyr.2023.07.021
81. Shen Y-S, Huang G-T, Chang-Chien C-L, Huang LH, Kuo C-H, Hu AH (2023) The impact of passenger electric vehicles on carbon reduction and environmental impact under the 2050 net zero policy in Taiwan. Energy Policy 183:113838. https://doi.org/10.1016/j.enpol.2023.113838
82. Hosein AS, Whale J, Simsek Y, Urmee T (2023) Exploring energy policy scenarios to transition to a low carbon economy by 2050: a case study on the Northern Territory of Australia. Energy Policy 180:113663. https://doi.org/10.1016/j.enpol.2023.113663
83. Ubolsook P, Podong C, Sedpho S, Jansanthea P (2024) Assessing the environmental impact of construction waste management in northern Thailand: an approach to estimate greenhouse gas emissions and cumulative energy demand. J Clean Prod 467:142961. https://doi.org/10.1016/j.jclepro.2024.142961
84. Hollinger AS, McAnallen DR, Brockett MT, DeLaney SC, Ma J, Rahn CD (s. d.) Cylindrical lithium-ion structural batteries for drones. https://doi.org/10.1002/er.4937
85. World energy consumption by energy source 2050 (s. d. a) Statista. Consulté 18 septembre 2024, à l'adresse https://www.statista.com/statistics/222066/projected-global-energy-consumption-by-source/
86. World energy consumption by energy source 2050 (s. d. b) Statista. Consulté 17 septembre 2024, à l'adresse https://www.statista.com/statistics/222066/projected-global-energy-consumption-by-source/

Index

F. Latrache et al. (eds.), *Digital Technology for an Innovative Energy Transition*, Green Energy and Technology, https://doi.org/10.1007/978-3-031-89657-6

MIX
Papier aus verantwortungsvollen Quellen
Paper from responsible sources
FSC® C105338

If you have any concerns about our products,
you can contact us on
ProductSafety@springernature.com

In case Publisher is established outside the EU,
the EU authorized representative is:
Springer Nature Customer Service Center GmbH
Europaplatz 3, 69115 Heidelberg, Germany

Printed by Libri Plureos GmbH
in Hamburg, Germany